17th World Renewable Energy Congress 2016

E3S Web of Conferences Volume 23 (2017)

Manama, Kingdom of Bahrain
4 - 8 December 2016

Editors:

W.E. Alnaser
A.A. Sayigh

ISBN: 978-1-5108-5227-3

Printed from e-media with permission by:

Curran Associates, Inc.
57 Morehouse Lane
Red Hook, NY 12571

Some format issues inherent in the e-media version may also appear in this print version.

Printed by Curran Associates, Inc. (2017)

For additional information, please contact EDP Sciences – Web of Conferences
at the address below.

EDP Sciences – Web of Conferences
17, Avenue du Hoggar
Parc d'Activité de Courtabœuf
BP 112
F-91944 Les Ulis Cedex A
France

Phone: +33 (0) 1 69 18 75 75
Fax: +33 (0) 1 69 28 84 91

contact-edps@webofconferences.org

Additional copies of this publication are available from:

Curran Associates, Inc.
57 Morehouse Lane
Red Hook, NY 12571 USA
Phone: 845-758-0400
Fax: 845-758-2633
Email: curran@proceedings.com
Web: www.proceedings.com

TABLE OF CONTENTS

edp sciences

MATEC Web of Conferences

Statement of Peer review

In submitting conference proceedings to MATEC Web of Conferences, the editors of the proceedings certify to the Publisher that they adhere to the **Policy on Publishing Integrity** of the journal in order to safeguard good scientific practice in publishing.

1. All articles have been subjected to peer review administered by the proceedings editors.
2. Reviews have been conducted by expert referees, who have been requested to provide unbiased and constructive comments aimed, whenever possible, at improving the work.
3. Proceedings editors have taken all reasonable steps to ensure the quality of the materials they publish and their decision to accept or reject a paper for publication has been based only on the merits of the work and the relevance to the journal.

Title, date and place of the conference World Renewable Energy Congress – 17, 3-8 December 2016

Proceedings editor(s): [signature] (Prof. ALI SAYIGH)

Date and editor's signature

DOI: 10.1051/e3sconf/20172300001

INTRODUCTION AND SUMMARY

The initial Congress registration was 260 participants and later reached over 300 participants from 42 countries. There were 151 – submitted abstracts down loaded into USB and handed out with the Congress package to each participants.
In the morning session: Three presentations were given: The first one was by Dr Naji Al-Mutairi, Deputy Director General of KISR: **Shagaya Renewable Energy Initiative – Kuwait.**
Then Dr Abdul Sattar Al-Taie, Executive Director, Qatar National Research Fund: **Empowering Research on Energy Security, Sustainability and Clean Environment in Qatar.** The third presentation was by Prof Ali Sayigh, Chairman of WREC and DG of WREN: **Renewable Energy Progress – Can the GCC be an Exporter of Electricity from Renewables?** The chairmen of the main plenary were Dr Waheeb Al-Naser – Vice President of Bahrain University and Dr Fuad Al-Ansari, Dean of College of Engineering, University of Bahrain

Dr Al-Mutairi – Kuwait

Dr Al-Taie – Qatar

Prof Sayigh - UK

This was followed by two parallel plenary sessions with eight speakers, among them were:
Mr Rainer Hinrichs-Rahlwes – Germany, Dr Haris Doukas – Greece, Dr Lawrence Kazmerski – USA, and Mr Winfried Hoffmann – Germany. Titles of their talk are in the programme.

Mr Hoffmann

Dr Doukas

Dr Kazmerski

Mr Hinrichs-Rahlwes

 DOI: 10.1051/e3sconf/20172300001

In the afternoon three sessions ran in parallel. Some of the speakers are featured below.

On the right hand side Dr Kazmerski presentation on: **Photovoltaic Technology: Where we are, how we got here, and where we are going?**

The Congress Official Opening

On Sunday 4 December at 18:00 hr. His Excellency Prof Dr Riyad Yousif Hamzah – The President of Bahrain University, H E Dr Abdulhussain Bin Ali Mirza, Minister of Electricity and Water Affairs, The Patron of the Congress arrived to Hotel Crowne Plaza Conference Centre. After that H E Sheikh Mohamad bin Khalifa bin Ahmed Al Khalifa, Minister of Oil and Gas arrived to the Congress Venue. The guests and participants assembled at the main theatre where the opening of **World Renewable Energy Congress – 17,** took place. After the reading from the Holy Quran and the Bahrain National Anthem, Prof Dr Riyad welcome the two ministers and guests and expressed his delight that the WREC-17 is taking place in Bahrain. He stressed the importance of renewable energy to Bahrain and rest of the GCC countries and added that more than 40 papers from staff of the University of Bahrain were submitted to the Congress.

DOI: 10.1051/e3sconf/20172300001

H E Prof Riyad Hamzah

H E Dr Abdulhussain Mirza

Prof Ali Sayigh

 DOI: 10.1051/e3sconf/20172300001

The next to speak was H E Dr Mirza who expressed the aim was for The Kingdom of Bahrain to lay the foundation for the use of Renewable Energy and his ministry will legistlate the use of renewable energy and abide by the UN Convention of Climate Change. The last speaker was Prof Ali Sayigh who thanked the University of Bahrain, President, Dean of College of Engineering and H E Dr Mirza for being the Patron of the Congress. He also thanked the people of Bahrain and the Government of Bahrain for the welcome and hospitality shown to all the participants. Prof Sayigh expressed his specific thanks and appreciation to ISESCO, KFAS and KISR and all of the Sponsors for their support.

H E Dr Abdulhussain Mirza, Ms Gada Fouad, H E Sheikh Mohamed Al-Khalifa, Prof Ali Sayigh and Dr Erwin Susanto Sadirsan.

At the Closing Session Dr Fuad Al-Ansari Read the Congress Recommendations:

During the last four days there were 42 countries present at the Congress and more than 250 participants presented 150 contributions in various fields of renewable energy including energy efficiency and energy conservation in buildings. The participants are recommending the following:

1- To speed up the legislation of using renewable energy in Bahrain
2- To encourage the public and officials to work together to improve energy conservation, increase energy efficiency through cultural awareness, smart design and the effective retrofitting of buildings.
3- To set up a Renewable Energy Electricity Target of at least 20% by the year 2030 in order to catch up with many other countries in the region. Also power purchase agreements of renewable Energy electricity can facilitate speedy and cost effective installations at around 10 Fils/kWh.
4- Request the University of Bahrain to initiate at least one renewable energy event each year.
5- The Kingdom of Bahrain should focus on Green Economy by encouraging investment, capacity building and by initiating appropriate legislation to achieve those goals.
6- A truly sustainable future for the Kingdom can be achieved quickly by integrating the massive use of renewable energy systems, energy efficiency and also by understanding the essential need for change.

H.E.Dr. Abdulhussain Bin Ali Mirza
Minister of Electricity & Water Affairs

DOI: 10.1051/e3sconf/20172300001

Prof. Riyad Yousif Hamzah
President of University of Bahrain

Dr Fuad Al-Ansari - Chair of Organizing Committee
Dr. Mohab Mangoud - Registration Chair
Dr. Fay Al Khalifa - Operation Chair/ Events Planning & Design
Ms. Ebtisam Bin Butti- Publicity
Ms. Nada Mahmoud - Events Planning & Design
Mrs. Reema Albaker - PR / Events Planning
Mr. Bassem Elmeligy - Media Designer

Prof. Ali Sayigh, UK – Executive Chairman & Chairman of the Technical Committee
Prof. Waheeb Al-Naser, Bahrain - **Member**
Mr. Rainer Hinrichs - Rahlwes, Germany - **Member**
Dr. Larry Kazmerski, USA - **Member**
Prof. Don Swift-Hook, UK - **Member**
Prof. Robert Critoph, UK - **Member**
Dr. Jazaer Dawody, Sweden- **Member**
Prof. Nader Al-Bastaki, Bahrain - **Member**
Prof. Riadh Al-Dabbagh, UAE- **Member**
Dr. Shaker Haji, Bahrain- **Member**

Members of WREN COUNCIL At WREC-17

DOI: 10.1051/e3sconf/20172300001

EU – Workshop organized by Dr Haris Doukas, Dr Mustapha Taoumi

DOI: 10.1051/e3sconf/20172300002

17th WORLD RENEWABLE ENERGY CONGRESS
4-6 December 2017, University of Bahrain

Speech of H E the President of Bahrain University Professor Riyad Yousif Hamzah at the Opening:

Your Excellency Dr. Abdul Hussain bin Ali Mirza,
Minister of Electricity and Water Affairs
Your Excellency Shaikh Mohammed bin Khalifa bin Ahmed Al Khalifa, Minister of Oil
Prof Ali Sayigh
Director General of WREN, Chairman World Renewable Energy Congress, and Chairman of Iraq Energy Institute
Distinguished Guests and International Delegates:

It gives me great pleasure to welcome you here today to the *17th World Renewable Energy Congress*.

I would like to thank for Your Excellency Dr. Abdul Hussain Bin Ali Mirza, for his leadership and vision, firstly, for making this event happen and secondly, for his ongoing commitment to renewable energy and a clean environment for the Kingdom of Bahrain.

I am delighted that Bahrain is hosting a global congress that addresses the global issues of security, sustainability and clean energy. I am also very pleased to see that we have such wide representation from across the world from industry, academia, government and environmental organisations. It is only through collaboration on a global scale that solutions to the current and emerging energy problems can be found in a quicker, better and more sustainable way than ever before.

Global oil-price volatility has affected economies across the Middle East. Having seen severe price declines in the past, many countries in the region may be tempted to wait for prices to rise again. However, this situation is different; governments need new energy and development strategies. The Gulf's drive to become a renewable energy hub offers a real option for economic development in the Arab world.

The Middle East is a hot spot for renewable energy, gaining great speed over the last 18 months. The region has tripled renewable energy investment over the last 12

DOI: 10.1051/e3sconf/20172300002

months, with more investment planned. The International Energy Agency reported in September that energy investment globally has reached $1.8 trillion last year; the scale of the problem and the scale of investment are clear for all.

The Gulf region (GCC Countries) already have high solar radiation, relatively low land costs, low-cost labor, favorable loan rates, and low tax regimes. These qualify the GCC Countries are uniquely positioned to have the world's new lowest price solar leader.

With many countries now having renewable energy targets and with such a large amount of new infrastructure necessary in order for countries to hit their proposed targets, the GCC Countries will require to develop expertise in a diverse range of roles, from development, manufacturing and on-going management. This will necessitate Academia to respond and work closely with regional and international experts, industry leaders and government agencies to develop the sustainable talent pool in order to meet the challenges ahead.

University of Bahrain (UoB) have already started to tackle regional sustainability issues, our recently approved transformation plan highlights that our signature research will be in the areas of renewable energy, water and food security. Desalination is vital for the Kingdom of Bahrain: it supplies about 90% of freshwater to our islands and consumes 30% of our primary energy. The topics of water supply and energy security are closely linked to those of food security and economic security in Bahrain and more broadly across the GCC Countries.

Two weeks ago, UoB held a 2 day symposium on these particular issues with 40 universities attending from the Gulf and UK, as a result several joint projects will commence to find sustainable solutions to the regional and global challenges that we face.

In 2017 UoB will also commence long term joint research projects in our signature research areas with the University of Oxford, University of Aston and University of Surrey. These projects will integrate industry and academic stakeholders from both the UK and from the Gulf region with a view to achieving implementation and impact from our joint research. We will continue to collaborate in order to the make best use of our world class resources, expertise and environment.

Finally, I wish for a productive and valuable congress that brings great minds together to identify specific challenges and more importantly to build collaborative links in order to develop policy, projects and solutions.

Thank you.

Prof Dr Riyad Yousif Hamzah, Conference Chairman, President, University of Bahrain Kingdom of Bahrain

DOI: 10.1051/e3sconf/20172301001

Experimental study of the effect of fully shading on the Solar PV module performance

Monadhil Al-chaderchi*, K. Sopain, M. A. Alghoul***, T. Salameh***

**Sustainable and Renewable Energy Engineering (SREE) Department, University of Sharjah*

***Solar Energy Institute (SREI), University Kabangsaam (UKM)*

****Energy and Building Center, Kuwait Institute for Scientific Research, Safat 13109, Kuwait.*

**** Center of Research Excellence in Renewable Energy (CoREEE),King Fahd University of Petroleum and Minerals(KFUPM),Dhahran31261,Saudi Arabia*

Abstract

Experimental tests were performed to study the effects of shading for different string inside the photovoltaic (PV) panels, power equipped with different number of diodes from the same manufacturer as of solar panel. The IV curve for all cases were recorded to see how the bypass diodes will reduce the effects of shading .The case for 3 by pass diode show the best performance of solar PV module under shading phenomena.

Key Words: PV solar Panel, Shading, Bypass diode, fully shading

1. Introduction and literature review

Photovoltaic renewable energy system are widely used as source of green energy because of the positive impact on the environment, direct electricity without conversion and low pollution to the atmosphere, as well as competitive with the traditional energy system, due to maturity of these technologies. Shading is one of the most important phenomena that affects on the performance of PV panels, the shading can happen by different reasons, some of these reasons are related to atmospheric condition such as dust and surrounding conditions such as human being , constructions, birds, dropping of birds, plants and between panels in arrays.

Shading is categorized in to two main types; (i) full shading (ii) partial shading. Most of researchers were performed tests and simulations to study the effect of partially and fully shading on the performance of PV panels.

In the study [1], the *I–V* and *P–V* characteristics of a PV array under a nonuniform insolation due to partial shading were presented based on modeling and simulation scheme performed by MATLAB. [2] they developed A generalized PV module model n MATLAB based on the mathematical equation of PV cell. The model is used to simulated partially shaded cells in the PV module and also for partial shaded modules in the PV string under various weather conditions. The impact of shading on array characteristics using MATLAB model of PV array were observed by the study of [3]. Different insolations incident were considered on different models. The output of the MATLAB model with the theoretical values were compared. The goal of this study is to design and develop simulation of solar panels connected in series integrated with a boost converter and MPPT algorithm taking into account the effects of partial shading. In the study [4], MATLAB/Simulink was used to design and simulate a complete off-grid PV module based power generation system. The value of standard solar irradiance about 1 KW/m^2 was used for Bangladesh. The simulation model was examined the performance of all components (blocks) for the off-grid PV module system. Adaptive inertial weight particle swarm optimization (AIWPSO) algorithm was proposed to solve the low convergence speed and search accuracy of the Particle swarm optimization (PSO) under partially shading by the study of [5].[6] they studied the impact of using bypass diodes on the solar cell performance under shading at a module level. Algorithms based on MATLAB was designed to study the effects of non-overlapping bypass diode configuration in a randomly shaded solar module. In the study [7], the behavior of a photovoltaic device (cell or module) under partial shading conditions were reviewed and analyzed by using MATLAB. [8 - 9] they investigated the harmful effects of partial shading of series and parallel connected Solar PV modules and compare their performance.

In this research paper, the I-V characteristic curve and power performance of 30 W PV panel under fully different shading cells condition will study for different number of bypass diode, mainly two and four diodes. The PV analyzer device will use to measure I-V and power curves for this panel.

2. Experimental setup and testing

The experiments were tested under real conditions of irradiance G (W/m^2) and temperature T (°C), where the I-V and power curves for PV module measured by using the PV analyzer (profitest), the electrical characteristic of PV panel is shown in Table 1. All the components used during the measurement are shown in figure (1). Different configurations of fully shading cases were

DOI: 10.1051/e3sconf/20172301001

tested for 30 watt PV panel with two or four bypass diodes per panel. The fully shading was applied on the cells by using cardboard.

Table 1 Electrical characteristic of PV panel used in the testing under STC.

Open circuit voltage (V_{oc})	22.5 V
Short circuit current (I_{sc})	2 A
Voltage at maximum power point (V_{mp})	18 V
Current at maimum power point (I_{mp})	1.67 A
Maximum power (P_{mp})	30 W
# of cells ad type	36 Monocrystalline

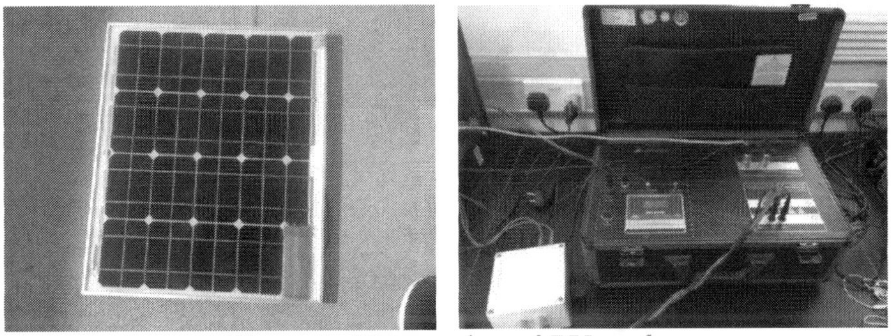
Figure 1: PV panels and PV analyzer

3. Results and discussion

Figure (2) shows the effect of shading size on both maximum power and fill factor, both of them are inversely proportional with the shading size as shown from the figure, that can explain how is the fully shading reduces the maximum power production from the PV panel, this reduction in the

DOI: 10.1051/e3sconf/20172301001

maximum power due to the drop in the short circuit current and hence the fill factor.

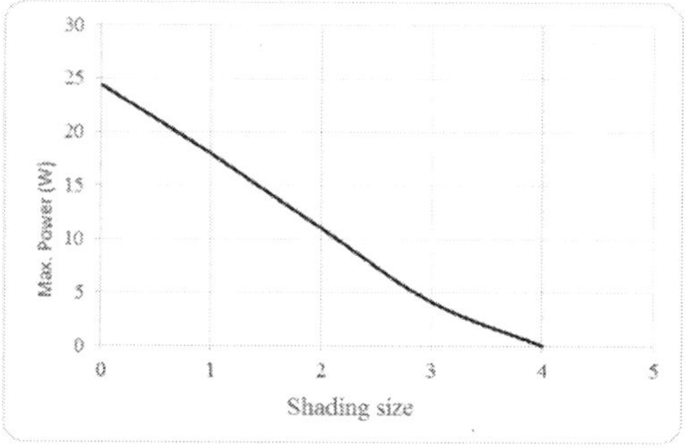

(a)

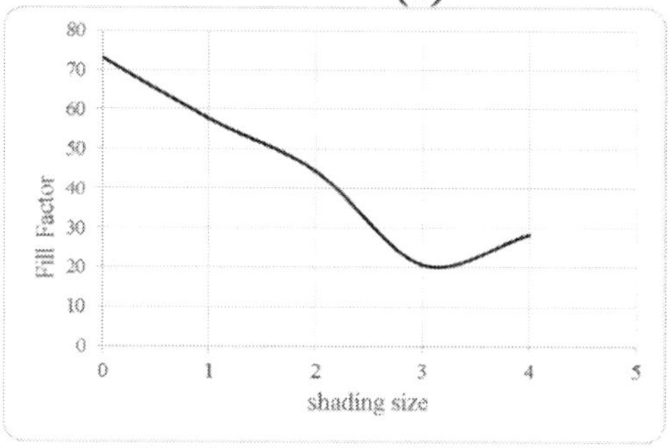

(b)

Figure 2: (a) Fill factor versus shading size (b) Maximum power versus shading size

The characteristic I - V curve for 30 W PV panel is shown in figure (3) for unshaded (reference) case and different shaded cases (1 cell, 2 cells, 3 cells and 4 cells) for both 2 and 4 bypass diodes. The effect of the number of shaded cells is obviously clear on the characteristic I - V curve of PV panel, the drop in the current for different shaded cells from the string (from left or right sides i.e. from first or fourth column) is typically same for all shaded configurations, the reference case has the highest short circuit current while the four shaded cells has the worst current (zero current), it worth to mention that the difference between the two and four bypass diode is clearly appears in the case of two cells from different string.

Figure (4) shows the power curves for unshaded (reference) case and different shaded cases for the 30 W panel size. The drop of the maximum power for each individual shaded case was corresponding to the drop of the short circuit current for each individual shaded case as shown in figure (3). The value of

 DOI: 10.1051/e3sconf/20172301001

the maximum power is determined by the value of voltage corresponding to maximum power for each individual shaded case. Similar to the characteristic I - V curves, the reference case has the highest power value (24.4 W) while the four shaded cells has the worst power value (0.031 W). More ever, for the cases of 1 cell shaded, 2 cells shaded from different strings and 3 cells shaded from different strings, the 4 bypass diodes saved more power than the 2 bypass diodes by 28%, 45% and 17% respectively. It is also clear that increasing the numbers of bypass diodes will improve the performance of PV panel under fully shaded cells condition, take in the account the price of these bypass diodes with respect to the price of PV panel, in order to keep acceptable energy production cost ($ / KWh).

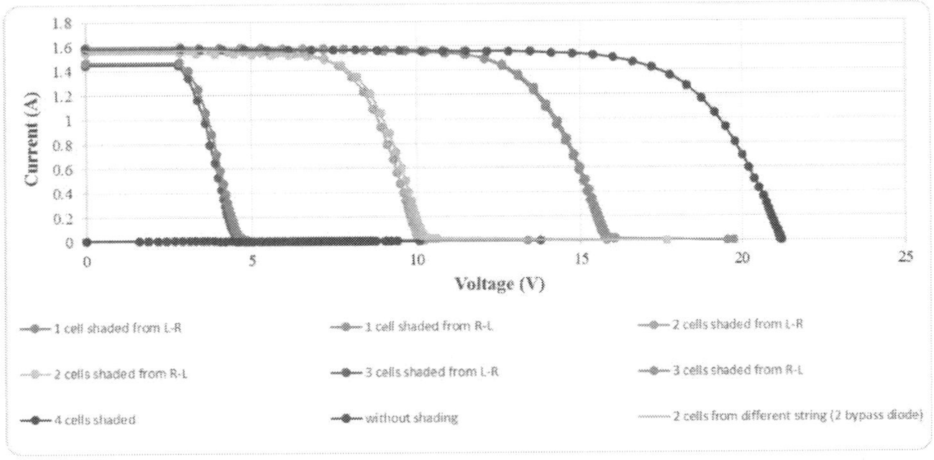

Figure (3) I – V characteristic curves for unshaded (reference) and shaded cases for 2 and 4 bypass doides.

DOI: 10.1051/e3sconf/20172301001

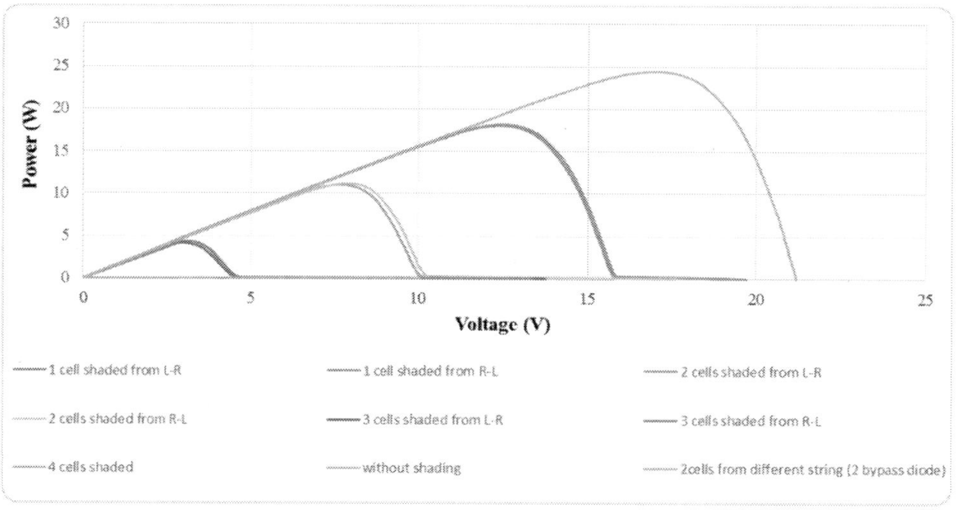

Figure (4) power curves for unshaded (reference) and shaded cases for 2 and 4 by pass doides.

4. Conclusion

The profitest PV analyzer device was used to measure the effect of fully shaded cells on the performance of PV panel equipped by two or four bypass diodes. The drop in the short circuit current and power of the 30 W PV panel depend on the number and the position of the shaded cells. The four bypass diodes save more power than the case of two bypass diodes taking in the account the cost of energy production.

References:

[1] Hiren Patel and Vivek Agarwal, " MATLAB-Based Modeling to Study the Effects of Partial Shading on PV Array Characteristics," *IEEE Trans. EnergyConvers.*, vol. 23, no. 1, pp. 302–310, March. 2008.

[2] S. Sheik Mohammed, D. Devaraj and T. P. Imthias Ahamed, "The modeling and performance analysis of PV modules under partial shaded condition," Indian Journal of Science and Technology, Vol. 9, no. 16, April. 2016

[3] Abhinav Kumar, Krishan Arora," observing the impact of shading on array characteristics using MATLAB model of p v array," vol. 8, no. 1, June. 2016.

[4] Protap Kumar Mahanta*, Khokan Debnath and Md.Habibur Rahman, "Modeling and Simulation of a PV Module Based Power System Using MATLAB/Simulink," Dhaka Univ. J. Sci. Vol. 62, no. 2, pp. 127-132, July 2014.

[5] Xiaoling Yuan, Dianfei Yang, Haoming Liu, " MPPT of PV System under Partial Shading Condition based on Adaptive Inertia Weight Particle Swarm Optimization Algorithm," The 5th Annual IEEE International Conference on Cyber Technology in Automation, Control and Intelligent Systems, June 8-12, 2015, Shenyang, China.

[6] Srinivasa Vemuru, Priyanka Singh and Mohammed Niamat, "Modeling Impact of Bypass Diodes on Photovoltaic Cell Performance Under Partial Shading,"

[7]O. G. García , J. C. Hernández, F". Jurado, "Senior Member, "Assessment of shading effects in photovoltaic modules.

[8] Ramaprabha Ramabadran, Badrilal Mathur, "Effect of Shading on Series and Parallel Connected Solar PV Modules Modern," Applied Science,Vol. 3, No. 10 October 2009.

[9] R.Ramaprabha ,B.L.Mathur, "Impact of Partial Shading on Solar PV Module Containing Series Connected Cells," International Journal of Recent Trends in Engineering, Vol 2, No. 7, November 2009.

DOI: 10.1051/e3sconf/20172301002

SMART PV GRID TO REINFORCE THE ELECTRICAL NETWORK

Mohamed Y. AL-Hamad
Power Trade Senior Executive
Market Operations
GCC Interconnection Authority
K. S. A.
mhamad@gccia.com.sa

Isa S. Qamber
Dept. Electrical & Electronics Eng.
Professor of Electrical Eng. & Energy
University of Bahrain, P. O. Box 32038
Kingdom of Bahrain
iqamber@uob.edu.bh

Abstract:

Photovoltaic (PV) became the new competitive energy resources of the planet and needs to be engaged in grid to break up the congestion in both Distribution and Transmission systems. The objective of this research is to reduce the load flow through the distribution and transmission equipment by 20%. This reduction will help in relief networks loaded equipment's in all networks. Many projects are starting to develop in the GCC countries and need to be organized to achieve maximum benefits from involving the Renewable Energy Sources (RES) in the network.

The GCC countries have a good location for solar energy with high intensity of the solar radiation and clear sky along the year. The opportunities of the solar energy is to utilize and create a sustainable energy resource for this region.

Moreover, the target of this research is to engage the PV technology in such a way to lower the over loaded equipment and increases the electricity demand at the consumer's side.

Keywords:

Photovoltaic, Smart Grid, Peak load, Renewable energy, Energy Managements.

Introduction:

Over the last 10 years solar technology grew fast worldwide. The construction cost of the PV has dropped more than 70%. The GCC countries' noon time peak load demand is the most expensive hours, which is recorded during the day during summer periods from 13:00hrs to 14:30 hrs. The cost during the system peak duration goes very high, interim of generation costs. Also, these hours are the most stressed hours for transmission and distribution systems which are used to give consumers the required power for these hours [1].

 DOI: 10.1051/e3sconf/20172301002

Bayindir and others [2] explore the need of transforming the power grid into a more reliable, secure, efficient and clean network. The authors mentioned that the Smart Grids (SGs) take a large step to develop the global power utilities industry [2]. At present time the new concept of SGs and the functional relationship between the different subsystems and technologies. The SGs help to utilize the involvement of new Renewable Energy Sources for the benefit of the customers by reducing the costs of operation and network reinforcements. The utility sectors of the GCC networks are not created like these markets, where the demand response and development of the smart customer sector could achieve the utilization of the new concepts partially.

More utilization could be achieved by introducing the Markets in GCC region as proposed by Al-Hamad and Qamber [3, 4]. They observed that all member states have the opportunity of power trading in a short-term contract market, such as monthly contracts. All member states have the opportunity for power trading in the spot market.

The present paper focuses solely on the use of PV system in the smart grid and shows its importance and utility. PV is used worldwide in many applications, from the major cities in the developed countries to primary rural villages in developing states. The PV system is illustrated as a means to generate electricity by utilizing directly from the sunlight. The PV is a cost-effective alternative in rural areas where extending a utility power line is very expensive. The PV has less operating cost and takes less maintenance, generate electricity without polluting the environment and without creating disturbance. The PV can be integrated into a grid with other types of generators [5]. These are considered as advantages of the PV.

Single PV/Solar cells are connected electrically to form PV modules. The solar cells can be connected in a certain way to power a wide kind of electrical equipment. The components taken to complete a PV system may include batteries, charge controller, inverter, dis-connectors or fuses, ground circuit and wiring.

Integrating PV into the residential areas will help develop Green Energy Market in GCC countries. The PV helps in putting back the formal facade, roofing materials and keeping off the cost of support structures. This organization is called the Building-Integrated PV (BIPV). These systems include crystalline modules integrated into roofing systems and practiced as 'eyebrows' over windows. As well, these systems practiced glass-on-glass modules used in skylights and view walls.

DOI: 10.1051/e3sconf/20172301002

In many applications, PV is used in Utility Systems including large centralized generation, transmission and distribution support including housing, commercial, and agricultural demand. In addition, in the more modern system, it is used in the Demand-side management (DSM).

DSM systems have a special value of energy between high demand and low demand, where high demand for the power present the most expensive period. Transmission and distribution support has a value because of the need of more infrastructure and investment to transfer the energy, where utilities can install a PV near to overloaded equipment to eliminate or delay the need for costly upgrades.

Renewable Energy Targets of GCC Countries:

GCC country's renewable energy targets illustrated in Table (1), presents decent renewable energy targets. The targets [6] varied from 5% to 20% until 2030.

Table (1) Renewable Energy Targets of GCC Countries

GCC State	Target of Renewable Energy	Year
Bahrain	5% of the Capacity	2020
Kuwait	5% of Generation	2020
Kuwait	15% of Generation	2030
Oman	10% of Generation	2020
Qatar	20% of the Capacity	2030
Saudi Arabia	9.5 GW	2020
Saudi Arabia	54 GW	2040
Abu Dhabi	7% of Capacity	2020

DOI: 10.1051/e3sconf/20172301002

Dubai	5 GW of Capacity	2030

The average target is 5% of the member state's consumption, where the growth of consumption assumed at 5% per annum. The requirement of the renewable energy is around 34,180 GWh for GCC States for the year 2020. Considering the deployment of solar energy to meet the 5% renewable energy target would mean installing a deployment of active production from Renewable Energy Sources in GCC countries around 20,000 MW by the year 2020.

Retail Electricity Prices:

The electricity prices in the GCC countries have always been a subject of debate due to the subsidies at upstream level and hence cannot be exactly quantified. While the GCC states are actively working on having a cost reflective tariffs in place, the existing average retail electricity tariffs as presented in Table (2).

Table (2) Average Retail Electricity Tariffs in GCC

Retail Electricity Tariff in US ¢/kWh			
GCC State	Residential	Commercial	Industrial
Bahrain* [7]	4.11	5.17	5.98
Kuwait* [8]	3.87	8.30	3.32
Oman [9]	3.25	5.19	4.16
Qatar [10]	2.38	3.20	4.12
Saudi Arabia [11]	2.22	5.69	4.80
UAE [12]	8.36	4.36	4.99

*Proposed Tariffs for 2017

PV Unit Components:

The main PV system components are Solar Cells Array, Charge Controller, Battery, DC-DC converter, DC-AC inverter, and AC load connected to electrical power grid are presented in [13].

A Charge Controller is used to govern the flow of electricity from the PV modules to the battery as well as both DC-DC converter and DC-AC inverter. It keeps the battery fully charged without overcharging it. Then, when the controller senses that the battery is fully loaded, it breaks the flow of direction from the PV modules. The controller allows charge to flow from the modules into battery, the load or both when the load is getting power.

The battery is used to charge/discharge the electricity, which will be utilized at night-time or when the modules are not generating sufficient power to meet load demand. Deep-cycle batteries are utilized in this scheme [14]. Depth of discharge in conjunction with battery capacity is a fundamental parameter in the design of battery bank of PV system and the effect of their multiplication represents the energy which can be pulled from a shelling. A deep-cycle battery is designed to gradually discharge and recharge up to 80% of their capacity hundreds of times. The shallow - cycle battery should not be used in PV system because it can discharge only 20% of their capacity. Thus, if extracting much more of its capacity and for several times, the battery will be damaged and no longer be able to hold a commission. In order to reach the same usable capacity, a shallow-cycle battery must have a larger capacity than a deep-cycle battery. PV systems may accept to be sized to lay in sufficient batteries to meet energy during the night.

The Inverter also may be used in PV system to convert the DC electricity produced by PV modules or batteries into AC electricity. The Inverter cost is affected by various factors such as, the quality of electricity it needs to grow, the incoming DC voltage, the load required, the amount of extra surge power the AC loads need for little periods and whether the inverter has any extra features like meters and indicator lights.

PV Array Arrangement:

With respect to the PV Array Arrangement there are three systems for connecting PV arrays, which are series connection, parallel connection and combination of series & parallel connection. The voltage for the PV module in series connection is

DOI: 10.1051/e3sconf/20172301002

additive while the current/ampere stays the same. The voltage stays the same in the parallel connection where the current is additive.

Normally, One PV panel available in the market generates 200W. The PV array arrangement proposed is a combination of 8 x 7 panels. The array combination will generate 11200W. The power factor of the electrical load connected in the main electrical network in the Kingdom of Bahrain is 0.85. The apparent power for the PV array proposed for 0.85 power factor is 13.167kVA [15].

The inverter and charging/discharging of battery efficiency is found to be close to 80% [14]. Counting the losses of inverter and battery efficiency, the total power transferred to the grid is 10.54 kVA.

PV Array in Electricity Distribution System:

The modern electricity distribution system starts from the principal sub-station and reaches by the secondary service enter the customer's meter socket. It conducts electricity from the transmission organization and presents it to consumers.

This paper proposes to use PV system in the city network. In the Gulf region domestic load increases in summer time due to wide usage of air-conditioners. City network requires reinforcement due to increase of domestic cargo. Increase in load becomes a critical issue if the substation transformer increases more than 80% of its capacity and ambient temperature more than 40°C [15].

The transformer load ability versus ambient temperature is shown in Fig (1), which is established in the IEEE standard [13, 16]. The normal lifetime of the transformer (meaning normal solid insulation life) is 180,000 hours or 20.6 years are based on continuous operation at a design hotspot temperature of 110°C.

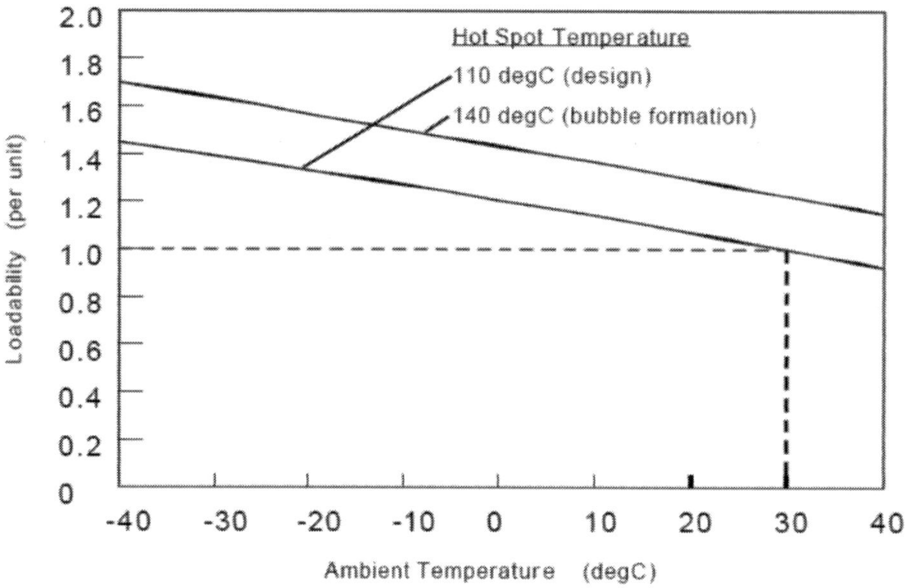

Fig (1) Transformer load ability versus ambient temperature

Table (3) shows nonlinear formula relating the rate of loss of life, which increased with an addition in the 'hot spot' temperature. Hot spot temperature is shown by the 'winding temperature' gauge. When the ambient temperature reaches 45°C and the transformer capacity reaches a full rate, the loss of life estimated will reach 16 times of normal loss life.

It should be noted that the operation at the 'oil bubbles' condition is taken for a short period of time as the bubbles re-fade out when the oil cools.

In the Kingdom of Bahrain ambient temperature in summer rises normally over 40°C, and reaches 45°C in July and August. Therefore, transformer loading should not be more than 80% based on the IEEE standard to avoid loss of life [16].

Table (3) Rate of loss of life to other values of hot spot temperature

Hot Spot Temperature Rate of Loss of Life(°C)	*degrees Celsius relative to normal(°C)*
110 (design value)	1
117	2
124	4

 DOI: 10.1051/e3sconf/20172301002

131	8
139 (oil bubbles)	16
147	32

Reinforcement of Network Results:

The medium-voltage (MV) of electricity distribution system consists of many rings connected between different MV bus bars. Each ring contains a number of substations, which transfer the power from MV to the Low Voltage (LV). The ring load accepted for the Kingdom of Bahrain is 6MVA. Number of substations connected at the ring on average are 10 substations. On average, a substation has 5 or 7 Low Voltage feeders supplying the homes. Normally one feeder is supplying 5 homes.

Average number of houses connected to one substation of capacity 1500 kVA, 1000 kVA and 500 kVA are 30, 20 and 10 houses, respectively. If each house has PV unit, total load generated by the PV array for different types of substation will be 316 kVA, 211 kVA and 105 kVA, respectively. The load generated by the PV array as compared to the capacity of the transformer of distribution substation is equal to 21%. This measure of load generated is equivalent of a load required to relieve transformers from the 100% loading to the 80% loading level.

This technique can be applied to avoid reinforcement of the Low voltage feeder, whose loading exceeds 80% of its LV feeder capacity. Likewise, the 11kV feeder will be reinforced by the same measure of loading, which is 21% of the current burden. All equipment's in the transmission side will be also relieved by 20%. The Power generated by power station will be cut by 21%.

Impact of renewable energy on the tariffs as per our study, considering the international fuel prices, the generation costs in GCC countries for new power plants are in the range of US ¢ 4.2 to US ¢15/kWh considering long-term, short-term and peak scenarios. The range of prices calculated for conventional power along with the solar tariffs discovered through a transparent bidding process in GCC States is illustrated in Fig (2).

 DOI: 10.1051/e3sconf/20172301002

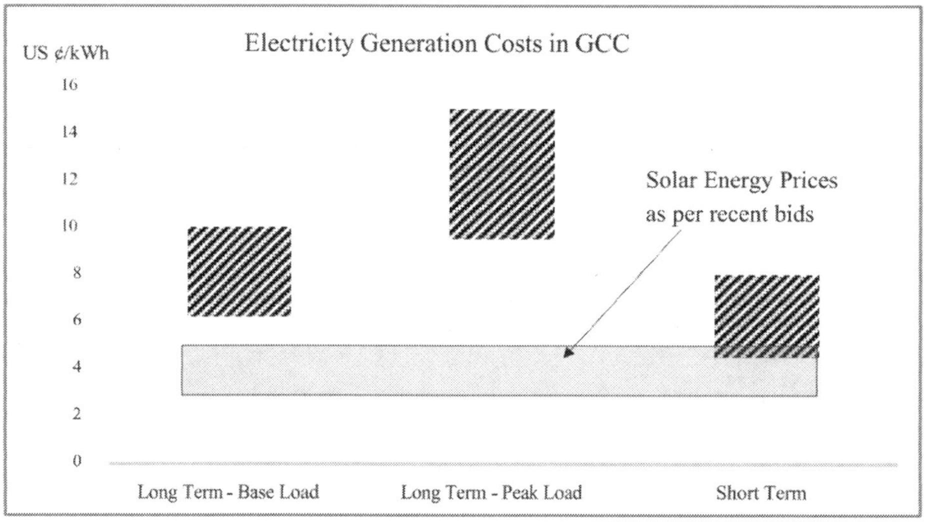

Fig (2) Electricity Generation Cost & Solar Prices

Impact of Solar Prices on Retail Tariff in GCC:

While the generation costs of conventional power plants are more than USD 0.0242 per kWh, the average realization is less than that. However, the tariffs of solar energy projects in the recent bids appear to be closer [17] to the retail tariffs of the GCC States.

It can be seen from Fig (2) that the solar tariffs are already proving to be more economical than the conventional sources of power generation.

Considering the conservative solar tariff of USD 0.0242 per kWh [17], which was the highest tariff in the most recent tariff based bid for the solar power project. The increase in average retail tariff of the GCC states for a share of 5% of solar energy is presented in Fig (3).

DOI: 10.1051/e3sconf/20172301002

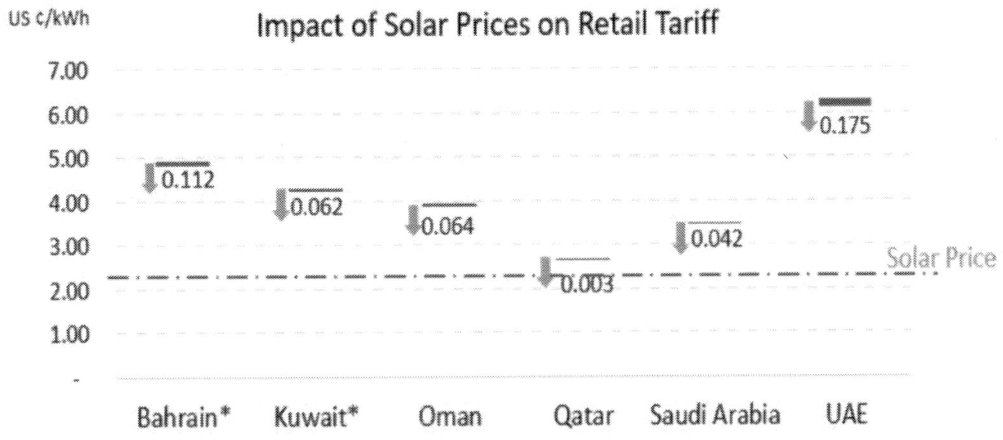

Fig (3) Impact of Solar Prices on Average Retail Tariff

Fig (3) demonstrates that while there is a marginal increase in average retail electricity tariff due to the addition of 5% solar energy, all member states have a reduction in average retail tariff. This will justify adding Renewable Energy Sources generation to the electricity utility industry.

Innovative policies and regulations that can be introduced at national level of renewable energy:

Some of the established policies for promotion of renewable energy for retail sale are wheeling of energy, banking, energy, renewable energy certificates and virtual power purchase agreements (PPA). Wheeling and banking policies have been carried out in many states.

Wheeling is the transfer of electric energy through transmission and distribution lines from the generator to the consumer. The same generated energy may not reach the consumer, but the electricity is passed to the consumer through energy displacement process.

For promotion of renewable energy, this wheeling (transmission) process needs to be simplified to the renewable energy generator and the consumer for scheduling. The existing base will hold good for the consumer to claim total or partial zip from

 DOI: 10.1051/e3sconf/20172301002

the renewable energy generated with small or no modifications to the existing metering infrastructure of the consumer.

Oftentimes, the highest-quality renewable resources are situated far from major electricity load centers and in those cases, wheeling polices can support transmission to deliver renewable generation to those load centers.

Suggested types of polices are as follows:

a) Wheeling Policy for wheeling of electricity. Wheeling needs to be digested by the trading parties in addition to the wheeling charge the distribution network for the transmission & the distribution network of the consumer and distribution network of the consumer. For the above computation, the wheeling charges for withdrawal of energy by the member nations have not been taken due to unavailability of data. Nevertheless, it should be mentioned that as a share of wheeling incentives for the renewable energy, the wheeling charges should be waived off by the member states to realize the scheme (renewable energy wheeling) more viable and attractive for the consumers. The utilities may also put up rebates on the transmission and distribution loss.
b) Banking/Metering Policy - Member State host the project should allow the banking & metering facility for the renewable energy generated which can be scheduled for consumption later by the consumer. The modes and terms & conditions for banking and metering of energy may be discussed with all the member states for agreeing terms.
c) Exchange Renewable Energy Sources with GCCIA - The energy may also be allowed to be exchanged through GCCIA with the member states in an agreed percentage with each member state.
d) Regional Renewable Energy Certificates - The Regional Renewable Energy Certificates system may as well be implemented at the GCC regional level. Nevertheless, the concept of wheeling and banking of energy will involve all the stakeholders in the growth of renewable energy in the area.

Conclusion:

DOI: 10.1051/e3sconf/20172301002

The Smart Grid could help GCC countries in addressing the high load and stress on the electricity network. The Photovoltaic systems could be used to add electricity generation units in the city, areas where there is no space for PV arrays. The house's roof can be used to install the PV units to handle a portion of local demand. Millions of dollars are spent annually in the GCC countries to support the network, which can be utilized in building the PV smart grid in residential areas and to satisfy the aim of the reinforcement issue.

The proposed scheme will encompass the following advantages:

i) Relieves the distribution network, transmission network and force generated by approximately 5% to 20% by the year 2030.
ii) All States will achieve their renewable energy targets, and allows the grid system to become more interoperable at a continual level to heighten protection and cost effectiveness "Smart grid".
iii) With subsidies being reduced, the average retail tariffs are now close to the solar generation tariffs.
iv) Cuts the cost of generation, transmission and distribution of electricity throughout the grid.

The present study recommendation is to make the investment in developing the GCC Smart Grid in smart way by utilizing the budget reduction expected from reinforcement schemes to build new PV system in the residential areas and create the Smart Grid in the GCC region.

DOI: 10.1051/e3sconf/20172301002

References:

1. M. Y. Al-Hamad, A. A. Al-Ebrahim, I. S. Qamber, "Opportunities for Power Trading through Diversity of Annual Maximum Load for GCC Countries", GCC Cigre 2013 Conference & Exhibition, Abu Dhabi, 18-20 November 2013.
2. R. Bayindir, I. Colak, G. Fulli, and K. Demirtas "Smart grid technologies and applications", Renewable and Sustainable Energy Reviews, Volume 66, December 2016, Pages 499-516.
3. M. Y. Al-Hamad, I. S. Qamber, "Activating GCC Power Market: Developing the Power Market and Involving the Electric Utilities", Paper No. A085, Cigre GCC Power 2014, 10-12 November 2014, Manama, Kingdom of Bahrain.
4. Isa S. Qamber, Mohammed Y. Al-Hamad, "Trading Opportunities Forecasted Benefits at Peak Load for GCC Countries", Electric Power and Water Desalination Conference, 22-23 February 2016, Doha, Qatar.
5. M. AL-Hamad "Maximum Power Transfer from Photovoltaic Power Cell to Drive three Phases Induction Motors", University of Bahrain, Master thesis, 2005.
6. Renewable Energy Market Analysis, the GCC Region, IRENA – 2016.
7. Electricity and Water Authority. Tariff and charges rates collected in May 2016, Retrieved from http://www.mew.gov.bh/default.asp?
8. Energy Bill Increase to Hike Rents, Prices. Article retrieved from http://news.kuwaittimes.net /website/energy-bill-increase-hike-rents-prices/
9. My e-Portal, Tariff – MEDC collected in May 2016, Retrieved from http://www.electricity. com.om/tabid/178/Default.aspx
10. Qatar General Electricity & Water Corporation, Retrieved from http://www.km.com.qa/ customerservice/pages/tariff.aspx
11. Electricity & Co-Generation Regulatory Authority, Retrieved from www.ecra.gov.sa.
12. RSB Notification accessed from ADDC website, Retrieved from http://www.addc.ae/images/New%20Tariff%20Structure.pdf
13. Mohamed Y. AL-Hamad and Qahtan Hassan A.Alfaris, "Smart PV Grid to Reinforce the Electrical Network", Jordan Cigre National Committee (JCNC), 9-11 November 2010, Alzarqa, Kingdom of Jordan.
14. Battery Company: Northern Arizona Wind & Sun, 4091 East Huntington Drive, Suite B, Flagstaff, AZ 86004, http://www.windsun.com/Batteries/Battery_FAQ.htm.
15. Mohamed Y. AL-Hamad "Managing Transformers & Feeders During Minor Overloading In Primary Substations to Avoid Load Shedding", POWER-GEN Middle East 2008, 17th February 2009, Bahrain International Exhibition Centre.
16. Dick DeBlasio "IEEE P2030 Smart Grid Interoperability Standards Development Kick-Off Meeting Hosted by INTEL Corporation", Santa Clara, California, June 3-5, 2009.
17. Sam Pothecary, Abu Dhabi: Three world record bids entered for Sweihan solar project, PC Magazine, 20th September 2016, Article, Retrieved from http://www.pv-magazine.com/news/details/beitrag/abu-dhabi--three-world-record-bids-entered-for-sweihan-solar-project_100026191/#ixzz4RChVGe23

Sustainable recycling technologies for Solar PV off-grid system

Bhavesh Uppal, Adish Tamboli, Nandan Wubhayavedantapuram

Global Production Engineering, Technical University Berlin,
Department Assembly Technology and Factory Management,
Pascalstr. 8-9, D – 10587, Berlin, Germany
Email: b.uppal@campus.tu-berlin.de, adishtambo@gmail.com, wsnandan@gmail.com

Abstract

Policy makers throughout the world have accepted climate change as a repercussion of fossil fuel exploitation. This has led the governments to integrate renewable energy streams in their national energy mix. PV off-grid Systems have been at the forefront of this transition because of their permanently increasing efficiency and cost effectiveness. These systems are expected to produce large amount of different waste streams at the end of their lifetime. It is important that these waste streams should be recycled because of the lack of available resources. Our study found that separate researches have been carried out to increase the efficiencies of recycling of individual PV system components but there is a lack of a comprehensive methodical research which details efficient and sustainable recycling processes for the entire PV off-grid system. This paper reviews the current and future recycling technologies for PV off-grid systems and presents a scheme of the most sustainable recycling technologies which have the potential for adoption. Full Recovery End-of-Life Photovoltaic (FRELP) recycling technology can offer opportunities to sustainably recycle crystalline silicon PV modules. Electro-hydrometallurgical process & Vacuum technologies can be used for recovering lead from lead acid batteries with a high recovery rate. The metals in the WEEE can be recycled by using a combination of biometallurgical technology, vacuum metallurgical technology and other advanced metallurgical technologies (utrasonical, mechano-chemical technology) while the plastic components can be effectively recycled without separation by using compatibilizers. All these advanced technologies when used in combination with each other provide sustainable recycling options for growing PV off-grid systems waste. These promising technologies still need further improvement and require proper integration techniques before implementation.

Keywords: Recycling, PV off-grid system, Sustainability, Solar panels, Lead acid battery, WEEE

Introduction

Photovoltaic systems are becoming increasingly affordable for producing electricity. The total installed capacity of PV systems is estimated around 222 GW at the end of 2015 [1]. Crystalline-silicon PV technology dominates the market with 85–90% of the share because of its low cost and high efficiency [2]. Off-grid PV systems can be used for both domestic and low energy non-domestic purpose. Globally around 6 million homes were electrified with off-grid PV systems till 2015 [3]. These systems generally consist of PV module, lead acid battery and other electrical and electronic equipments. Many photovoltaic systems are at their end phase of life cycle and a lot of e-waste will be generated in the coming years. The average lifetime of crystalline silicon photovoltaic module is about 25–30 years. The global cumulative PV panel waste is expected to reach 60 million tons by the end of 2050 [1]. Average life of lead acid battery depends on its usage cycle. For any off-grid PV system till the end-of-life of PV module, generally 2-3 lead acid battery replacements are required. Inverters, charge controllers etc. if installed, operated and maintained properly may run through till the end-of-life of PV module. Our study found that separate researches have been carried out to increase efficiency of recycling of different PV system components. But there is no concrete methodical research available which will explain efficient and sustainable recycling processes for the entire PV off-grid system. In developing countries where off-grid system installations are increasing at a high rate, there is no industrial network available for recycling of PV off-grid system components in efficient and environmental friendly way. The specific combination of all sustainable and efficient recycling techniques for major off-grid system components is needed to analyze recycling efficiency and environmental impacts of whole system. This paper first evaluates possible recycling techniques for recovery of materials from major components of PV off-grid solar system. After the evaluation of technologies, a framework was developed which focused on the most efficient and sustainable processes for recycling of major PV off-grid system components.

Literature review

Extensive research has been carried out on recycling PV modules as first generation of PV modules are in their end-of-life phase. Latunussa et al. [4] mentioned that recycling of silicon PV modules can be carried out by physical/chemical treatment or combination of both. In their research, they focused on innovative process called FRELP for recycling of Si PV module. Granata et al. [5] studied in detail about various physical treatments for recycling of Si and CdTe photovoltaic modules. Their analysis presented that recycling by physical treatment with two blade rotors crushing followed by hammer crushing and eventually by a thermal treatment was the best option aiming to a direct recovery of glass. Ostrowski et al. [6] researched on various chemical treatment of recycling for crystalline silicon PV module. In their research, a series of etching processes were carried which includes etching of electric connectors, anti-reflective coating and n-p junction. Kang et al. [2] presented that 86% silicon can be recovered from any Si- PV module through most efficient

 DOI: 10.1051/e3sconf/20172301003

recycling techniques. Ashfaq et al. [7] analyzed that about 90% of the materials recovered from solar panels can be recycled into useful products.

Many developments are being carried out in recycling lead acid batteries by emphasizing new technological methods. Espinosa et al. [8] studied about the current methods for battery recycling like separation of components, pyrometallurgy and hydrometallurgy. Their research presented about the different technological process available for recycling the batteries. Jolly et al. [9] studied about the process of recycling lead acid batteries by separation process i.e. battery breaking & smelting for production of lead and polypropylene. Blanpain et al. [10] analysed the advantages and drawbacks of the different recycling technologies. Ferracin et al. [11] studied the lead recovery from lead acid batteries using an electro-hydrometallurgical process. Lin et al. [12] conducted experimental analysis of vacuum process for recycling of waste lead storage battery. Zhang et al. [13] studied about the practices in secondary lead recycling technologies (Electrowinning) and development of sustainable greener recycling methods, such as Paste-to-Paste recycling and hydrogen-lead oxide fuel cell method. Chen et al. [14] analysed the environmental impacts of lead acid batteries. Their research suggested the environmental risk assessment methodology for the lead acid batteries. Zabaniotou et al. [15] studied about the application of modern recycling technologies in accordance to the European and Greek legislation aiming at the recovery of lead; polypropylene and sulphuric acid from spend lead acid batteries.

Research has also been carried out to understand the amount of E-waste or WEEE being produced every year, the components of E-waste, the methods to recycle them and the problems associated with recycling E-waste. Widmer et al. [16] presented the global outlook of the problems associated with handling large amounts of E-waste. Their research reviewed the legal frameworks of different countries to tackle e-waste management issue. Ongondo et al. [17] reviewed the global WEEE in the year 2010 in different countries. Their paper discussed on the common priority areas for WEEE across the globe and the future WEEE generation, treatment, prevention and regulation activity. Zhang et al. [18] reviewed the latest recycling technologies to extract metals from WEEE. They elaborated on the shortfalls of each technology from technical and environmental perspective and discussed the future development of recycling technologies for extraction of metals from WEEE. Makenji et al. [19] examined in detail the mechanical recycling of plastics from WEEE. The effectiveness of mechanical separation techniques like floating, hydro-cyclone, air tabling, near infra-red and electrostatic sorting for recycling plastics from WEEE was presented in detail. Vazquez et al. [20] studied the effect of the use of campatibilizers on the recycling of plastic waste from WEEE. Their study suggested an improvement in the mechanical properties of preliminary WEEE plastics through use of campatibilizers. Chi et al. [21] reviewed low cost informal practices of e-waste recycling with a special focus on China and the problems associated with informal e-waste recycling techniques. Huang et al. [22] studied the health hazards for residents around e-waste recycling sites. Their research suggested that the incremental lifetime cancer risk of five carcinogenic metals Cr, Co, Ni, As and Cd was higher than acceptable range.

Method

Selection and classification of the research methods were divided into three steps: The first step was to conduct a literature review by a comprehensive search on Sciencedirect and online web database with keywords "recycle" and "technology" and "crystalline silicon module" and "WEEE" and "Lead acid battery". The abstract of the relevant papers were studied and the literatures were classified depending on the component of recycling. In second step, Recycling technologies were studied for crystalline silicon PV modules, lead acid batteries and WEEE. These technologies were evaluated based on sustainability in terms of economic, environmental and social perspectives. In the last step, the most promising technologies that address efficient and sustainable recycling ways were selected and presented in a scheme. Finally, future recommendations for improvements were suggested.

Photovoltaic Module

Crystalline silicon-type PV module consists of glass, aluminum frame, EVA coating, solar cells and terminal box [5]. Following table explains various physical /chemical treatments or their combination for recovery of various materials such as aluminum, glass, silicon, silver, copper, Ethylene vinyl acetate, plastics etc.

Table 1: Various Processes developed for Recycling of Crystalline PV panels [4]

Studied Process	Outcome of the process
Pyrolysis (in a conveyer belt furnace or in a fluidized bed reactor)	Separation up to 80% of wafers and almost 100% of glass

 DOI: 10.1051/e3sconf/20172301003

EVA removed by dissolution in organic solvents	Silicon cell can be separated without any damage from single cell module
Pyrolysis process of EVA at different heating rates under different oxidizing atmosphere	The pyrolysis behavior of EVA is strongly influenced by the content of acetate in the EVA
Thermal and chemical process applied to a large sample of module	The process achieved a separation yield of about 76% of the cells
Two-step heating process of EVA followed by chemical processes with acid and alkali	85% of copper and 62% of the silicon were separated
Dissolution of EVA by organic solvents and treatment of the PV cell by chemical etching	The process allowed to recover up to 86% of the silicon with very high purity
Two processes based on two blade rotors crushing followed by thermal treatment and two blade rotors crushing followed by hammer crushing	Recover 80–85% of the glass
FRELP (Full Recovery End Life Photovoltaic	Up to 95% recovery of silicon

Full Recovery End Life Photovoltaic (FRELP):

A recent research project has been financed by the EU and in partnership with PV Cycle Italy, titled "Full Recovery End of Life Photovoltaic project–FRELP", aiming at maximizing the recycling of the different materials of a silicon PV panel. The FRELP project had the objective of developing an innovative recycling process (successively defined as 'FRELP process) for Crystalline Si PV waste aiming at maximizing the recovery of all the material. This method implies combination of various methods with innovative approach (explained in table 1) to recover solar panel component materials [2]. At first all waste PV panels are collected at most optimized location by consideration of transport. Transport is generally ignored in recycling calculations. FRELP aims to choose optimized location for recycling, collecting and landfill sites. Automated system with robots is utilized to dismantle PV waste panel. In this phase, aluminum frame, cables, photovoltaic cell, glass, plastic parts are separated. Glass is separated by mechanical detachment with prior treatment by short and medium

 DOI: 10.1051/e3sconf/20172301003

wave infra-red to PV panels. By using sieving and optical based separation system, glass is separated from impurities. Incineration process is used for recovery of PV cell materials such as aluminum and silicon. Output of incineration process is treated with acid (HNO_3) leaching process. This process separates metal and silicon from incineration ash. With leaching process, up to 95% of silicon can be recovered. Lastly, electrolysis is used to recover silver and copper from leaching process solution and waste sludge is collected to landfill it [2].

Lead Acid Batteries

Lead Acid Batteries (LABs) are the most commonly used storage unit in the off-grid solar PV systems due to low cost and high energy density. About 97% of all lead-acid batteries are recycled today and have highest recycling rate compared to other consumer products [23]. The recycling of LABs includes recovery of valuable materials like lead, polypropylene, sodium sulphate crystals and lead pastes and helps to reduce the environmental effects [12]. Generally, discarded LABs are the major resource of secondary lead; more than 85% in the total amount of secondary lead is from LABs [10]. For recycling purpose, analyses of major components of the LABs are considered. The current recycling practices are considered for evaluation of the recyclability of LABs in off-grid solar PV systems. The recyclability rate of the batteries depends on the type of the process used. In addition, environmental aspects are also considered to address the sustainable recycling techniques. The methods are as follows:

(1) Pyrometallurgical Process: Pyrometallurgical process is the most common method that is currently in practice for recycling of LABs. The LABs are smelted in the furnace to produce antimonial lead bullion and soda slags or mattes. Many developments were done in this process which includes the utilization of different type of furnaces and preheating equipment [13]. Recent technological methods in the smelting process of LAB's are QSL process, Isasmelt Paste Smelting process, Kivcet method, Kaldo smelting process and Rotary or Reverberatory furnace smelting. The recyclability rate of the lead acid battery by Pyrometallurgical method is 90%. Although, the process is efficient it has severe environmental effects due to SO_2 emissions and lead emissions from fuming at elevated temperatures [12].

(2) Electro-hydrometallurgical Process: In recent years, considering the environmental impact by recycling of LABs by pyrometallurgy process, a cleaner technology, electro-hydrometallurgical technique was developed. The process employs alternative techniques instead of smelting like dissolving the

DOI: 10.1051/e3sconf/20172301003

lead sludge in acidic or basic media. In this process, the lead acid batteries are crushed and sulphuric acid, plastic, metallic and non-metallic portions are separated. The plastic components are reprocessed again and the metallic components are re-melted. The battery sludge which contains major lead compounds undergoes desulphurization process. After desulphurization process, the leaching process is carried out using a suitable electrolyte to solubilize the lead compounds remaining in the residue. Finally, the lead is then recovered from solution by electrowinning technique using insoluble anodes [11].

(3) Vacuum Process: One of the recent developments in LABs recycling includes the utilization of vacuum technologies. The process includes the separation of the LABs into 3 components: (1) Plastic shell and cover; (2) Lead grids, (3) Lead pastes [12]. Initially, the grids are melted under vacuum to generate lead bullions by separating the oxidized component of the lead. The lead bullion produced by this process is of high purity. The lead pastes are also reduced in vacuum to generate lead bullion. The plastic components are recycled using pyrolysis process to generate pyrolysis oil and pyrolysis gas. As per the experimental research carried out by Deqiang et al., the lead bullion recovery ratio is 96.29% from grids and 98.98% from lead pastes. Thus, the vacuum technologies control emissions, metal loss and energy consumption.

Waste Electrical and Electronic Equipment

The WEEE of a typical off-grid PV system consists of but not limited to inverters, charge controllers, cables, switches, fuses, disconnects, MCB's, protection devices etc. Traditional WEEE disposing techniques include disassembly, classification, sorting and inefficient material extraction processes. For recycling metals, the traditional methods are incineration, separation through hydraulic bed shaking and acid leaching process. Incineration involves burning WEEE directly. It is a tedious process with different sub-processes and treatment techniques to extract the material. Through hydraulic bed shaking only crude copper particles can be extracted. However, this process generates a lot of waste water and residues. This can lead to secondary pollution if not handled carefully. The use of acids in acid leaching process of extraction can cause major environmental hazards and safety issues if not handled cautiously. These techniques do not provide value addition as precious metals are not extracted efficiently leading to large waste and bio hazards [18]. When different plastic wastes are accumulated together, it becomes economically unviable to separate them and recycle. This is because sorting techniques are labour intensive in their present state [19]. Recycling process can be considered successful only when it is easy to perform, is effective in recovering the maximum amount of material, is environmentally friendly and does not cause any health hazards. Over the years new and improved recycling technologies have been developed which encompass some of the above characteristics. Table 2 lists all kinds of improved technologies for recycling metals from WEEE along with their advantages and disadvantages with regards to environmental and technological aspect:

Table 2: Recycling technologies for WEEE [18]

Recycling technology	State of recovery metals	Species and effect of recovery metals	Advantages of environmental and technological aspect	Disadvantages of environmental and technological aspect
Pyrometallurgical technology	Pure solid metals	Almost all metals; high recovery rate	Totally-commercial; owing some dust chamber and exhaust gas treatment plant	High investment; fine particles pollution; High energy consumption; difficult to recover precious metals
Mild extracting technology	Solution	Almost all metals; recovery rate related to the reagents and reaction condition	Low toxicity; simple and easy accessibility; relative low environment damage	High cost of some reagent; some reagents could result in waste water pollution
Biometallurgical Technology	Solution	For few specific metals; considerable recovery rate for Cu, Zn, Au, etc.	Environment-friendly; low cost of investment	For few specific metals (Cu, Zn, Au); no industrialization
Electrochemical technology	Pure solid metals	Only for specific metals; high recovery rate	High recovery efficiency; low investment; mature technology	Waste water and residue pollution; low industrialization;
Supercritical technology	Solid mixture of metals	Almost all metals; high recovery rate	High recovery efficiency; low cost of investment; low environment damage	Failure to recycle single metal; waste oil and waste gas pollution; low industrialization
Vacuum metallurgical technology	Solid single metal	Only for high vapour pressure metals; high recovery rate	Environment-friendly; short technological process	Relative high cost of investment; low industrialization; immature technology

DOI: 10.1051/e3sconf/20172301003

Other metallurgical technologies (ultrasonical, mechanochemical technology)	Solid mixture of metals/ solution	Need to combined with other technologies to recycle metal	Environment-friendly; low cost of investment	Failure to recycle single metal; no industrialization

Plastic from WEEE manly consists of thermoplastic copolymers like ABS (Acrylonitrile-Butadiene-Styrene), HIPS (High Impact Polystyrene), PC (Polycarbonate) etc. Separating these plastics is a labor intensive and tedious task. Automated methods like spectroscopic techniques are effective but they are expensive and difficult in handling. Thus, it is beneficial if the entire plastic waste can be treated together. This type of recycling can be achieved using compatabilizers by making a blend. Compatibilization by addition is the most popular technique because of the ease to process. Compatibilizers like SEBS (Styrene-Ethylene-Butylene-Styrene), SBS (Styrene-Butadiene-Styrene) and SBR (Styrene-Butadiene-Rubber) are used for the blends of ABS and HIPS and their selection is based on the similarity of their structure with the plastics. Miscibility of the compatibilizer with the plastic blend is the main factor that is responsible for the choosing them. Through this method, compatibilization blend can be obtained with double ultimate strength and modulus with regards to the physical blend. This technique also improves the mechanical properties of initial plastic waste [20].

Discussion

Though, extensive research has been carried out on the recycling of individual components of a solar PV system, a research on the whole PV off-grid system recycling is needed. In this study, a sustainable recycling model for solar PV off-grid systems was developed. The model creation was based on the selection of most environmentally friendly and economically feasible recycling practices for the individual components of the system. Fig1. shows the schematic view of sustainable recycling technologies for PV off-grid system.

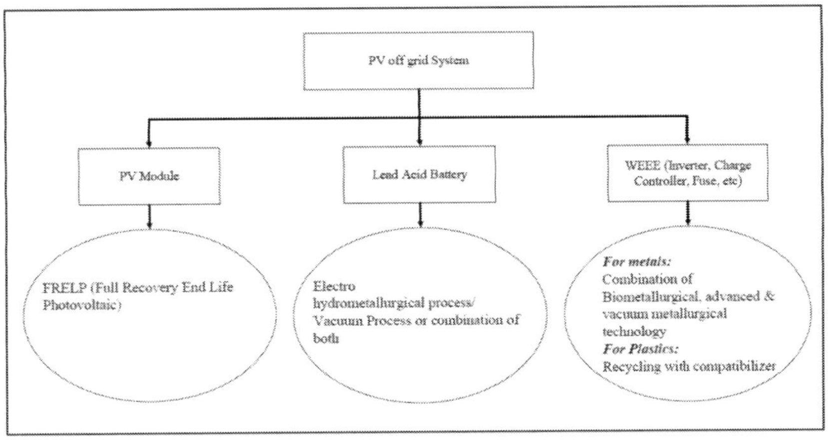

 DOI: 10.1051/e3sconf/20172301003

Fig. 1. Sustainable recycling technologies for Solar PV off-grid systems

For PV module, FRELP is considered as the most sustainable recycling technique from the available recycling methods. Latunussa et al. performed Life Cycle Assessment (LCA) method to analyze impacts of recycling of silicon PV modules with FRELP method. LCA analysis focuses mainly sub processes such as transport of the PV waste to the site, the incineration processes, and the further metal recovery from the bottom ash. Result of LCA study on FRELP method states that this method causes less environmental impact with considerable reduction of energy consumption and low CO2 emissions. FRELP process targets full recovery of entire solar panel component materials and their exploitation for Re-use. Up to 98% by weight of components of solar panel can be converted into salable product [24]. In case of lead acid batteries, Electro-hydrometallurgical process & Vacuum technologies offer better recycling techniques than the traditional Pyrometallurgical process due to reduced emissions and higher lead recovery rates. However, there are still constraints in the electro-hydrometallurgical process which includes additional costs for the chemicals used for electrolytes and deterioration of anodes in electrowinning process. On the other side, vacuum technologies offer more opportunities to avoid metal loss, environmentally cleaner technology with less emissions and consuming less energy for the processing of lead acid batteries. The choice of technology also depends on the other components of the PV off-grid system components like WEEE elements for recycling, where the plastic components from the lead acid battery can be combined with the plastic waste of WEEE. For WEEE, Biometallurgical technology, vacuum metallurgical technology and advanced metallurgical technologies (ultrasonical, mechanochemical technology) provide the right solutions for environmentally friendly recycling with a considerable recovery rate of the material. However, these technologies are either material selective or can recover metals with high vapour pressure or needs to be combined with other recycling technologies. On the positive side, these technologies have a low investment cost except for vacuum metallurgical technology, thus should be encouraged to be easily industrialized and refined. Since WEEE is an extremely complicated system, it is difficult for a single technology to provide all the solutions. Use of compatibilzers like SEBS, SBS, and SBR offer tremendous opportunities to avoid large plastic waste handling and improve the initial material properties. These technologies will be further developed hence combining these technologies to complement each other provides a good sustainable option for recycling in the future.

Conclusion

The environmental effects of fossil fuels have led the governments all over the world to change their policies to integrate renewable energies in the energy mix. Climate change is real, so more and more people are opting for solar energy as a means of sustainable electricity supply. This has resulted to a great increase in the number of PV off-grid systems being sold and this trend will continue in the future so long as the technology becomes cheaper, more efficient and easily available. The first generation of PV off -grid systems are at an end of their life span. They will generate a lot of waste and such waste streams will only increase in the future. Traditional methods of recycling e-waste do not provide all the answers to solve this growing problem. These methods are energy intensive, expensive, inefficient and create environmental and health hazards. Current and future recycling technologies for PV off-grid systems are presented in this review. PV off-grid systems are complicated and contain a lot of different parts. For solar modules, through FRELP 95% of silicon can be recovered and 98% by weight of components can be converted into products that can be sold. LAB recycling through Electro-hydrometallurgical process & Vacuum technologies offer eco-friendly recycling methods with excellent recovery rates like 98% lead recovery. Technologies like vacuum metallurgical technology, ultrasonical technology offer great metal recovery solutions from WEEE and can be industrialized if nurtured properly. All these advanced technologies when used in combination with each other provide sustainable recycling methods for growing PV-off grid systems waste. It is important to continuously research on these technologies to further improve their efficiencies and reduce cost so that they can be successfully realized in the future. More work should be done on ways of smoothly integrating these technologies with each other. Thus, it can be said that future recycling technologies for PV off-grid systems will meet the environmental, economic and social needs successfully.

References

[1] IRENA, IEA-PVPS (2016): End-of-life management solar photovoltaic panels. International Renewable Energy Agency and the International Energy Agency Photovoltaic Power Systems (ISBN 978-92-95111-99-8). Available online at http://www.irena.org/DocumentDownloads/Publications/IRENA_IEAPVPS_End-of-Life_Solar_PV_Panels_2016.pdf.

[2] Kang, Sukmin; Yoo, Sungyeol; Lee, Jina; Boo, Bonghyun; Ryu, Hojin (2012): Experimental investigations for recycling of silicon and glass from waste photovoltaic modules. In *Renewable Energy* 47, pp. 152–159. DOI: 10.1016/j.renene.2012.04.030.

[3] IRENA (2015): Off-grid renewable energy systems: status and methodological issues. International Renewable Energy Agency. Available online at http://www.irena.org/DocumentDownloads/Publications/IRENA_Off-grid_Renewable_Systems_WP_2015.pdf.

[4] Latunussa, Cynthia E.L.; Ardente, Fulvio; Blengini, Gian Andrea; Mancini, Lucia (2016): Life Cycle Assessment of an innovative recycling process for crystalline silicon photovoltaic panels. In *Solar Energy Materials and Solar Cells. DOI:* 10.1016/j.solmat.2016.03.020.

[5] Granata, G.; Pagnanelli, F.; Moscardini, E.; Havlik, T.; Toro, L. (2014): Recycling of photovoltaic panels by physical operations. In *Solar Energy Materials and Solar Cells* 123, pp. 239–248. DOI: 10.1016/j.solmat.2014.01.012.

[6] Klugmann-Radziemska, Ewa; Ostrowski, Piotr (2010): Chemical treatment of crystalline silicon solar cells as a method of recovering pure silicon from photovoltaic modules. In *Renewable Energy* 35 (8), pp. 1751–1759. DOI: 10.1016/j.renene.2009.11.031.

[7] Ashfaq, Haroon; Hussain, Ikhlaq; Giri, Ajay: Comparative analysis of old, recycled and new PV modules. In *Journal of King Saud University - Engineering Sciences. DOI:* 10.1016/j.jksues.2014.08.004.

[8] Espinosa, Denise Crocce Romano; Bernardes, Andréa Moura; Tenório, Jorge Alberto Soares (2004): An overview on the current processes for the recycling of batteries. In *Journal of Power Sources* 135 (1–2), pp. 311–319. DOI: 10.1016/j.jpowsour.2004.03.083.

[9] Jolly, R.; Rhin, C. (1994): Recycling of Materials in Industry The recycling of lead-acid batteries: production of lead and polypropylene. In *Resources, Conservation and Recycling* 10 (1), pp. 137–143. DOI: 10.1016/0921-3449(94)90046-9.

[10] Blanpain, Bart; Arnout, Sander; Chintinne, Mathias; R. Swinbourne, Douglas (2014): Chapter 8 - Lead Recycling. In : Handbook of Recycling. Boston: Elsevier, pp. 95–111.

[11] Ferracin, Luiz C.; Chácon-Sanhueza, Abel E.; Davoglio, Rogério A.; Rocha, Luis O.; Caffeu, Daniele J.; Fontanetti, Adilson R. et al. (2002): Lead recovery from a typical Brazilian sludge of exhausted lead-acid batteries using an electrohydrometallurgical process. In *Hydrometallurgy* 65 (2–3), pp. 137–144. DOI: 10.1016/S0304-386X(02)00087-7.

[12] Lin, Deqiang; Qiu, Keqiang (2011): Recycling of waste lead storage battery by vacuum methods. In *Waste Management* 31 (7), pp. 1547–1552. DOI: 10.1016/j.wasman.2011.02.019.

[13] Zhang, Wei; Yang, Jiakuan; Wu, Xu; Hu, Yuchen; Yu, Wenhao; Wang, Junxiong et al. (2016): A critical review on secondary lead recycling technology and its prospect. In *Renewable and Sustainable Energy Reviews* 61, pp. 108–122. DOI: 10.1016/j.rser.2016.03.046.

[14] Zhang, Jing; Chen, Chuanmin; Zhang, Xueying; Liu, Songtao (2016): Study on the Environmental Risk Assessment of Lead-Acid Batteries. In *Selected Proceedings of the Tenth International Conference on Waste Management and Technology* 31, pp. 873–879. DOI: 10.1016/j.proenv.2016.02.103.

[15] Zabaniotou, A.; Kouskoumvekaki, E.; Sanopoulos, D. (1999): Recycling of spent lead/acid batteries: the case of Greece. In *Resources, Conservation and Recycling* 25 (3–4), pp. 301–317. DOI: 10.1016/S0921-3449(98)00071-8.

[16] Widmer, Rolf; Oswald-Krapf, Heidi; Sinha-Khetriwal, Deepali; Schnellmann, Max; Böni, Heinz (2005): Global perspectives on e-waste. In *Environmental and Social Impacts of Electronic Waste Recycling* 25 (5), pp. 436–458. DOI: 10.1016/j.eiar.2005.04.001.

[17] Ongondo, F. O.; Williams, I. D.; Cherrett, T. J. (2011): How are WEEE doing? A global review of the management of electrical and electronic wastes. In *Waste Management* 31 (4), pp. 714–730. DOI: 10.1016/j.wasman.2010.10.023.

[18] Zhang, Lingen; Xu, Zhenming (2016): A review of current progress of recycling technologies for metals from waste electrical and electronic equipment. In *Journal of Cleaner Production* 127, pp. 19–36. DOI: 10.1016/j.jclepro.2016.04.

[19] Makenji, K.; Savage, M. (2012): 10 - Mechanical methods of recycling plastics from WEEE. In : Waste Electrical and Electronic Equipment (WEEE) Handbook : Woodhead Publishing Series in Electronic and Optical Materials: Woodhead Publishing, pp. 212–238.

[20] Vazquez, Yamila V.; Barbosa, Silvia E. (2016): Recycling of mixed plastic waste from electrical and electronic equipment. Added value by compatibilization. In *Waste Management* 53, pp. 196–203. DOI: 10.1016/j.wasman.2016.04.022.004.

[21] Chi, Xinwen; Streicher-Porte, Martin; Wang, Mark Y.L.; Reuter, Markus A. (2011): Informal electronic waste recycling: A sector review with special focus on China. In *Waste Management* 31 (4), pp. 731–742. DOI: 10.1016/j.wasman.2010.11.006.

[22] Huang, Chun-Li; Bao, Lian-Jun; Luo, Pei; Wang, Zhao-Yi; Li, Shao-Meng; Zeng, Eddy Y. (2016): Potential health risk for residents around a typical e-waste recycling zone via

DOI: 10.1051/e3sconf/20172301003

inhalation of size-fractionated particle-bound heavy metals. In *Journal of Hazardous Materials* 317, pp. 449–456. DOI: 10.1016/j.jhazmat.2016.05.081.

[23] BCI (2016): How a Lead Battery is Recycled. Battery Council International. Available online at http://batterycouncil.org/?page=Battery_Recycling.

[24] SASIL (2010): FRELP Project. SASIL. Available online at http://www.sasil-life.com/index.php?option=com_content&view=article&id=85&Itemid=83&lang=en.

Uncertainty Model for Total Solar Irradiance Estimation on Australian Rooftops

Hassan Al-Saadi, Rastko Zivanovic & Said Al-Sarawi

School of Electrical and Electronic Engineering, The University of Adelaide

Adelaide SA 5000, Australia.

hassan.al-saadi, rastko.zivanovic, said.alsarawi}@adelaide.edu.au

Abstract.

The installations of solar panels on Australian rooftops have been in rise for the last few years, especially in the urban areas. This motivates academic researchers, distribution network operators and engineers to accurately address the level of uncertainty resulting from grid-connected solar panels. The main source of uncertainty is the intermittent nature of radiation, therefore, this paper presents a new model to estimate the total radiation incident on a tilted solar panel. Where a probability distribution factorizes clearness index, the model is driven upon clearness index with special attention being paid for Australia with the utilization of best-fit-correlation for diffuse fraction. The assessment of the model validity is achieved with the adoption of four goodness-of-fit techniques. In addition, the Quasi Monte Carlo and sparse grid methods are used as sampling and uncertainty computation tools, respectively. High resolution data resolution of solar irradiations for Adelaide city were used for this assessment, with an outcome indicating a satisfactory agreement between actual data variation and model.

Keywords- Solar irradiation prediction, tilted surface, goodness of fit, diffuse fraction correlation.

Introduction

The current prevalence of solar rooftops in Australia incentivizes academic researchers, engineers and power network operators to investigate in detail solar engineering applications. As part of this tendency, solar irradiance prediction has been given a tremendous interest and revealed different approaches and methods. The importance of such studies lies in the fact that, to date, a small portion of Photovoltaic (PV) rooftops are equipped with real-time communication, measurement and control devices, resulting in increased cost [1]. Nonetheless, the majority of PV installations are not observable and controllable. While the availability of PV measurement and control devices is subject to high costs, it is difficult to collect the required specification information from each unit. Meanwhile, solar energy reaching the ground can be predicted with high accuracy

 DOI: 10.1051/e3sconf/20172301004

by the utilization of statistical estimation functions at a given time for a predefined location, as discussed hereinafter. Therefore, obtaining a prior accurate knowledge on solar radiation can help in improving the quality of feed-in energy, reducing the ancillary costs of weather forecast equipment, mitigating the negative impacts of high PV penetration, especially in residential areas, storage quantifying and a better integration of PVs into the so-called "smart grid".

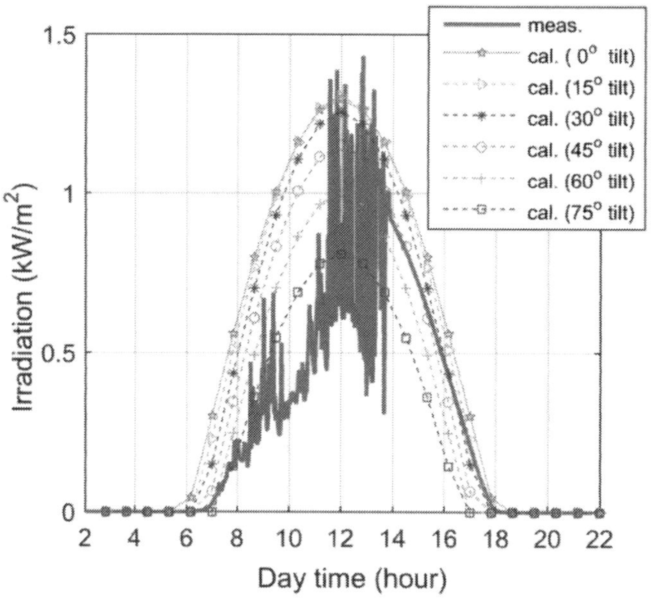

Figure 1 Measured data of one day in January 2016, compared with calculations of expected values of probabilistic irradiation model in hourly evaluations.

The major step to overcome the adversity of uncertainty level is the use of several irradiance representation models in which strengths and weaknesses of each model can be examined as well as their complexity. In this paper, a new model for uncertainty prediction of solar irradiation incident on a tilted surface is presented. Where such model is yet to be satisfactorily modelled, this model is built upon a given characterization of clearness index and diffuse fraction of radiation while taking into consideration the Australian climate conditions. Since the index and the fraction are proven to have a functional relationship [2], the properties of clearness index can dictate the level of uncertainty in the total incident radiation. In this regard, the probabilistic evaluation of uncertainty, given probability density function (PDF) similar to the one given in [3], can specify the amount of the ground-reaching radiation.
The contribution of this work is to provide a simple and easy to use model that can be incorporated or adapted into other models to stimulate interdisciplinary research and system engineering applications. The uncertainty modelling of irradiation is presented in the following section with the emphasis on total irradiation estimation with the embedment of solar variation. In Section 3, four different statistical tools, namely root mean square error (RMSE), mean bias error (MBE), Kolmogorov-Smirnov test (KST) and correlation coefficient (R^2) are used as part of this model assessment. Then, the implementation of the presented model is carried out and assessed with high resolution solar radiation data for two different seasonal days at several day times in Adelaide are discussed in Section 4. The last section presents a summary of this research outcomes.

Irradiation Uncertainty Modelling

Total Irradiation

The total radiation incident on a tilted surface, I_T, with the assumption of anisotropic sky, can be rewritten as:

$$I_T = \frac{k_t}{I_g} \cdot [\left(I_b I_o + I_d\, I_b\right) R_b + I_d (I_o - I_b)\left(\frac{1+\cos\beta}{2}\right)\left[1 + \sqrt{\frac{I_b}{I_g}} \sin^3\frac{\beta}{2}\right]] + I_g\, \rho\, \left(\frac{1-\cos\beta}{2}\right). \quad (1)$$

In this model, circumsolar diffuse, horizon brightening as well as anisotropy factor are incorporated into the isotropic model resulting in the so called "HDKR" model, named after the initials of Hay, Davies, Klucher and Reindl researchers [4]. The I_o, I_g, I_b and I_d are the exteritorial, global, beam and diffuse irradiances, respectively; R_b is Ratio of beam radiation on tilted surface to that on horizontal surface. The tilt

DOI: 10.1051/e3sconf/20172301004

angle is denoted by β and the reflectance ground factor is ρ. k_t is the hourly clearness index.

Solar Variation

For the variations of solar radiation, two vital factors have been identified to be representing the clearness of the sky: clearness index, k_t, and diffuse fraction, k_d. These quantities are highly correlated with different functions established on the subject, such as piecewise linear [2], piecewise 4th-order polynomial [5], piecewise 3rd-order polynomial [6], piecewise 2nd-order polynomial [7], logistic [8]. The logistic function is proven to be the best-fit based on Australian meteorological conditions. Later, the logistic function is further complicated adding more sensitive factors, for example, the effects of apparent solar time, persistence of global radiation level and daily clearness index. Providing the value of k_t, the k_d can be computed using the logistic function as:

$$k_d = \frac{1}{1+e^{-B(B\prime - k_t)}}, \tag{2}$$

where B and B' are specified parameters for the logistic based diffuse fraction that can be computed as demonstrated in [8].

The behavior of k_t is assumed to follow a special distribution with various approaches are proposed given the characteristics of the place, such as time standard meridian, number of the days in a year, longitude, latitude, altitude, and ground reflectance, further details can be found in [9]. In this paper, we choose to utilize the single Gamma distribution proposed by Holland and Huget [3] where the probability density is characterized by time and locations.

Uncertainty Model

As the discussed hereinbefore, it is possible to form an uncertainty model when the uncertainty behavior of k_t is known. Therefore, after rewritten (1) to be a function of k_t and k_d, the hourly total irradiation on a surface with tilt angle, β, can be expressed as:

$$I_T = \Big[R_b(n,t) + k_d\big[-R_b(n,t) + k_t R_b(n,t) - k_d k_t R_b(n,t) + c(k_d) - k_t c(k_d) + k_d k_t c(k_d)\big] + \rho \left(\frac{1-\cos\beta}{2}\right)\Big] \times H_o(n) \times r_d(n,t), \tag{3}$$

where

 DOI: 10.1051/e3sconf/20172301004

$$c(k_d) = \left(\frac{1+\cos\beta}{2}\right)\left[1+\sqrt{1-k_d}\,sin^3\frac{\beta}{2}\right].$$

The predefined quantities that function of n, which is the number of days, and t, day time, are shown with two brackets next to them while the uncertain quantity, k_t, is just indexed by t. In addition, the values of H_o are just indexed by n (*cf.* [4] & [3] for extra information).

For the initial assessment of the model's results, Figure 1 shows the hourly-calculated total irradiations (kW/m^2) for different tilt angles in a day for the city of Adelaide in comparison with the measured (one-minute) solar exposure data taken from the data portal webpage [10]. Herein, we utilize the Sparse Grid Technique [11] for the computations of the hourly expectation of I_T. 0 to 1 is the chosen domain for k_t as it is conventionally assumed. This is as the k_t is the ratio of terrestrial to extraterrestrial radiation idealy this should have a maximum value of 1, however if more influential factors are included such as the edges of the clouds the ratio could reach to 1.4 [9]. Regarding the figure and although the objectivity of assessing the model graphically is doubtable, a preliminary indication of the effectiveness of the present model can be drawn from these expectations.

Model's Assessment

The quality assessment of the modeled data is performed using goodness-of-fit techniques such as Root Mean Square Error (RMSE), Mean Bias Error (MBE), Coefficient of determination R^2 and Kolmogorov-Smirnov (KS) as in the following subsections. The usefulness of these metrics is to test the linear relationship between modelled and measured data.

Root mean square error

RMSE delivers a positive value with a tendency towards zero for ideal models. In the assessment of uncertainty models, the normalized version of RMSE with respect to predicted values is, herein, employed, as defined in [12] with objective justifications explained. Then, the short-term performance is evaluated as a smaller value means a better performance. The RMES value can be obtained using:

$$\text{RMSE} = \sqrt{\frac{1}{n}\sum_{i=1}^{n}\frac{Y_{i,calc.}-Y_{i,meas.}}{Y_{i,calc.}}}. \qquad (4)$$

where Y represents a set of data with the subscripts "$calc.$" and "$meas.$" represent calculated values from the model and measured values taken from meteorological stations. The subscript i denotes the i-th value of an hourly solar irradiation; n is the number of the hours from different days in a month across involved years.

DOI: 10.1051/e3sconf/20172301004

Mean bias error

The mean bias error (MBE) gives an indication of the average bias of the model. The value of MBE could be negative indicating the underestimation of tested model, as the desirable values are the low positive ones. Despite the popularity of this test, it is intuitively obvious that an overestimation may cancel an underestimation for individual observations, which is a well-recognized limitation of the MBS. According to [13], the formula of MBE with the normalization in respect to the calculated data can be expressed as:

$$\text{MBE} = \frac{1}{n}\sum_{i=1}^{n}\frac{Y_{i,calc.}-Y_{i,meas.}}{Y_{i,calc.}}, \quad (5)$$

The descriptions of the subscripts i, n, $calc.$ and $meas.$ are mentioned in the previous subsection.

Kolmogorov-Smirnov test

The Kolmogorov-Smirnov test (KST) involves the use of cumulative density functions (CDF), F, of both model and empirical data where the large difference is computed as follows:

$$\text{KST} = \sup_{1\le i\le n}\left|F_{i,meas.} - F_{i,calc.}\right|, \quad (6)$$

In (6), $F_{i,meas.}$ represents the CDF of empirical probability during i-th observation while $F_{i,calc.}$ stands for the CDF of thereortical model probability at i-th sample. The smallest value of KST indicates the goodness of fit of the proposed model.

The correlation coefficient (R^2)

The correlation coefficient, R^2, is the determination of linear relations between calculated and measured observations. In this test, the variance of errors to the variance of the calculated data is compared, examining the quality of the model. It can be computed as follows:

$$R^2 = 1 - \frac{\sum_{i=1}^{n}\left(Y_{i,meas.}-Y_{i,calc.}\right)^2}{\sum_{i=1}^{n}\left(Y_{i,meas.}-\bar{Y}_{meas.}\right)^2}, \quad (7)$$

DOI: 10.1051/e3sconf/20172301004

The descriptions of variable Y and the subscripts i, n, $calc.$ and $meas.$ are mentioned in the previous subsection; and $\bar{Y}_{i,meas.} = \sum_{i=1}^{n} Y_{i,meas.}$. The desirable result of this test is that whenever its value is close to a positive integer one.

Implementation and Discussion

As the proposed uncertainty model involves hourly clearness index calculation, the computation of monthly average daily clearness index is necessary providing only monthly average solar irradiation. Therefore, two monthly quantities are used for Jan. and July, 8.8 $Mj.m^{-2}$ and 24.4 $Mj.m^{-2}$ respectively. This is as we intend, in addition, to test the seasonal variations in the performance of the model. β is chosen to be $30°$ and $k_{t\,min}$ and $k_{t\,max}$ to be 0 and 1.4pu, respectively. Moreover, ground reflectance is 0.2pu. It is of importance to mention that for data sampling from the proposed model the Quasi Monte Carlo Method is utilized with the number of samplings is equal to the empirical data involved. In the meanwhile, the 1-min dataset of solar radiation can be obtained from Australian Bureau of Meteorology with the support of Australian Solar Institute (works under Australian Renewable Energy Agency - ARENA). The years considered in the current study are 95, 96, 97, 04, 05, 06, 07, 08, 10, 11, 14, 15, 16 for assessment of the model in January and 95, 96, 97, 04, 05, 06, 07, 08, 10, 11, 14, for the month July, resulting, for each hour, 24180 and 20460 measurements, respectively. The work is implemented using Matlab and MS Excel and run on Intel® i7-2600 4-core @ 3.4GHz processor.

In the implementation and assessment of the model, average hourly solar radiation is achieved for each consecutive minutes, which is the most common convention for characterizing solar radiations in short-time series analysis. From the presented simulation and result, it is clear that the presented model covers the most possible outcomes in solar radiation modelling in Australia. The statistics employing goodness-of-fit techniques mentioned in the previous section are shown in the

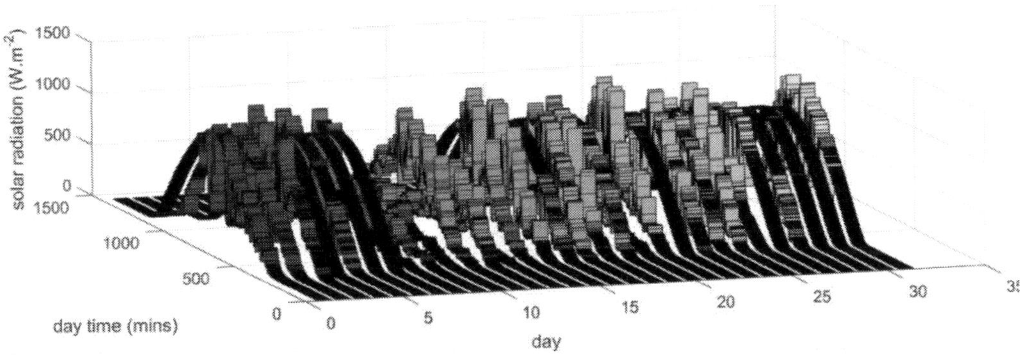

Figure 2 Empirical data of beam and diffuse solar radiations combined in Adelaide.

DOI: 10.1051/e3sconf/20172301004

Tables 1 and 2. In these two tables, model assessment accounted for seasonal variations are present where summer in Table 1 and winter in Table 2.

The statistical data from the assessment show various results as some are significantly noteworthy. The bolded results are distinguished among the rest to be desirable. It is clear that the presented modeling results are not consistent for individual day times, as in Table 1 & 2 (a), (b) or for n, as in Table 1 & 2 (c) has been given. However, there is a tendency that the model performs batter overall when $n = 30$, according to RMSE. In terms of MBE, the statistics obtained for certain days distributed around the month, (10, 15, 20, 25, 30), show an acceptable fit between measured data and modelled ones.

The outcome of the presented results are consistent with other studies such as in [14] where ∓10% is considered satisfactory. Based on the sign of MBE tests, no overestimation is noticed for the assessment of January while in July over-/under-estimations are obvious. KST and R^2 are implemented and compared with RMSE and MBE as, in Table 1(a), KST shows similar assessment with RMSE and MBE. This is while in Table 2(b) the KST and R^2 are in agreement with MBE at 12:00PM.

DOI: 10.1051/e3sconf/20172301004

Table 1. Statistics obtained employing RMSE, MBE, R^2 and KST for: a) different times of a day when $n = 15$ of January, b) different times of a day when $n = 30$ of January, c) an hour around noon of different n in January.

(a)

Day time	RMSE	MBE	KST	R^2
09:00AM	6.1535	-0.9744	0.0164	-1.9793
10:00AM	**3.7191**	**-0.6683**	**0.0105**	-1.1501
11:00AM	4.7896	-0.7512	0.0139	-1.0727
12:00PM	4.5223	-0.7629	0.0160	**-0.9765**

(b)

Day time	RMSE	MBE	KST	R^2
09:00AM	1.7170	**-0.1121**	0.0174	**-0.9286**
10:00AM	**1.0822**	-0.1416	0.0127	-1.1836
11:00AM	1.4021	-0.2211	**0.0121**	-1.0813
12:00PM	1.3748	-0.3022	0.0151	-1.2591

(c)

n	RMSE	MBE	KST	R^2
5	33.3050	-7.0233	0.0208	-5.9079
10	8.5701	-1.6641	0.0175	-2.0678
15	3.9784	-0.6463	0.0157	-0.8369
20	2.3555	-0.3349	**0.0149**	-0.5994
25	1.6203	**-0.2616**	0.0150	**-0.8457**
30	**1.3748**	-0.3022	0.0151	-1.2591

Table 2. Statistics obtained employing RMSE, MBE, R^2 and KST for: a) different times of a day when n = 15 of July, b) different times of a day when n = 30 of July., c) an hour around noon of different n in July.

(a)

Day time	RMSE	MBE	KST	R^2
09:00AM	5.4950	**-0.5668**	0.0470	-24.9514
10:00AM	8.4597	-0.8117	0.0442	-13.3852
11:00AM	**4.9798**	-0.6498	0.0375	-8.9466
12:00PM	8.1474	-0.8518	**0.0352**	**-7.2341**

(b)

Day time	RMSE	MBE	KST	R^2
09:00AM	0.8828	0.6268	0.0452	-28.1024
10:00AM	1.1153	0.5053	0.0432	-14.8841
11:00AM	**0.7712**	0.4496	0.0347	-10.3655
12:00PM	1.0730	**0.3727**	**0.0333**	**-9.3545**

(c)

n	RMSE	MBE	KST	R^2
5	294.0569	-45.9685	0.0394	-19.5802
10	41.0301	-6.4490	0.0362	-11.8846
15	10.2086	-1.2191	0.0353	-7.7100
20	3.7727	**-0.0592**	0.0348	**-6.3111**
25	1.8125	0.2873	0.0341	-6.9740
30	**1.0730**	0.3727	**0.0333**	-9.3545

Overall, the objectiveness of the goodness-of-fit techniques proves the acceptable quality of the proposed model. However, due to the limited range of empirical data to within few years a few "non-desirable" values of RMSE, MBE, R^2 and KST can be seen. Indicating that for a more accurate modelling there is a need to have fully evaluate solar prediction model for more years.

Conclusion

This paper presents a new uncertainty model for the estimation of total solar radiation reaching Australian rooftops. The uncertainty is characterized by the PDF of clearness index when the diffuse fraction is functionally correlated through the utilization of logistic relation. Four assessment techniques are employed to examine the effectiveness of the model with the involvement of high resolution data recorded for Adelaide.

The assessment covers the day hours for two different days taken seasonal variations into consideration. In this assessment, different day times give different outcomes when an hour around noon and three before noon. However, there is an agreement with root mean square error (RMSE), mean bias error (MBE) and Kolmogorov-Smirnov test (KST) in evaluating the model during the hour around 10:00AM when the number of the day is 15. While the agreement is apparent for MBE, KST and the correlation coefficient (R^2) when 12:00PM at 30 as a day's number in the model. Although the agreements do not exactly occur in all times and days, other assessment results are still within the acceptable range comparing with other work, mentioned in the main scripts.

The considerations that should be paid for future work are more assessments on other Australian cities. Moreover, Expanding the time domain with different day times could yield more reliable outcomes of the proposed model.

DOI: 10.1051/e3sconf/20172301004

References

[1] L. Cibulka, M. Brown, L. Miller and A. V. Meier, "User requirements and research needs for renewable generation forecasting tools that will meet the needs of the caiso and utilities for 2020," in *A White Paper Report Prepared by CIEE*, 2012.

[2] J. F. Orgill and K. G. T. Hollands, "Correlation equation for hourly diffuse radiation on a horizontal surface," *Sol. Energy,* vol. 19, no. 4, pp. 357-359, Dec. 1977.

[3] K. G. T. Hollands and R. G. Huget, "A probability density function for the clearness index, with applications," *Solar Energy,* vol. 30, no. 3, p. 195–209, Dec. 1983.

[4] J. A. Duffie and W. A. Beckman, Solar engineering of thermal processes, vol. 3, New York: Wiley, pp. ???, 2013.

[5] J. Chandrasekaran and S. Kumar, "Hourly diffuse fraction correlation at a tropical location," *Sol. Energy,* vol. 53, no. 6, pp. 505-510, Dec. 1994.

[6] S. Karatasou, M. Santamouri and V. Geros, "Analysis of experimental data on diffuse solar radiation in Athens, Greece, for building applications," *Int. J. Sustain. Energy,* vol. 23, no. (1-2), pp. 1-11, Mar. 2003.

[7] J. Soares, A. P. Oliveira, M. Z. Božnar, P. Mlakar, J. F. Escobedo and A. J. Machado, "Modeling hourly diffuse solar-radiation in the city of São Paulo using a neural-network technique," *Appl. Energy,* vol. 79, no. 2, pp. 201-214, Oct. 2004.

[8] J. Boland, B. Ridley and B. Brown, "Models of diffuse solar radiation," *Renew. Energy,* vol. 33, no. 4, pp. 575-584, Apr. 2008.

[9] K. T. Hollands and H. Suehrcke, "A three-state model for the probability distribution of instantaneous solar radiation, with applications." *Solar Energy,* vol. 96, pp. 103-112, Oct. 2013.

[10] [Online]. Available: http://reg.bom.gov.au/climate/reg/oneminsolar/index.shtml, last accessed 12th of Dec, 2016.

[11] T. Gerstner and M. Griebel, "Numerical Integration Using Sparse Grids," *Numerical Algorithms,* vol. 18, no. 3-4, pp. 209-232, Jan. 1998.

[12] L. Martín, L. F. Zarzalejo, J. Polo, A. Navarro, R. Marchante and M. Cony, "Prediction of global solar irradiance based on time series analysis: application to solar thermal power plants energy production planning," *Solar Energy,* vol. 84, no. 10, pp. 1772-1781, 2010.

[13] T. E. Hoff, R. Perez, J. Kleissl, D. Renne and J. Stein, "Reporting of irradiance modeling relative prediction errors," *Progress in Photovoltaics: Research and Applications,* vol. 21, no. 7, pp. 1514-1519, 2013.

[14] Robaa, S. M. "Evaluation of sunshine duration from cloud data in Egypt," Energy 33.5 (2008): 785-795.

[15] B. Ridley, J. Boland and P. Lauret, "Modelling of diffuse solar fraction with multiple predictors," *Renew. Energy,* vol. 35, no. 2, pp. 478-483, Feb. 2010.

Investigating the Effect of Titanium Dioxide (TiO2) Pollution on the Performance of the Mono-crystalline Solar Module

Zeki Ahmed Darwish[1,2], K. Sopian[1*], Hussein A Kazem[3], M.A. Alghoul[4] and Hussain Alawadhi[5]

[1]*Solar Energy Research Institute (SERI), Universiti Kebangsaan Malaysia, Bangi 43600 Selangor, Malaysia*

[2]*Ministry of Education, United Arab Emirates, Dubai PO. Box 3962*

[3]*Faculty of Engineering, Sohar University PO Box 44, Sohar, PCI 311 Oman*

[4]*Energy and Building Research Center, Kuwait Institute for Scientific Research, Safat 13109, Kuwait*
[4]*Center of Research Excellence in Renewable Energy (CoRERE), King Fahd University of Petroleum and Minerals (KFUPM), Dhahran 31261, Saudi Arabia*

[5]*Research Center for Advanced Materials, University of Sharjah, PO Box 27272 United Arab Emirates*

ABSTRACT

This paper presents a study of titanium oxide TiO_2 as one of the components of dust pollution affecting the PV performance. This pollutant can be found in various quantities in different locations around the world. The production of energy by different types of photovoltaic systems is very sensitive and depends on various environmental factors. Dust is one of the main contributing factors, yet the type of the dust is often neglected when studying the behaviour of the solar panel. In this experimental work we have studied the performance of the mono-crystalline solar module as affected by the density of TiO_2. The reduction of the PV module power caused by titanium dioxide under various mass densities was investigated. The results showed that the TiO_2 has a significant effect on the PV output power. The dust density varied between 0-125 $g.m^{-2}$. The corresponding reduction of the PV output power increased from 0 to 86.7%. This is based on various influencing parameters such as: short circuit current (I_{sc}), maximum current (I_m), open circuit voltage (V_{oc}), maximum voltage (V_m), maximum power (P_m) and efficiency (E). Two functions are proposed as a mathematical model in order to explain this behaviour, namely the exponential and Fourier functions. The coefficients of all general models are valid for this type of dust with a density value ranging from 0-125 $g.m^{-2}$.

Keywords: Titanium dioxide, dust effect, short circuit current, open circuit voltage, maximum power, fill factor

*Corresponding author. Tel.: +0060389118023; fax: +0060389118574. E-mail address: ksopian@ukm.edu.my (K. Sopian)

 DOI: 10.1051/e3sconf/20172301005

Introduction

Numerous technologies are currently used in order to harness solar radiation into suitable energy in the form of heat or electricity. Photovoltaic (PV) and solar thermal power (CSP) are the main technology systems used in order to convert light into electricity. An efficient application of this technology has a significant potential in both improving the economy and reducing pollution [1]. It has been widely acknowledged that photovoltaic power is closely linked to the use of solar energy for the production of electricity on a global scale. One of the main aspects to consider prior to introducing this type of technology to the market is represented by the efficiency of the solar cell. Extensive research is currently being conducted in order to improve the efficiency of the solar for future market use. Furthermore, a gradual improvement has been observed in relation to the efficiency of the monocrystalline silicon solar cell on an annual basis, from only 15% in the 1950s to 17% in the 1970s and a potential to increase up to 28% nowadays. According to Zhao et al. [2] photovoltaic performance is highly dependent on external conditions such as wind, temperature, irradiation and dust pollution. Furthermore, dust is also known to affect photovoltaic performance considering that it may prevent radiations from reaching the photovoltaic module [3]. Previous studies about the effects of dust on PV and solar thermal panels date back to over seventy years ago and focus on the thermal cells and the effect of dust on the reflectance of mirrors [4]. For instance, Sayigh who is credited as a pioneer in the Gulf region conducted a detailed investigation on the effect of dust on solar flat-plate collectors. The experiments included seven flat-plate collectors arranged in pairs and subjected to various inclinations of 0°, 30° and 60° respectively, with the seventh collector inclined vertically at 90°. In the case of each pair, one collector was cleaned regularly while the other was left undisturbed [5]. The amount of solar energy absorbed by the unclean plates was computed and subsequently compared to the values obtained in the case of the clean plates. Within this context, a dusty collection of approximately 2.5 $g/m^2/day$ was recorded between the months of April and June. Moreover, Sayigh et al. [6] analysed the effect of dust accumulation on tilted glass plates in Kuwait. Within the framework of their study, a plate transmittance reduction from 64% to 17% was observed in the case of tilt angles ranging from 0° to 60° following 38 days of exposure. As indicated by the horizontal collector, the useful energy gain was further reduced by 30% after 3 days of dust accumulation. Monto and Rohit classified the research exploring the effects of dust on the PV performance into two phases: the first phase appraising

 DOI: 10.1051/e3sconf/20172301005

the research conducted between 1940 and 1990 and the second phase appraising the research conducted between 1990 and 2010. A critical assessment of the research conducted during phase one was performed in order to evaluate the dust (pollution) level and composition at the location of the solar collector over the course of the experimental study. Very little investigation was conducted on the typical characteristics of dust itself, such as type, composition, accumulation rate and fill factor. The second phase refers to the salient limitations which still need to be overcome within the framework of this research prior to the integration thereof in the PV system design, such as optical and electro-chemical properties [7]. Although many types of dust and pollutants have been formerly studied, there are still certain types whose effects are yet to be explored. Kaldellis et al. [8] conducted a systematic experimental study of the different type of pollutions such as red soil, limestone and ash. The results indicated that the highest reduction was caused by the deposition of red soil, limestone and ash, respectively. Hussein et al. [9] studied the effects of air pollution on the performance of the PV module using red soil, ash, sand, calcium carbonate and silica, and it was found that the ash pollutant was the type of dust exerting the most significant impact on the PV module voltage. The highest value of PV voltage reduction of 25% was reported when ash was used. A comprehensive review of power loss due to soiling and the impact of dust on the use of solar energy was introduced by various researchers [10, 11, 12, 13]. In recent years, human activities have caused serious environmental problems, such as the ozone hole, urban smoke and increased atmospheric pollution. A limited amount of information is provided about the side effects of air pollution, and in particular about the degraded photovoltaic performance due to the deposition of solid particles [14]. The effects associated with dust pollution are strongly correlated with the environment in which the PV system is installed and the local conditions thereof, such as climate. For this reason, there is no general model to be applicable in all cases. Previous studies have indicated titanium oxide as one of the dust components in the United Arab Emirates and four other locations [15]. This study focuses on the effect of titanium oxide on the mono crystalline module performance and proposes a simple mathematical model for the short circuit current (I_{sc}), maximum current (I_m), open circuit voltage (V_{oc}), maximum voltage (V_m), maximum power (P_m) and efficiency (E).

 DOI: 10.1051/e3sconf/20172301005

Methodology

This section of the paper will detail the methodology used in order to investigate the effect of titanium oxide on the PV panel. Firstly, the set-up of the system which includes the solar simulator, data-acquisition, solar panel and parameters will be presented. This is subsequently followed by some samples being examined using a Maya2000Pro spectrometer in order to measure the transmittance and reflectivity coefficients. These are then examined using an electron microscope (SEM) in order to identify the size of the particle. Finally, an experimental verification was conducted in order to assess the relationship between the density of mass and the decrease in current, voltage, power and efficiency.

Solar simulator

The solar simulator consists of 23 units of halogen tungsten lamps with an output power of 500W. The lamps are distributed into nine columns with each column (1, 3, 5, 7and 9) containing 3 lamps and with columns (2, 4, 6, 8) containing 2 lamps. The design of a sun simulator was based on a previous research on the development of the solar simulator [16, 17]. The lamps consume 11.5 kW and the intensity of the solar irradiance has been controlled by nine voltage controllers with a capacity of 2KVA, 8Amp. The results were obtained from one module at 583 W/m^2 with an error of 3.89% in record reading and a distance of 120cm between the solar simulator and the PV module.

PV Module

A monocrystalline solar module with 40 solar cells was used in this research. Table 1 illustrates the module specifications at the technical standard test conditions of AM=1. 5, S=1000W/m^2 and T_c=25 ^{0}C.

Table 1: Technical characteristics of the PV module

Copex Solar: Model M100	
Peak Power	100 W_P
Power Tolerance	±5W
Voltage (V_{mp})	19.6V
Current (Imp)	5.11 A

DOI: 10.1051/e3sconf/20172301005

Open Circuit Voltage (V_{oc})	24.8 V
Short Circuit Current (I_{sc})	5.56 A
Weight	11 Kg
Dimensions	1335× 547× 40 mm

Data acquisition

The 3711A is a programmable DC electronic load, capable of supporting up to 360 W, 360V and 30A. The software for this device was developed by Ross Bunker (Copyright@2012 Gratings, Inc.) in order to read the main electrical parameters of the solar module such as the short-circuit $I_{sc,}$ the open-circuit voltage $V_{oc,}$ the maximum power $P_{m,}$ the fill factor FF, the shunt resistance R_{sh} and the series resistance R_s. In the final setup the data acquisition was connected to the computer and also directly to the PV panel, which was located under the solar simulator with a zero tilt angle. The test started with the study of a clean PV module followed by the gradual addition of 15g of TiO_2 until reaching 90g.

Dust properties

Figs. 1a, 1b and 1c show three regions of the light spectrum representing the ultraviolet ray's region (UV), visible ray's region (VIS) and infrared ray's region (IR), respectively. The percentage of reflectance, transmittance and absorbance was measured using Maya 2000Pro spectrometers. The percentage value of reflectance in the wavelength range from (376.8-816) nm is 84.9%, while the percentage of transmittance and observance is equal to 15% and 0.1%, respectively. A scanning electron microscope (SEM) was used in order to examine the TiO_2 sample distributed on the module and it was noted that the particles were non-homogenous. The diameter from different regions was then measured and it was found to range between (0.37-0.68) µm using the (Image J Software), as indicated in Fig .2.

DOI: 10.1051/e3sconf/20172301005

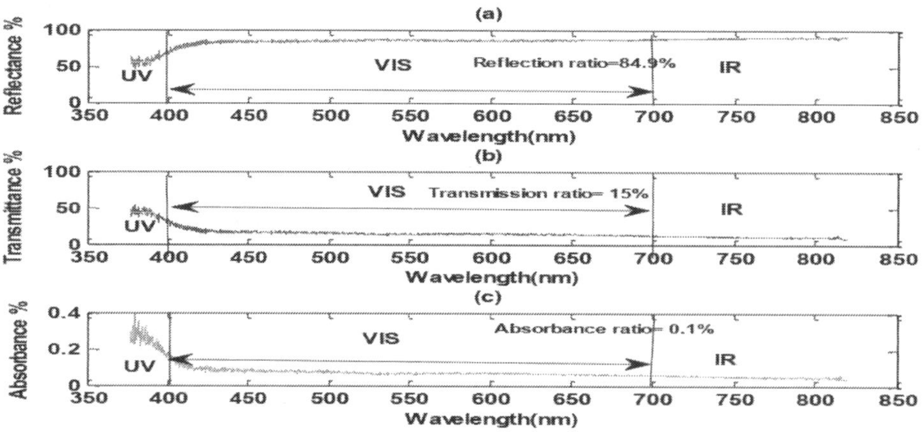

Fig.1 Spectroscopy of TiO_2 (a) Reflectance (b) Transmittance and (c) Absorbance

Fig.2. SEM images of TiO_2

Results and discussion

Effect TiO_2 on I-V and P-V characteristics

A I-V measurement is the best tool to provide more information on the overall performance of a PV module. Figs. 3a and 3b present the current – voltage (I-V) and power – voltage (P-V) curves for the PV module under two conditions, namely clean and dusty conditions. It was found that titanium oxide deposition has a significant effect on the short circuit current of the panel. The current loss due to TiO_2 was 85.4% where the dust density increased from 0-125 g/m^2. When the open circuit voltage of the dusty case was slightly lower than that of the clean case, the loss was 11.08%. A 52.3W drop in the maximum power as a result of the titanium

 DOI: 10.1051/e3sconf/20172301005

oxide deposition was observed. The output of the maximum current (I_m) and maximum voltage (V_m) was also affected by the titanium accumulation. It was found that the current decreased from (3.011-0.44) A while the voltage decreased from (20.08-18.36) V. These results reflect on the efficiency of the PV performance, showing that the reduction in the efficiency was (19.2-2.4) %. The studies carried out in Senegal have shown that the maximum power output loss can range from 18% to 78% respectively for the polycrystalline (ps-Si) and the monocrystalline module (mc-Si) where the modules were not cleaned following exposure to a Sahelian environment over the course of one year [18]. The outdoor studies conducted in Egypt have shown that the reduction of P_m as a result of dust deposition could reach 35% after two months without cleaning the PV modules [19].

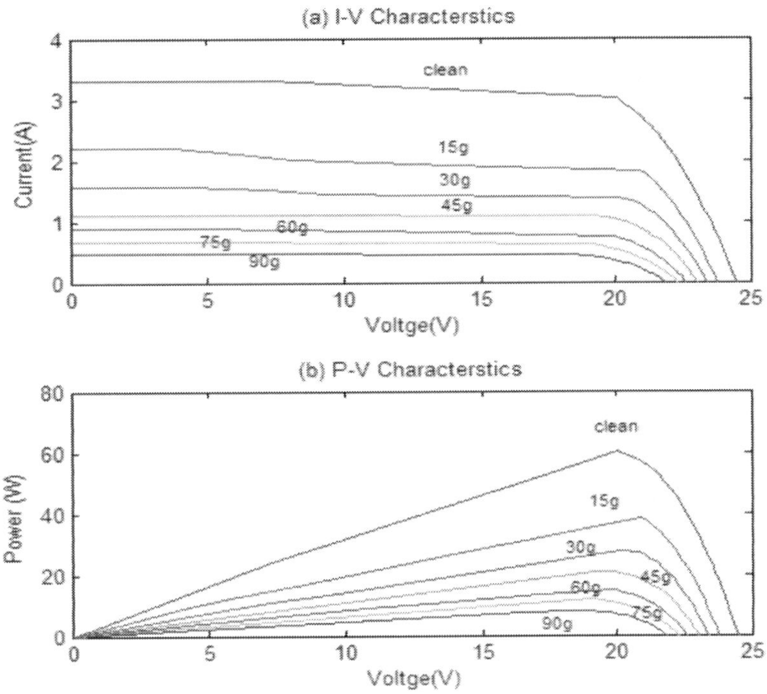

Fig.3. Comparison between the I-V and P-V characteristics of the PV module under clean and different density values of the masses for TiO_2.

DOI: 10.1051/e3sconf/20172301005

Effect of TiO_2 on the FF, R_s, R_{sh} and R_{max}

The fill factor (FF) represents the squareness of the I-V curve and is mainly related to the resistive losses in a solar cell. Theoretically, in perfect cases, its value can be 100%, corresponding to the square I-V curve. But it is not a suitable practice to square I-V. There are many losses which reduce the value of the fill factor which do not depend directly on the density mass of pollution. Dust obscures the amount of solar radiation that reaches the solar cells and it was observed that the increase in the amount of dust does not affect the value of the fill factor. As demonstrated in Fig .4 some strong fluctuations were observed in the FF, the maximum value was 0.80 and the minimum value was 0.72 at 62.5 g/m^2 and at 20g/m^2 respectively, but in the clean case this was 0.74 and 0.77 at 125g/m^2. This result is similar to the findings of Vorster who studied the relation between solar radiation and the fill factor using the PV concentrator array [20]. The fill factor can be mathematically written as a function of the open circuit voltage normalised (v_{oc}) [21]

$$FF = \frac{v_{oc} - \ln(v_{oc} + 0.72)}{v_{oc} + 1}$$

(1)where v_{oc} is $V_{OC}/(\frac{kT}{q})$

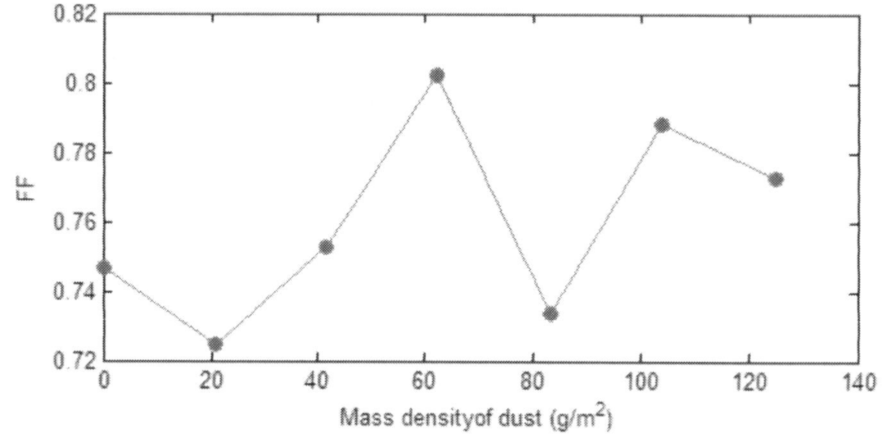

Fig.4 Fill factor versus density of mass

DOI: 10.1051/e3sconf/20172301005

In an ideal case only, the value of (FF=1) in relation to the I-V curve of a solar cell is square, but the cell I-V curve actually behaves as exponential. Thus in the best possible case, the value of FF is less than one. These types of losses are classified into electrical losses and arise from the parasitic resistance, series (R_s) and shunt resistance (R_{sh}) of the cell. Such losses cannot be evaded or cannot be reduced beyond their fundamental limit. The series resistance represents the sum of the resistance values of all the components that come in the path of the current, while the shunt resistance is due to the leakage across the P-N junction. The maximum resistance (R_m) represents the lesser value of resistance that yields maximum power (P_m). The equation for the R_m of the module can be written as $R_m=V_m/I_m$. It was found that the series resistance, shunt resistance and maximum resistance were increased from (1.4-7.7) Ω, (43.77-341) Ω, (6.6-41.6) Ω respectively, as shown in Figs. 5a, 5b and 5c. Nand et al. [22] noticed that the series resistance increases with a higher dust factor and shadow effects.

DOI: 10.1051/e3sconf/20172301005

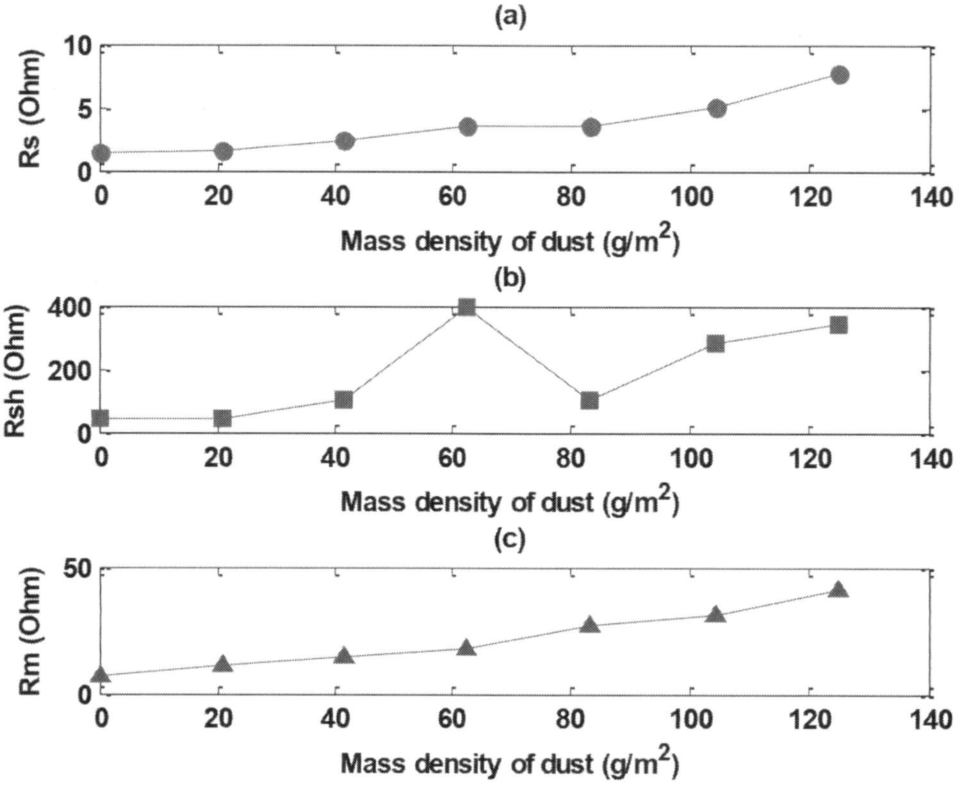

Fig.5 The effect of TiO_2 on the (a) series resistance (b) shunt resistance and (c) maximum resistance.

Performance prediction model

Based on the experimental study, six electrical characteristics of a PV module have been selected and represented mathematically as follows: exponential function used to represent the short circuit current (I_{sc}), maximum current (I_m), maximum power (P_m) and efficiency (E), the Fourier function used to represent the open circuit voltage (V_{oc}) and maximum voltage (V_m). It was found in the literature that the researcher used the low-degree polynomial in order to simulate the dust deposition density so as to simulate the dust deposition on current, power efficiency etc. These correlations are valid for small values for dust deposition

DOI: 10.1051/e3sconf/20172301005

density. This problem was solved by using a polynomial of a higher degree, e.g. of the 7th degree [23]. In order to reduce the number of coefficients which change from one case to another it was found that the exponential and Fourier functions are more efficient to represent the electrical parameters. Table 2 shows the function which was chosen to represent these parameters in this study.

Table 2: Implicit equations for electrical characteristics I, V, P and E with coefficients

General equation	Coefficients& Goodness of fit
Isc = a*exp(b*x) + c*exp(d*x)	a = 0.2761 (-0.2095, 0.7617); b = -0.04348 (-0.09889, 0.01193); c =0.7241 (0.2363, 1.212); d =-0.01258 (-0.01825, -0.006906); SSE: 0.0001874; R-square: 0.9997; Adjusted R-square: 0.9993; RMSE: 0.007903; x- density of mass (g/m^2)
Im = a*exp(b*x)	a = 0.9554 (0.8559, 1.055); b = -0.01653(-0.01967, -0.01339); SSE: 0.009588; R-square: 0.9818.; Adjusted R-square: 0.9782; RMSE: 0.04379; x- density of mass (g/m^2)
Voc=a0+a1*cos(x*w) + b1*sin(x*w)	a0 = 2.48e+005 (-6.742e+012, 6.742e+012) a1 = -2.48e+005 (-6.742e+012, 6.742e+012) b1 = 277.3 (-3.768e+009, 3.768e+009) w = -3.897e-006 (-52.96, 52.96), SSE: 4.586e-005; R-square: 0.9948; Adjusted R-square: 0.9896, RMSE: 0.00391
Vm = a0 + a1*cos(x*w) + b1*sin(x*w) + a2*cos(2*x*w) + b2*sin(2*x*w)	a0 = 0.9772 (0.8828, 1.072), a1 = 0.01008 (-0.1092, 0.1294) b1 = 0.0383 (-0.07907, 0.1557), a2 = 0.009359 (-0.151, 0.1698) b2 =0.03457 (-0.1006, 0.1698), w = 0.04179 (-0.00399, 0.08757), SSE: 0.0002826; R-square: 0.9726; Adjusted R-square: 0.8359,

DOI: 10.1051/e3sconf/20172301005

	RMSE: 0.01681; x- density of mass (g/m^2)
Pm= a*exp(b*x) + c*exp(d*x)	a = 0.1677 (0.05908, 0.2764), b = -0.1192 (-0.3147, 0.07634) c = 0.8323 (0.7261, 0.9385), d = -0.0144 (-0.01617, -0.01264) SSE: 0.000165; R-square: 0.9997; Adjusted R-square: 0.9994 RMSE: 0.007416; x-density of mass (g/m^2)
E = a*exp(b*x) + c*exp(d*x)	a = 0.6816 (0.5735, 0.7897), b = 0.002083 (0.0008552, 0.003311), c = -0.6816 (-0.7877, -0.5755), d =-0.03034 (-0.03701, -0.02367), SSE: 9.499e-005, R-square: 0.9998, Adjusted R-square: 0.9997, RMSE: 0.005627. x- density of mass (g/m^2)

Conclusion

In this paper the effect of the titanium oxide accumulation on the performance of the PV module has been investigated. Titanium oxide is one of the dust components in the United Arab Emirates at different rates and in various locations. It was found that TiO_2 has a significant effect on the short circuit current of the panel while the open circuit voltage is slightly lower than that the clean case, thus reflecting the maximum power and efficiency reduction. The parasitic resistance shunt and series resistances increased with the increasing density of mass in addition to the maximum resistance, which represents the maximum power value of the panel. Two functions were selected, namely the Fourier and the exponential functions with coefficients to represent six electrical characteristics.

Acknowledgments

The authors would like to honestly thank the National University of Malaysia for the provision of the grant TRGS/1/2014/UKM/01/11/1 to support this work. The authors are deeply grateful to Mr. Khamis Saeed Alkindi for consenting to the construction of the laboratory in his farm. Finally, thanks go to Mr. Mohammed

Shameer, the supervisor of the Research Center for Advanced Materials, University of Sharjah for his for assistance in the analysis of dust samples using an electron microscope (SEM).

References

[1] S. E. Colesca and C. N. Ciocoiu, "An overview of the Romanian renewable energy sector," *Renewable and sustainable energy reviews,* vol. 24, pp. 149-158, 2013.
[2] J. Zhao, A. Wang, P. Campbell, and M. A. Green, "A 19.8% efficient honeycomb multicrystalline silicon solar cell with improved light trapping," *IEEE Transactions on Electron Devices,* vol. 46, pp. 1978-1983, 1999.
[3] V. Tyagi, N. A. Rahim, N. Rahim, A. Jeyraj, and L. Selvaraj, "Progress in solar PV technology: research and achievement," *Renewable and Sustainable Energy Reviews,* vol. 20, pp. 443-461, 2013.
[4] S. Ghazi, A. Sayigh, and K. Ip, "Dust effect on flat surfaces–A review paper," *Renewable and Sustainable Energy Reviews,* vol. 33, pp. 742-751, 2014.
[5] A. Sayigh, "Effect of dust on flat plate collectors," in *Sun: Mankind's Future Source of Energy,* 1978, pp. 960-964.
[6] A. Sayigh, S. Al-Jandal, and H. Ahmed, "Dust effect on solar flat surfaces devices in Kuwait," in *Proceedings of the workshop on the physics of non-conventional energy sources and materials science for energy,* 1985, pp. 353-367.
[7] M. Mani and R. Pillai, "Impact of dust on solar photovoltaic (PV) performance: Research status, challenges and recommendations," *Renewable and Sustainable Energy Reviews,* vol. 14, pp. 3124-3131, 2010.
[8] J. Kaldellis, P. Fragos, and M. Kapsali, "Systematic experimental study of the pollution deposition impact on the energy yield of photovoltaic installations," *Renewable Energy,* vol. 36, pp. 2717-2724, 2011.
[9] T. Khatib, H. Kazem, K. Sopian, F. Buttinger, W. Elmenreich, and A. S. Albusaidi, "Effect of dust deposition on the performance of multi-crystalline photovoltaic modules based on experimental measurements," *International Journal of Renewable Energy Research (IJRER),* vol. 3, pp. 850-853, 2013.
[10] T. Sarver, A. Al-Qaraghuli, and L. L. Kazmerski, "A comprehensive review of the impact of dust on the use of solar energy: History, investigations, results, literature, and mitigation approaches," *Renewable and Sustainable Energy Reviews,* vol. 22, pp. 698-733, 2013.
[11] S. C. Costa, A. S. A. Diniz, and L. L. Kazmerski, "Dust and soiling issues and impacts relating to solar energy systems: Literature review update for 2012–2015," *Renewable and Sustainable Energy Reviews,* vol. 63, pp. 33-61, 2016.
[12] A. Ndiaye, A. Charki, A. Kobi, C. M. Kébé, P. A. Ndiaye, and V. Sambou, "Degradations of silicon photovoltaic modules: A literature review," *Solar Energy,* vol. 96, pp. 140-151, 2013.

DOI: 10.1051/e3sconf/20172301005

[13] J. K. Kaldellis, A. Kokala, and M. Kapsali, "Natural air pollution deposition impact on the efficiency of PV panels in urban environment," *Fresenius Environmental Bulletin,* vol. 19, pp. 2864-2872, 2010.

[14] M. R. Maghami, H. Hizam, C. Gomes, M. A. Radzi, M. I. Rezadad, and S. Hajighorbani, "Power loss due to soiling on solar panel: A review," *Renewable and Sustainable Energy Reviews,* vol. 59, pp. 1307-1316, 2016.

[15] Z. A. Darwish, H. A. Kazem, K. Sopian, M. Al-Goul, and H. Alawadhi, "Effect of dust pollutant type on photovoltaic performance," Renewable and Sustainable Energy Reviews, vol. 41, pp. 735-744, 2015.

[16] H. Garg, A. Shukla, I. Madhuri, R. Agnihotri, and S. Chakravertty, "Development of a simple low-cost solar simulator for indoor collector testing," *Applied energy,* vol. 21, pp. 43-54, 1985.

[17] Z. A. Darwish, K. Sopian, H. Alawadhi, H. A. Kazem, and M. Alghoul, "The impact of Calcium Carbonate on the photovoltaic performance: An indoor experimental study," *International Journal of Applied Engineering Research,* vol. 11, pp. 2091-2097, 2016.

[18] A. Ndiaye, C. M. Kébé, P. A. Ndiaye, A. Charki, A. Kobi, and V. Sambou, "Impact of dust on the photovoltaic (PV) modules characteristics after an exposition year in Sahelian environment: he case of Senegal," International Journal of Physical Sciences, vol. 8, pp.1166-1173, 2013.

[19] M. Ibrahim, B. Zinsser, H. El-Sherif, E. Hamouda, G. Makrides, G. E. Georghiou, et al., "Advanced Photovoltaic test park in Egypt for investigating the performance of different module and cell technologies," in Proceedings of the 24th Symposium Photovoltaic Solar Energy, Staffelstien, Germany, 2009.

[20] F. Vorster, E. Van Dyk, and A. Leitch, "Investigation on the IV characteristics of a high concentration, photovoltaic array," in *Photovoltaic Specialists Conference, 2002. Record of the Twenty-Ninth IEEE*, 2002, pp. 1604-1607.

[21] M. A. Green, "Accuracy of analytical expressions for solar cell fill factors," *Solar Cells,* vol. 7, pp. 37-340, 1982.

[22] N. Kishor, M. G. Villalva, S. R. Mohanty, and E. Ruppert, "Modeling of PV module with consideration of environmental factors," in *ISGT Europe*, 2010, pp. 1-5

[23] Z. A. Darwish, H. A. Kazem, K. Sopian, M. Alghoul, and H. Alawadhi, "Experimental investigation of dust pollutants and the impact of environmental parameters on PV performance: an experimental study," *Environment, Development and Sustainability,* pp. 1-20.

E3S Web of Conferences 23, 01006 (2017)
World Renewable Energy Congress-17
DOI: 10.1051/e3sconf/20172301006

ASSESSING THE ACCURACY OF GLOBE THERMOMETER METHOD IN PREDICTING OUTDOOR MEAN RADIANT TEMPERATURE UNDER MALAYSIA TROPICAL MICROCLIMATE

N.G. Khrit[a] , M.A. Alghoul*, [b,c], K. Sopian*, [a], A.A. Lahimer [a], and O.K. Elayeb[a]

[a] *Solar Energy Research Institute, Universiti Kebangsaan Malaysia, 43600 Bangi, Selangor, Malaysia*

[b] *Energy and Building Research Center, Kuwait Institute for Scientific Research, P.O. Box 24885, Safat 13109, Kuwait*

[c] *Center of Research Excellence in Renewable Energy (CoRe-RE), Research Institute, King Fahd University of Petroleum and Minerals (KFUPM), Dhahran 31261, Saudi Arabia*

**Corresponding authors: dr.alghoul@gmail.com (M.A. Alghoul), ksopian@ukm.edu.my (K. Sopian)*

Abstract:

Assessing outdoor human thermal comfort and urban climate quality require experimental investigation of microclimatic conditions and their variations in open urban spaces. For this, it is essential to provide quantitative information on air temperature, humidity, wind velocity and mean radiant temperature. These parameters can be quantified directly except mean radiant temperature (T_{mrt}). The most accurate method to quantify T_{mrt} is integral radiation measurements (3-D short-wave and long-wave) which require using expensive radiometer instruments. To overcome this limitation the well-known globe thermometer method was suggested to calculate T_{mrt}. The aim of this study was to assess the possibility of using indoor globe thermometer method in predicting outdoor mean radiant temperature under Malaysia tropical microclimate. Globe thermometer method using small and large sizes of black-painted copper globes (50mm, 150mm) were used to estimate T_{mrt} and compare it with the reference T_{mrt} estimated by integral radiation method. The results revealed that the globe thermometer method considerably overestimated T_{mrt} during the middle of the day and slightly underestimated it in the morning and late evening. The difference between the two methods was obvious when the amount of incoming solar radiation was high. The results also showed that the effect of globe size on the estimated T_{mrt} is mostly small. Though, the estimated T_{mrt} by the small globe showed a relatively large amount of scattering caused by rapid changes in radiation and wind speed.

1. INTRODUCTION

Mean radiant temperature (T_{mrt}) is one of the key variables in the calculation of human energy balance and the assessment of thermal comfort through thermal indices. The concept of T_{mrt} considers radiative heat exchange between a human body and its surrounding which causes an energy gain –or loss– of a person depending on their surface temperatures and emissivity. It is defined as the uniform temperature of a hypothetical spherical surface surrounding the subject (emissivity

 DOI: 10.1051/e3sconf/20172301006

ε=1) that would result in the same net radiation energy exchange with the subject as the actual, complex radiative environment [1].

The determination of T_{mrt} is important for the assessment of human bioclimatology and heat stress in the urban environments. In such environments the radiation fluxes densities are highly influenced by urban structures that absorb, emit, reflect radiative energy in the shortwave and longwave spectrum [2]. This results in spatial and temporal variation in radiation fluxes densities, as well as radiant non-uniformity [2, 3]. Quantitative information of short- and long wave radiation fluxes is essential for an accurate estimation of the T_{mrt} [4]. For this detailed integral radiation measurements of radiation fluxes have been developed and used as a way to determine T_{mrt} [2-5]. The method offers significant advantages in terms of accuracy and fast response; however, it requires complex and costly equipment. Any attempt to use the method for further understanding of T_{mrt} from temporal and spatial scales (i.e., multi-site field measurements) is impractical and costly [6].

A simple and inexpensive method for estimating T_{mrt} is globe thermometer method [7]. It was developed and is commonly used for indoor environments [7, 8]. However, in outdoor environments with high intensity and variability of climate data, the accuracy of globe thermometer is sometimes unacceptable [5, 6, 9]. Despite of that, it is still attracting the attention as an alternative and low cost method for further evaluation in outdoor environment [2, 3, 5, 10, 11].

So far, limited researches in the literature compared globe thermometer method with the integral radiation measurements (radiometers) for the estimation of T_{mrt} under sub-tropical [12, 13] and tropical climate zones [5]. Kántor et al. (2014) compared measurements of standard 150mm-black globe thermometer with integral radiation measurements in order to improve the assessment of outdoor thermal comfort in subtropical Taiwan [13]. Tan et al. (2013) used a smaller (38 mm in diameter) gray-painted acrylic globe to validate the globe method by six-directional measurements in Singapore and suggested a modified T_{mrt} equation to obtain better values of T_{mrt} [5]. Therefore, the aim of this study is to examine the accuracy of two black-painted hollow copper globes (50mm and 150mm) in estimating T_{mrt} in the tropical outdoor urban setting of Malaysia when compared with the integral radiation measurements as a reference method. Also the effect of the globe diameter on the estimated T_{mrt} is investigated to determine whether the two globe thermometers are equally suitable for estimating T_{mrt} in tropical urban setting, or if each size has different estimation accuracy of the T_{mrt}.

DOI: 10.1051/e3sconf/20172301006

2. MATERIALS AND METHOD

2.1. Measurement site and microclimate station unit

The measurement site is located at a pedestrian walkway at the main campus of the National University of Malaysia, Bangi, Malaysia (2°.55'N, 101°.46'E). The main campus is spatially constrained by abutting landscape and tropical rainforest that necessitated the densification of the built environment at the university. The floor of the pedestrian walkway is made of flat concrete bricks and the adjacent buildings facades are made of red brick and grey concrete. The measurement site and the microclimate station unit are shown in Fig 1(a) and Fig 1(b), respectively.

Fig. 1(a-b): (a) The measurement site at the pedestrian walkway, (b) Microclimate station unit for measuring three-dimensional short and long-wave radiation fluxes, besides two globe thermometers of 50mm and 150mm diameters.

As shown in Fig 1(b), a microclimate station was equipped with a set of instruments as defined in Table 1. This includes sensors to measure air temperature, relative humidity, and wind speed. Three net radiometers, each consists of two pyranometers and two pyrgeometers, were set up on the station stand and oriented in east-west, north-south, upward and downward directions in order to measure the three-dimensional short and long-wave radiation fluxes. Two black-painted copper globe thermometers of 50mm and 150mm in diameters were installed within the space enclosed by the radiometers. The instruments were fixed at the same measuring point as much as possible and at a height of approximately 1.1m above the ground. The three-cup anemometer was set at height of 1.5m above the ground. The recording interval was set to 2-min-averages of 10s scan for radiation fluxes, and 2-min-averages of 30s scan for other microclimatic data.

DOI: 10.1051/e3sconf/20172301006

Table 1: Measured parameters and instruments

Quantity	Instrument
Air temperature, T_a	Skye rht+ PT100 sensor
50-mm globe temperature, T_{g1}	Delta ohm, TP3276.2, PT100 sensor.
150-mm globe temperature, T_{g2}	Delta ohm, TP3275, PT100 sensor.
Air speed, V_a	Delta ohm, Probe with omnidirectional hot wire
Relative humidity, RH	Skye rht+
Shortwave and longwave radiation, K, L	Delta OHM, LP NET 14

3. MEAN RADIANT TEMPERATURE ESTIMATION METHODS

3.1. Integral (3-dimensional) radiation measurement method

This method provides the most accurate estimate of the T_{mrt} [6, 9]. A mean radiant flux density (S_{str}) have to be calculated first to determine a part of the radiation that is significant to a person [1]. S_{str} is a summation of the measurements of radiation fluxes from all six directions multiplied by the angular factors between a person and the surrounding surfaces, according to this equation [1, 9].

$$S_{str} = \alpha_k \sum_{i=1}^{6} K_i F_i + \varepsilon_p \sum_{i=1}^{6} L_i F_i \qquad (1)$$

K_i = the short-wave radiation fluxes (i = 1–6)
L_i = the long-wave radiation fluxes (i = 1–6)
α_k = the absorption coefficient for short-wave radiation (standard value 0.7)
ε_p = the emissivity of the human body (standard value 0.97)
F_i = the angular factors between a person and the surrounding surfaces (i = 1–6).

For a standing –or walking– person, F_i is commonly set to 0.22 for radiation fluxes from the horizons and 0.06 for upward and downward radiation fluxes. For a sphere, F_i is 0.167 for all six directions.

 DOI: 10.1051/e3sconf/20172301006

The mean radiant temperature can be calculated in °C using equation (2) [1, 9]:

$$T_{mrt} = \left[\frac{S_{str}}{\varepsilon_p \sigma}\right]^{0.25} - 273.15 \quad (2)$$

Where, σ is the Stefan–Boltzmann constant (5.67×10^{-8} Wm^{-2} K^{-4})

3.2. Globe thermometer method

The method requires measurements of globe temperature in combination with measurements of air temperature and wind speed. Its theory was explained thoroughly by Kuehn et al., (1970) [14]. In the theory, when a globe thermometer is in equilibrium, the heat gained (or lost) by radiation is equal to the heat lost (or gained) through convection. Globe temperature represents a weighted average between radiant and ambient temperatures [5, 6, 9]. The formula to calculate the T_{mrt} is given by [7]:

$$T_{mrt} = \left[(T_g + 273.15)^4 + \frac{1.06 \times 10^8\, V_a^{0.6}}{\varepsilon\, D^{0.4}} \times (T_g - T_a)\right]^{0.25} - 273.15 \quad (3)$$

Where T_g is the globe temperature (°C), T_a is the air temperature (°C), V_a is the air speed (m/s), ε is the globe emissivity (0.95 for a black globe) and D is the globe diameter (m). The weighting parameter $1.06 \times 10^8\, V_a^{0.6}$ represents the globe's mean convection coefficient.

4. RESULT AND DISCUSSION

This case study was carried out in a greenery open urban space; the obtained results were based on one-day measurement in warm-humid conditions and intense solar radiation. The measurements were conducted on 8th October 2016 from 09:30 to 21:00 local time. The weather in this day brought hot humid conditions with a maximum air temperature of 33.9°C, average air temperature of 30.7°C and average wind speed of 1.29 m/s.

4.1. Three-dimensional short and long-wave radiation fluxes

The temporal course of shortwave fluxes from the four cardinal points K_{east}, K_{west}, K_{north} and K_{south}, as well as downward/incoming and upward/outgoing shortwave

DOI: 10.1051/e3sconf/20172301006

fluxes K↓ and K↑ are shown in Fig. 2 (a). The high variation in shortwave radiation fluxes can be explained by rapid changes in weather conditions from clear to cloudy conditions. This resulted in a sharp fluctuation in K↓ with a maximum value of 1042 W/m^2 recorded at 11:42. The upward shortwave radiation K↑ followed the same pattern of K↓ but with much less amounts (dependent on angle of incidence and surface albedo). Its maximum value (161.1W/m^2) was recorded at the same time that the maximum K↓ was recorded. In the morning when the site was sunny from east, K_{east} reached its maximum record of 450 W/m^2 at 09:54. When the site was sunny from west, K_{west} exhibited sharp increase (462 W/m^2) around 17:35.

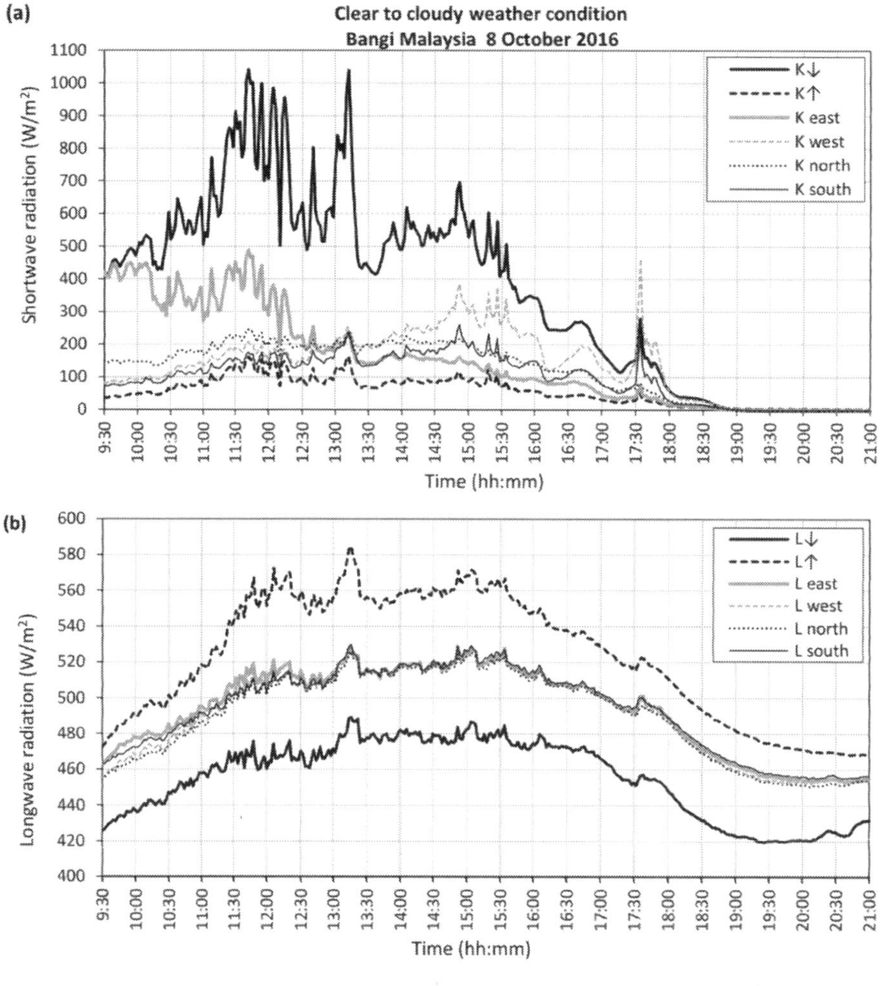

Fig. 2: (a-b): Three-dimensional shortwave and longwave radiation fluxes.

DOI: 10.1051/e3sconf/20172301006

As shown in Fig. 2 (b), longwave radiation fluxes from the four cardinals, L_{east}, L_{west}, L_{north} and L_{south} were close and ranged from 450 to 530 W/m^2, reaching their maximum in the afternoon. Longwave radiation from the sky $L\downarrow$ was lowest during the day, with a maximum record of 489 W/m^2 at 13:14. The upward long-wave radiation $L\uparrow$ was the highest amongst all longwave fluxes, with a maximum of 585 W/m^2 at 13:14. This was due to low albedo and high emissivity of the concrete floor, causing higher absorption of solar radiation and increased surface temperature and thus higher emittance of $L\uparrow$.

4.2. Validation of globe thermometer in a tropical outdoor setting

In this study, two globe thermometers of different diameters (50mm and 150mm) were used; each consisted of a black-painted hollow copper sphere with temperature sensor in its centre. Temperatures registered for small globe T_{g1} and for large globe T_{g2}, as well as air temperature T_a and wind speed V_a were used to estimate mean radiant temperatures $T_{mrt\,(g1)}$ and $T_{mrt\,(g2)}$ according to equation (3). Integral radiation measurements from the net radiometers were used to obtain the reference mean radiant temperature, $T_{mrt\,(rad.)}$ for a standing person. As shown in Fig. 3, the rapid change in solar radiation caused a high variation in $T_{mrt\,(rad.)}$, and may also be reflected in the globe thermometers measurements. The maximum $T_{mrt\,(rad.)}$ was 70.6 °C and occurred at which the maximum $K\downarrow$ occurred, that is at 11:42.

The estimated $T_{mrt\,(g1)}$ and $T_{mrt\,(g2)}$ from globe thermometer measurements considerably overestimated T_{mrt} during the middle of the day; however they tended to underestimate it in the morning and late evening. The difference between the two methods was obviously high when the amount of incoming solar radiation was high. The highest difference between $T_{mrt\,(rad.)}$ and $T_{mrt\,(g1)}$ was 31.7°C, which occurred at 11:36, while the highest difference between $T_{mrt\,(rad.)}$ and $T_{mrt\,(g2)}$ was 26.9°C, which occurred at the same time. The general form of this finding has been confirmed in earlier studies [9, 11, 12, 15-17]. The high absorption of the black globe to short-wave radiation causes T_{mrt} to be overestimated. The maximum $T_{mrt\,(g1)}$ by the small globe was 95.2°C, while the maximum $T_{mrt\,(g2)}$ by the large globe thermometer was 91.5°C at 11:42. The difference between $T_{mrt\,(g1)}$ and $T_{mrt\,(g2)}$ is mostly small during the day.

Also as shown in Fig. 3, the difference in globe size results in obviously different globe temperatures. The large globe temperature T_{g2} was higher than the small globe temperature T_{g1} throughout daytime, except in the morning and late evening, with a highest difference of 4.7°C, which occurred at 13:14.

DOI: 10.1051/e3sconf/20172301006

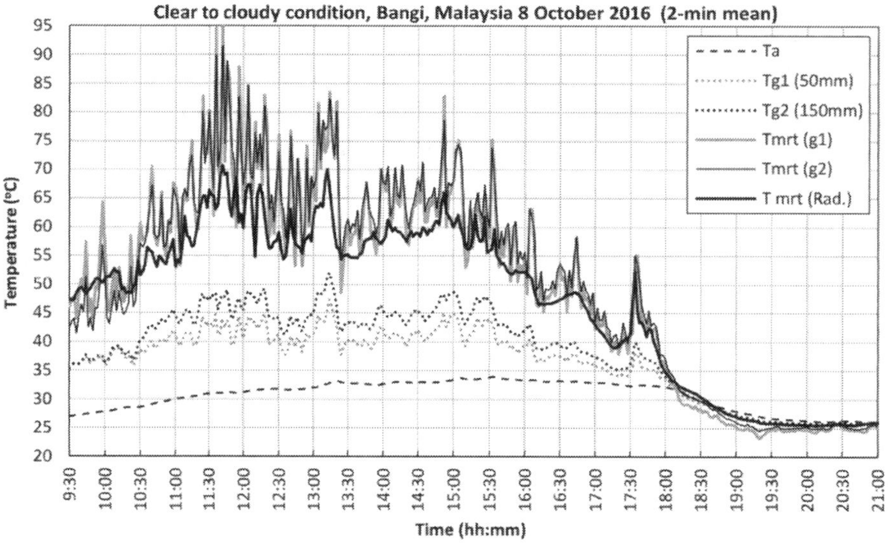

Fig. 3. Temperatures of the 50mm globe (T_{g1}), 150mm globe (T_{g2}) and air temperature (T_a). Estimated T_{mrt} using integral radiation measurements, $T_{mrt\,(rad.)}$ and estimated T_{mrt} using globe thermometer of 50mm diameter, $T_{mrt\,(g1)}$ and 150mm diameter, $T_{mrt\,(g2)}$.

4.3. Effect of the globe size and outdoor condition on the response time

In Fig. 3, there was a large amount of scattering in T_{mrt} data. This was because of the delay in the globes' response to rapid changes in radiation and wind speed. The response of the standard black globe is too long; it takes about 15 min to reach equilibrium [12, 16]. Thus, in case of rapidly changing outdoor conditions, the globe equilibrium is never reached, resulting in scatter in T_{mrt} data [1, 9].

It is also seen that the small globe thermometer has a faster response than the large one. As shown in Fig. 4 (a), this resulted in a rather larger scatter in the estimated $T_{mrt\,(g1)}$ by the small globe (lower R^2). This finding agreed with results of other researchers [6, 9, 12, 15-17], who reported that downsized globe thermometers (less than 50 mm in diameters) have faster response time (5-10 min) but also give a relatively larger scatter in T_{mrt} values. The results in Fig. 4 (a) indicate that the accuracy of $T_{mrt\,(g1)}$ and $T_{mrt\,(g2)}$ was reduced considerably when the amount of shortwave radiation increased. This can be clarified in Fig. 4 (b) by plotting the difference between $T_{mrt\,(g1)}$ and $T_{mrt\,(rad.)}$ and between $T_{mrt\,(g2)}$ and $T_{mrt\,(rad.)}$ against $K\downarrow$. As shown, the black colour of the globes causes the influence of shortwave radiation to be overestimated, and thus decreased accuracy of T_{mrt}.

DOI: 10.1051/e3sconf/20172301006

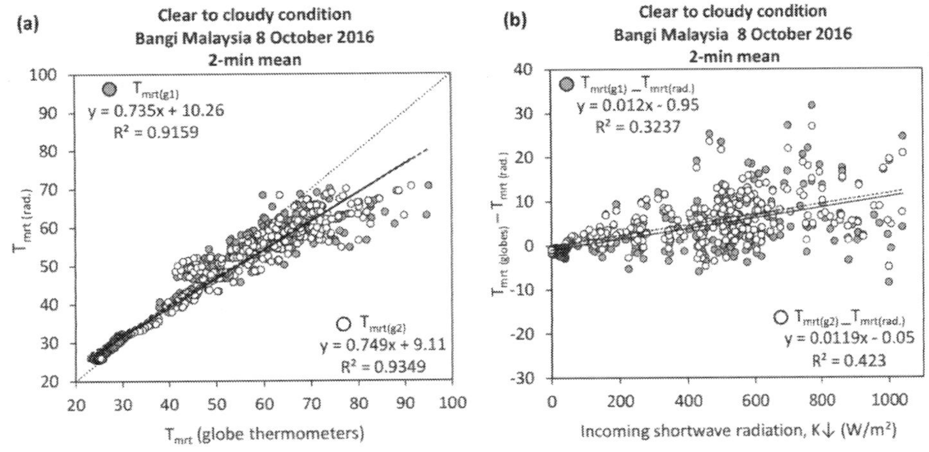

Fig. 4 (a-b); (a) T_{mrt} determined by integral radiation measurements, $T_{mrt\ (rad.)}$ *versus* that estimated by 50mm copper globe thermometer, $T_{mrt\ (g1)}$ and 150mm copper globe thermometer, $T_{mrt\ (g2)}$; (b) $T_{mrt\ (g1)}$ and $T_{mrt\ (g2)}$ differences to $T_{mrt\ (rad.)}$ *versus* incoming short-wave radiation, K↓.

In Fig. 5, $T_{mrt\ (g1)}$ and $T_{mrt\ (g2)}$ differences to $T_{mrt\ (rad.)}$ were plotted against wind speed. The results indicate that wind speed fluctuation had a slightly higher influence on the small globe, causing a higher scattering in the calculated T_{mrt} values (lower R^2).

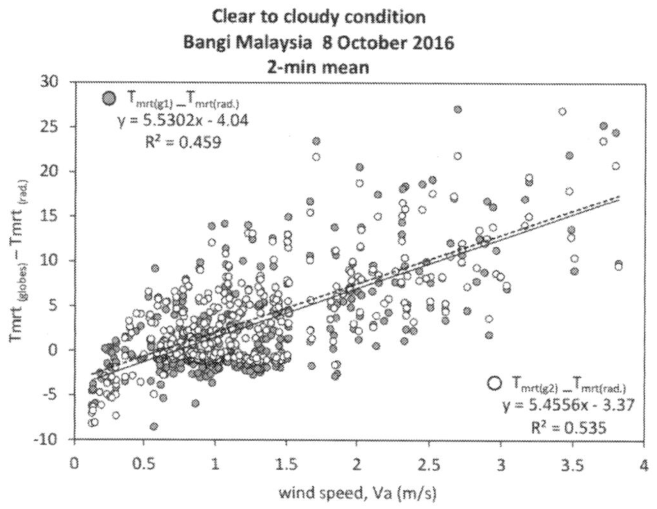

Fig 5: $T_{mrt\,(g1)}$ and $T_{mrt\,(g2)}$ differences to $T_{mrt\,(rad.)}$ *versus* wind speed

5. CONCLUSION

In this study globe thermometer method was used to estimate T_{mrt}, using two black-painted copper globes of 50mm and 150mm in diameters. Estimated T_{mrt} results were compared with integral radiation measurements as a reference method. The study showed that, in tropical outdoor urban setting, the globe thermometer method overestimated T_{mrt} during the middle of the day. In morning and late evening the difference between T_{mrt} data from the two methods became relatively small. The globe temperatures of the large and small globes increased as the amount of radiation increased, resulting in a reduced accuracy of the estimated T_{mrt}.

The results also showed that the obtained T_{mrt} values from the large and small globes fluctuated greatly with time; their differences from the reference T_{mrt} resulted in a large scatter in the function of radiation and wind speed. Furthermore, results showed that the small globe has a shorter delay to changes in outdoor conditions than the large globe. However, it gives slightly higher scatter in T_{mrt} values than the large globe, particularly during rapid changes in radiation and wind speed.

REFERENCES

[1] Matzarakis A, Rutz F, Mayer H. Modelling radiation fluxes in simple and complex environments: basics of the RayMan model. International Journal of Biometeorology. 2010;54:131-9.
[2] Gulyás Á, Unger J, Matzarakis A. Assessment of the microclimatic and human comfort conditions in a complex urban environment: modelling and measurements. Building and Environment. 2006;41:1713-22.
[3] Ali-Toudert F, Djenane M, Bensalem R, Mayer H. Outdoor thermal comfort in the old desert city of Beni-Isguen, Algeria. Climate Research. 2005;28:243-56.
[4] Höppe P. A new procedure to determine the mean radiant temperature outdoors. Wetter und Leben. 1992;44:147-51.
[5] Tan CL, Wong NH, Jusuf SK. Outdoor mean radiant temperature estimation in the tropical urban environment. Building and Environment. 2013;64:118-29.
[6] Johansson E, Thorsson S, Emmanuel R, Krüger E. Instruments and methods in outdoor thermal comfort studies–The need for standardization. Urban Climate. 2014;10:346-66.
[7] ISO I. 7726, Ergonomics of the Thermal Environment, Instruments for Measuring Physical Quantities. Geneva: International Standard Organization. 1998.
[8] Alfano FRdA, Dell'Isola M, Palella BI, Riccio G, Russi A. On the measurement of the mean radiant temperature and its influence on the indoor thermal environment assessment. Building and Environment. 2013;63:79-88.
[9] Thorsson S, Lindberg F, Eliasson I, Holmer B. Different methods for estimating the mean radiant temperature in an outdoor urban setting. International Journal of Climatology. 2007;27:1983-93.
[10] Kántor N, Unger J, Gulyás Á. Subjective estimations of thermal environment in recreational urban spaces—Part 2: international comparison. International Journal of Biometeorology. 2012;56:1089-101.
[11] Nikolopoulou M, Lykoudis S. Thermal comfort in outdoor urban spaces: analysis across different European countries. Building and Environment. 2006;41:1455-70.
[12] Wang S, 王上. Development of globe thermometers for measuring radiation and air speed. HKU Theses Online (HKUTO). 2015.
[13] Kántor N, Kovács A, Lin T-P. Looking for simple correction functions between the mean radiant temperature from the "standard black globe" and the "six-directional" techniques in Taiwan. Theoretical and Applied Climatology. 2015;121:99-111.
[14] Kuehn L, Stubbs R, Weaver R. Theory of the globe thermometer. Journal of applied physiology. 1970;29:750-7.
[15] Spagnolo J, De Dear R. A field study of thermal comfort in outdoor and semi-outdoor environments in subtropical Sydney Australia. Building and Environment. 2003;38:721-38.
[16] Nikolopoulou M, Baker N, Steemers K. Improvements to the globe thermometer for outdoor use. Architectural Science Review. 1999;42:27-34.
[17] Handbook A. Fundamentals. American Society of Heating, Refrigerating and Air Conditioning Engineers, Atlanta. 2001;111.

 DOI: 10.1051/e3sconf/20172301007

Effect of Different Solar Radiation Data Sources on the Variation of Techno-Economic Feasibility of PV Power System

*M.A. Alghoul [a, b,] *, Amer Ali [c], F.V. Kannanaikal [a], N. Amin [d], A.A. Aljaafar [c], Mohammed Kadhim [c] and K. Sopian [d]*
[a] Energy and Building Research Center, Kuwait Institute for Scientific Research, Safat 13109, Kuwait
[b] Center of Research Excellence in Renewable Energy (CoRe-RE), Research Institute, King Fahd University of Petroleum and Minerals (KFUPM), Dhahran 31261, Saudi Arabia
[c] The School of Applied Physics, Faculty of Science and Technology, Universiti Kebangsaan Malaysia
[d] Solar Energy Research Institute, Universiti Kebangsaan Malaysia, 43600 Bangi, Selangor, Malaysia
**Corresponding author e-mail: dr.alghoul@gmail.com*

Abstract: The aim of this study is to evaluate the variation in techno-economic feasibility of PV power system under different data sources of solar radiation. HOMER simulation tool is used to predict the techno-economic feasibility parameters of PV power system in Baghdad city, Iraq located at (33.3128° N, 44.3615° E) as a case study. Four data sources of solar radiation, different annual capacity shortages percentage (0, 2.5, 5, and 7.5), and wide range of daily load profile (10-100 kWh/day) are implemented. The analyzed parameters of the techno-economic feasibility are COE ($/kWh), PV array power capacity (kW), PV electrical production (kWh/year), No. of batteries and battery lifetime (year). The main results of the study revealed the followings: (1) solar radiation from different data sources caused observed to significant variation in the values of the techno-economic feasibility parameters; therefore, careful attention must be paid to ensure the use of an accurate solar input data; (2) Average solar radiation from different data sources can be recommended as a reasonable input data; (3) it is observed that as the size and of PV power system increases, the effect of different data sources of solar radiation increases and causes significant variation in the values of the techno-economic feasibility parameters.

Keywords: PV power systems, climate databases available online, effect of solar radiation from different climate databases, variation of techno-economic feasibility parameters

INTRODUCTION

In the literature, numerous researches discussed the techno-economic feasibility of hybrid renewable power systems [1-6] or PV power systems [7-10] using HOMER simulation tool. Techno-economic feasibility studies in the literature variated in their input data resources due to many climate databases available online. However, effect of different data sources of solar radiation on the techno-economic feasibility of PV systems are still a missing link in the literature despite there are many research areas in the literature investigated the effect of different data sources on the outcome of their studies [11-14]. Moreover, previous techno-economic feasibility studies usually dealt with a case study (a specific load for a respective location). Generalizing the outcome versus wide range of daily loads instead of a specific load is most effective. Therefore, the aim of this study is to enrich the literature with new research on the effect of different data sources of solar radiation on techno-economic feasibility of the PV power system versus wide range of daily loads. To success the verification, four data sources of solar radiation, different annual capacity shortages percentages (0, 2.5, 5, and 7.5) and wide range of daily load profile (10-100) kWh/day are implemented under Baghdad city location.

2. Materials and methods

Baghdad, Iraq's capital located at 33° latitude and 44° longitude. The monthly solar radiation from different data sources for Baghdad city is shown in Table 1. The solar radiation data sources are: Nasa climate site (SRC1) [15], weather base climate site (SRC2) [16], meteorological department data (Iraq) SRC3: (1984-2004) [17] and meteorological department data: (1962-1992) (Iraq) SRC4 [18]. The variation of monthly solar radiation obtained from different data sources is plotted in figure 1.

DOI: 10.1051/e3sconf/20172301007

The techno-economic feasibility parameters such as COE ($/kWh), PV array power capacity (kW), PV electrical production (kWh/year), No. of batteries, Battery lifetime (year) were simulated using HOMER. Then, the values of these parameters are compared to measure the significance of variation due to solar radiation from different data sources. Block diagram of the study evolution methodology is shown in Fig.2.

Table 1: Monthly solar radiation from different data sources for Baghdad city

Month	Monthly Solar Radiation (kWh/m^2)			
	SRC1: Climate site (Nasa)	**SRC2: Climate site (Weather base)**	**SRC3: Meteorological department data (1984-2004)**	**SRC4: Meteorological department data (1962-1992)**
January	3.08	2.97	2.96	3.07
February	4.09	4.03	4.03	3.98
March	5.16	4.97	4.97	4.97
April	6.19	5.39	5.44	6.03
May	7.15	6.47	6.48	6.87
June	7.57	7.56	7.58	7.87
July	7.45	7	6.99	7.75
August	6.94	6.77	6.71	7.28
September	6.23	5.56	5.55	6.16
October	4.56	3.97	3.97	4.62
November	3.37	2.97	2.86	3.41
December	2.67	2.61	2.64	2.88

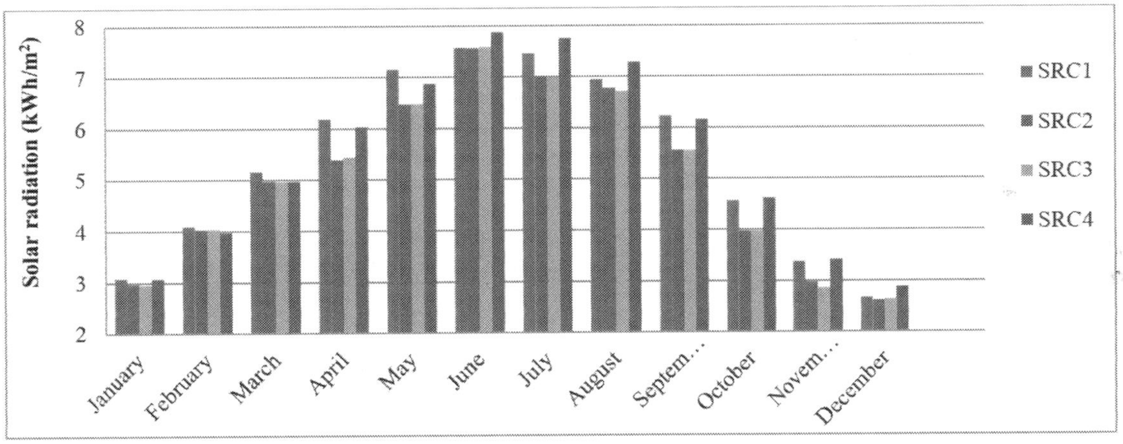

Fig 1: Variation of monthly solar radiation obtained from different data sources

DOI: 10.1051/e3sconf/20172301007

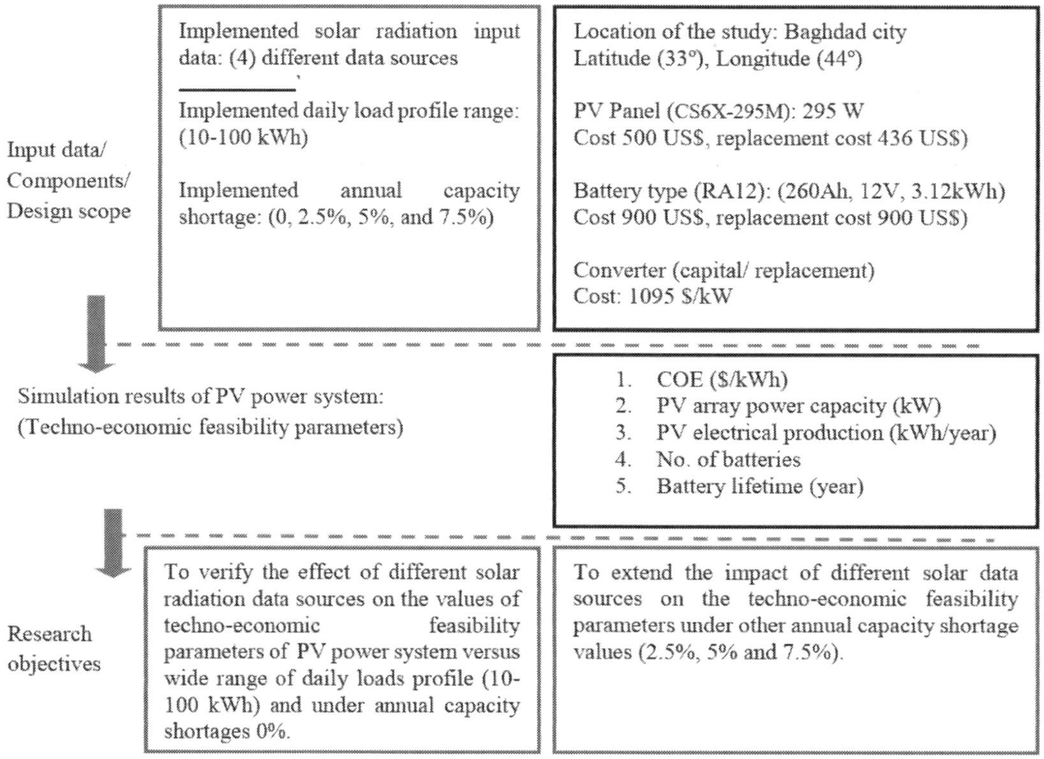

Fig. 2. Block diagram of the study evolution methodology

3. SIMULATION RESULTS AND DISCUSSION

Figure 3 illustrates the effect of multiple data sources of solar radiation on COE of PV power system versus wide range of daily load profile (10kWh-100kWh) for different annual capacity shortage. It can be seen that different data sources leaded to obvious variation in COE values for each capacity shortage. Comparison between the effects of these solar resources specifies that SRC3 and SRC4 are the most diverging resources at all capacity shortages with a maximum deviation of 22%. SRC1 and SRC2 vary at 0% capacity shortage but later at higher capacity shortage values they tend to converge. At the low load range, higher and unstable COE values can be witnessed meanwhile at medium and higher loads, stable COE values can be found. This is attributed to the fact that the adopted components are considered oversize for small loads. For example, the analysis showed that the adopted battery type (RA12, 260Ah, 12V, 3.12kWh) in this study is found oversize when designing PV system of small load.

Figure 4 illustrates the effect of multiple data sources of solar radiation (SRC1, SRC2, SRC3 and SRC4) on PV array power capacity versus wide range of daily load profile (10kWh-100kWh) for annual capacity shortage (0% and 5%). It can be perceived that different data sources of solar radiation lead to higher variation of PV array power at capacity shortage (0%) compare of that at capacity shortage (5%).

Figure 5 plots the effect of multiple data sources of solar radiation (SRC1, SRC2, SRC3 and SRC4) on PV electrical production versus wide range of daily load profile (10kWh-100kWh) for annual capacity shortage (0% and 5%). The results revealed that different data sources of solar radiation lead to significant variation of PV electrical production at annual capacity shortage (0%) while this variation vanished at capacity shortage (5%).

DOI: 10.1051/e3sconf/20172301007

Figure 6 displays the effect of multiple data sources of solar radiation (SRC1, SRC2, SRC3 and SRC4) on No. of batteries versus wide range of daily load profile (10kWh-100kWh) for annual capacity shortage (0% and 5%). It can be seen clearly that different data sources of solar radiation lead to significant variation in No of batteries at both capacity shortages.

Finally, effect of different data sources (SRC1, SRC2, SRC3 and SRC4) of solar radiation on battery lifetime at capacity shortage (0% and 5%) is shown in Fig 7(a-b) respectively. It is seen that different data sources of solar radiation leaded to significant variation in battery lifetime at capacity shortages 5% compare of that at capacity shortage 0%.

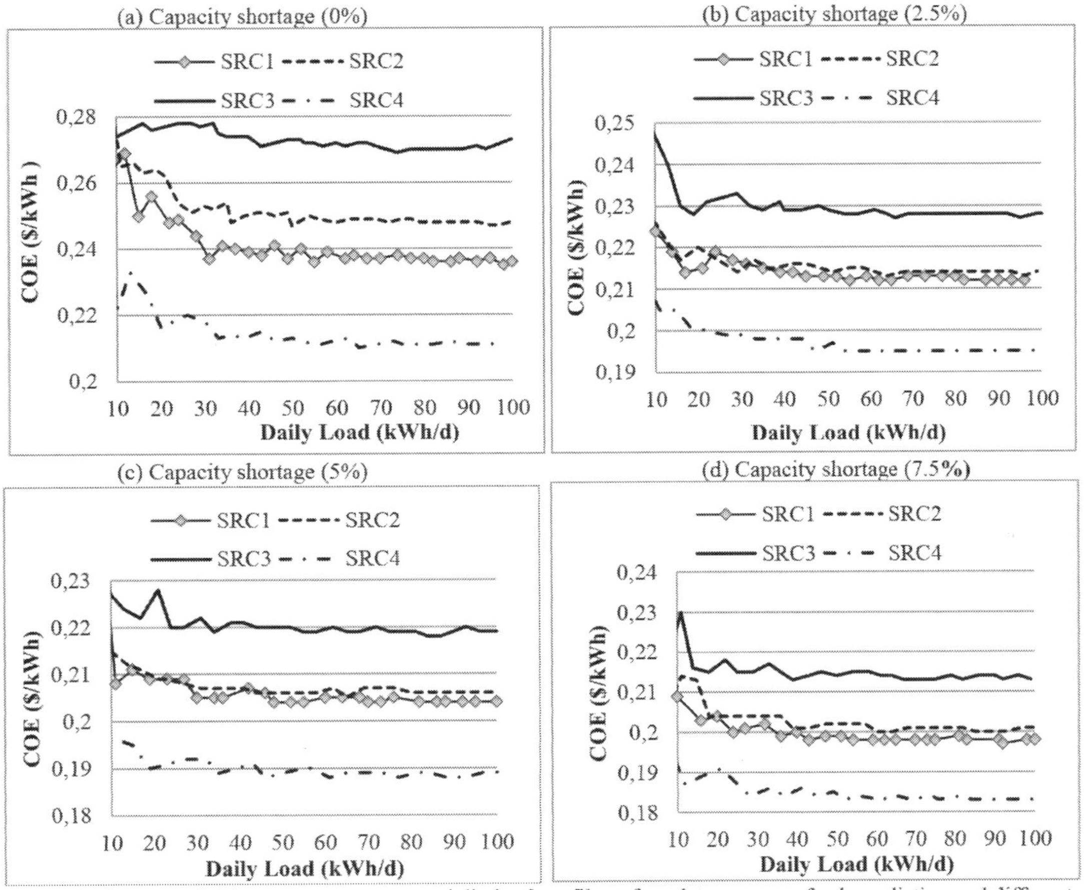

Fig 1 (a-d): COE of PV power system versus daily load profile at four data sources of solar radiation and different annual capacity shortage

DOI: 10.1051/e3sconf/20172301007

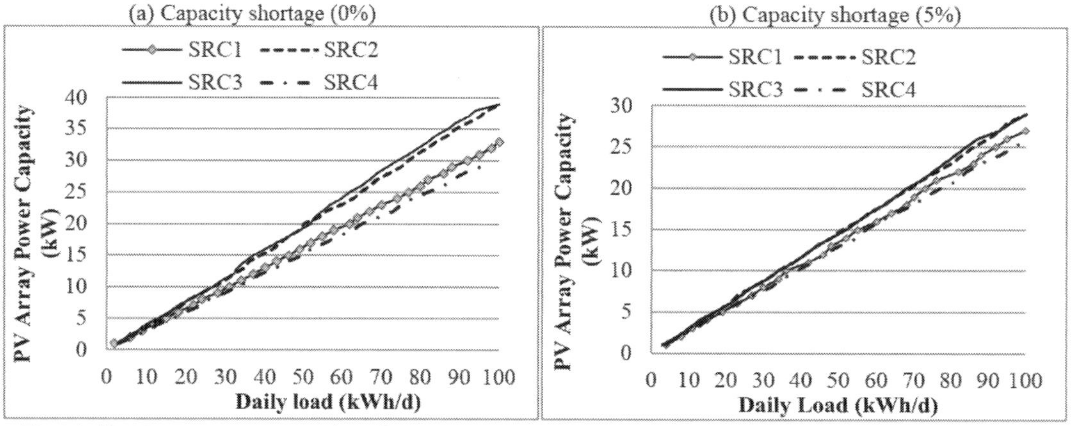

Fig 4 (a-b): PV array power capacity of PV system versus daily load profile at four data sources of solar radiation and two different annual capacity shortage

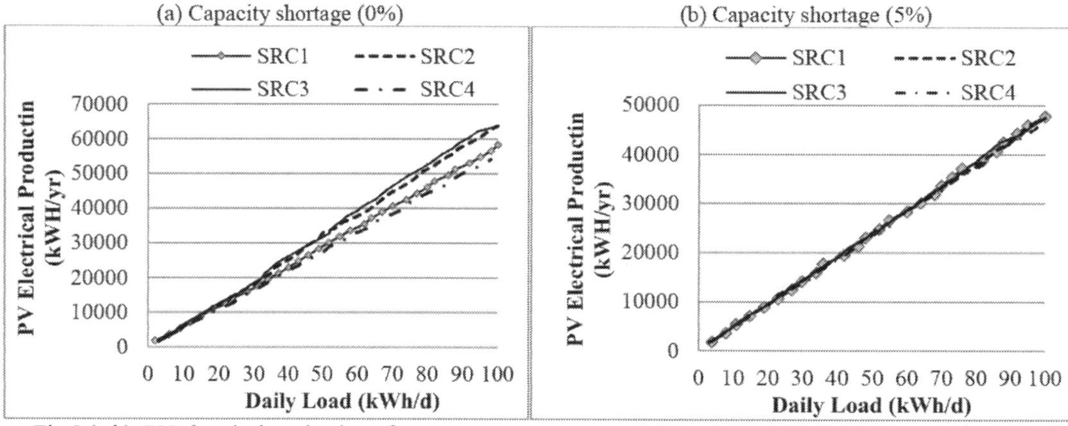

Fig 5 (a-b): PV electrical production of PV power system versus daily load profile at four data sources of solar radiation and two different annual capacity shortage

(a) Capacity shortage (0%)

(b) Capacity shortage (5%)

 DOI: 10.1051/e3sconf/20172301007

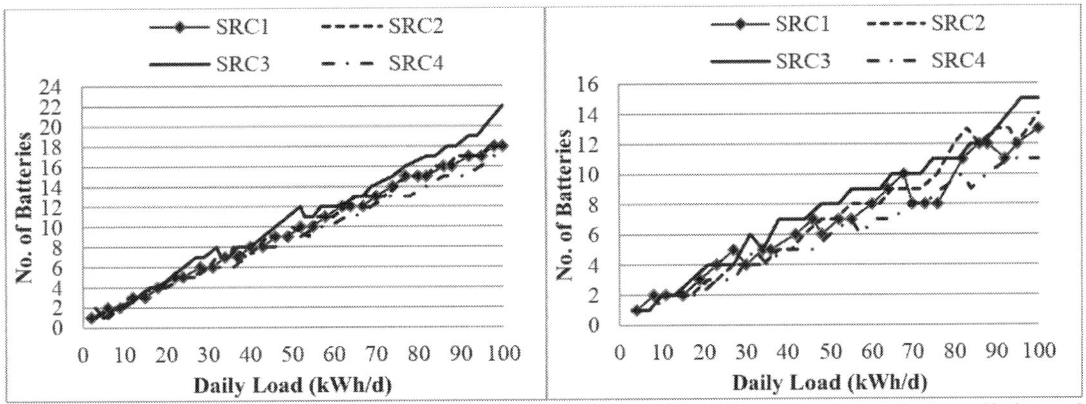

Fig 6 (a-b): No. of batteries of PV power system versus daily load profile at four data sources of solar radiation and two different capacity shortages

(a) Capacity shortage (0%) (b) Capacity shortage (5%)

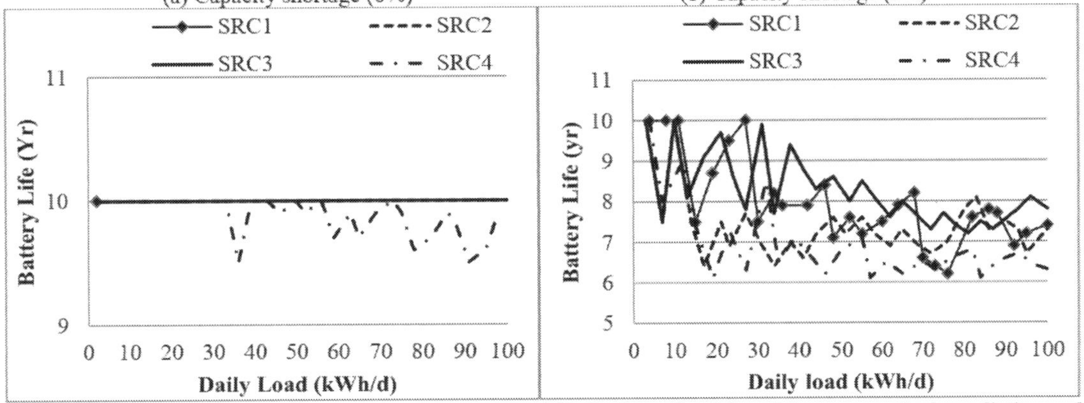

Fig. 7(a-b): Battery lifetime of PV power system versus daily load profile at four data sources of solar radiation and two different annual capacity shortage

4. CONCLUSION

The outcomes from this study are drawn as below:

- Solar radiation data from different data sources of Baghdad city caused variations in the values of the techno-economic feasibility parameters. So, careful attention must be paid to implement the proper input data of solar radiation in future works.
- This outcome of this study urge the authors to recommend average solar radiation from different data sources to be used as accurate input data for achieving the optimum techno-economic feasibility of PV power systems in future works.

Referencies:

[1] Braun. M., Girma. Z. 2013. Techno Economic Assessment and Optimization Study of Hybrid Power System Using Homer Software for Electrification of Rural District in Ethiopia. *International Journal of Renewable Energy Research*. 3(3).

DOI: 10.1051/e3sconf/20172301007

[2] Widatalla. A. M, Zinko. H. 2011. Designing a Photovoltaic Solar Energy System for a Commercial Building. Case Study: Rosa Park Hotel in Khartoum-Sudan. *World Renewable Energy Congress*. Vol. 11.

[3] Sopian, K., Zaharim, A., Ali, Y., Nopiah, Z.M., Razak, J.A. and Muhammad, N.S., 2008. Optimal operational strategy for hybrid renewable energy system using genetic algorithms. WSEAS Transactions on Mathematics, 7(4), pp.130-140. ISSN: 1109-2769.

[4] Shaahid, S.M & Elhadidy, M.A.2005. Economic analysis of hybrid photovoltaic–diesel–battery power systems for residential loads in hot regions—A step to clean future. *Renewable and Sustainable Energy Reviews* 11: 635–653.

[5] Shaahid, S.M. & Elhadidy, M.A. 2006. Technical and economic assessment of grid-independent hybrid photovoltaic–diesel–battery power systems for commercial loads in desert environments. *Renewable and Sustainable Energy Reviews 11: 1794–1810.*

[6] Malika Amini .2010. Renewable Energy Systems for Rural Health Clinics in Algeria: Homer Application. *Energy* 14: 1120–1130. http://homerenergy.com/webcast-downloads/Malika_Amini/health.pdf

[7] Amal A. Hassan , Abd El-Shafy A,Nafeh, Faten H. Fahmy &Mohamed A. El- Sayed. 2010. *Stand-Alone Photovoltaic System for an Emergency Health Clinic. International Conference on Renewable Energies and Power Quality. (ICREPQ'10).*

[8] I. M. Saleh Ibrahim Al-Jadi Photovoltaic in Libya Applications and evaluation. Proceedings of the International Conference on Renewable Energy for Developing Countries-20 http://www.udc.edu/docs/cere/PHOTOVOLTAIC%20IN%20LIBYA.pdf

[9] Al-Karaghouli, A., & Kazmerski, L. L. 2010. Optimization and life-cycle cost of health clinic PV system for a rural area in southern Iraq using HOMER Software.Solar Energy, 84(4), 710-714.

[10] Kazem, H.A., Ali, S.Q., Alwaeli, A.H., Mani, K. and Chaichan, M.T., (2013). Life-cycle cost analysis and optimization of health clinic PV system for a rural area in Oman. In Proceedings of the World Congress on Engineering, (Vol. 2, pp. 3-5). 2013, July , London, U.K.

[11] Rogers, J.D., 2009. Comparative Analysis Involving Wind Profile Data Sources. In Proceedings of the 20th AIAA Aerodynamic Decelerator Systems Technology Conference and Seminar. pp.1-13, 4 - 7 May 2009, Seattle, Washington,USA.

[12] Ben E. Lawsona, Michael D. Dayb, Michiala Bowenc, Rieks D. van Klinkend, Myron P. Zalucki 2010. The effect of data sources and quality on the predictive capacity of CLIMEX models: An assessment of Teleonemia scrupulosa and Octotoma scabripennis for the biocontrol of Lantana camara in Australia. Biological ControlVolume 52, Issue 1, January 2010, Pages 68–76

[13] Perez- Brunius et. Al 2007. Comparison of Upwelling indices off Baja California derived from three different wind data sources. CalCOFI Rep., Vol. 48, 2007, pp. 204-2014

[14] Jin, X., Hanesiak, J.M. and Barber, D.G., 2007. Time series of daily averaged cloud fractions over landfast first-year sea ice from multiple data sources. Journal of Applied Meteorology and Climatology, 46(11), pp.1818-1827.

[15] https://eosweb.larc.nasa.gov/sse/RETScreen/

[16] www.weatherbase.com

[17] Tadros, M. T. Y., Mustafa, M. A. M., & Abdel-Wahab, M. 2014. Estimation of the Global Horizontal Solar Radiation in Iraq. International Journal of Emerging Technology and Advanced Engineering Website: www.ijetae.com (ISSN 2250-2459, ISO 9001:2008 Certified Journal, Volume 4, Issue 8, August 2014)

[18] Mahdy, A. and Godu, H.A., 2010. Calculation and applications of net solar radiation in Iraq. Journal of alqadisiyah for pure science (quarterly), 15(1), pp.1-30.

E3S Web of Conferences 23, 02001 (2017)
World Renewable Energy Congress-17
DOI: 10.1051/e3sconf/20172302001

Sustainable Design Approach: A case study of BIM use

Wael Abdelhameed

Department of Architecture and Interior Design, University of Bahrain
wabdelhameed@uob.edu.bh; wael.abdelhameed@gmail.com

Abstract

Achieving sustainable design in areas such as energy-efficient design depends largely on the accuracy of the analysis performed after the design is completed with all its components and material details. There are different analysis approaches and methods that predict relevant values and metrics such as U value, energy use and energy savings. Although certain differences in the accuracy of these approaches and methods have been recorded, this research paper does not focus on such matter, where determining the reason for discrepancies between those approaches and methods is difficult, because all error sources act simultaneously. The research paper rather introduces an approach through which BIM, building information modelling, can be utilised during the initial phases of the designing process, by analysing the values and metrics of sustainable design before going into the design details of a building.

Managing all of the project drawings in a single file, BIM –building information modelling- is well known as one digital platform that offers a multidisciplinary detailed design –AEC model (Barison and Santos, 2010, Welle et.al., 2011). The paper presents in general BIM use in the early phases of the design process, in order to achieve certain required areas of sustainable design. The paper proceeds to introduce BIM use in specific areas such as site selection, wind velocity and building orientation, in terms of reaching the farther possible sustainable solution.

In the initial phases of designing, material details and building components are not fully specified or selected yet. The designer usually focuses on zoning, topology, circulations, and other design requirements. The proposed approach employs the strategies and analysis of BIM use during those initial design phases in order to have the analysis and results of each solution or alternative design. The stakeholders and designers would have a better effective decision making process with a full clarity of each alternative's consequences. The architect would settle down and proceed in the alternative design of the best sustainable analysis. In later design stages, using the sustainable types of materials such as insulation, cladding, etc., and applying sustainable building components such as doors, windows, etc. would add more improvements and enhancements in reaching better values and metrics.

The paper describes the methodology of this design approach through BIM strategies adopted in design creation. Case studies of architectural designs are used to highlight the details and benefits of this proposed approach.

DOI: 10.1051/e3sconf/20172302001

Sustainable Architectural Design

According to Wikipedia definition, the basic objectives of sustainability are to reduce consumption of non-renewable resources, minimize waste, and create healthy, productive environments.

Sustainability in architectural design seeks lower energy and water consumption; limit resource consumption through an emphasis on water recycling and solar energy; give priorities to non-toxic materials; minimize impact on climate change by reducing gas emissions; and emphasize quality and durability over price.

Sustainable architectural design is to design the built environment to respect the principles of social, economic and ecological sustainability. It attempts at reducing negative impacts on the environment in general, and the comfort of building occupants in particular. Its main objective, therefore, is to improve building performance in terms of the three main areas: society, economy and ecology.

Although the three main areas of sustainability are society, economy and ecology, this research paper is exclusively concerned with the ecological factors, the environmental factors and their impact on building performance at the initial design phases before the design is finalised. The first area, social factors, is viewed as immaterial values that would distinctly vary from one society to another. It therefore would preferably be evaluated by the architects based on the design location and the society values and customs. The second area, economic factors, has more obvious effect and impact in the final architectural design phases in selecting the finishing materials, more than in the initial phases of the design process itself.

In summary, the scope of this research paper is the environmental factors on building performance during the initial design phases. The paper focuses on small size projects where architects in the most cases are the only designers.

The Current Architectural Education and Practice

There are many factors and requirements that are taken into account during the architectural design process. Architectural design solution should satisfy the whole set of those interrelated factors and requirements. It is often necessary to develop an integrated solution for a whole set of interrelated requirements (Lawson, 2005).

The architect during conceptual designing process attempts to respect main factors namely: aesthetics, functions and economics. The architect seeks composing and proposing forms and spaces through required relationships and certain topologies in order to have a satisfactory design solution (Abdelhameed, 2004).

Other factors may influence the architectural design are: climate –weather temperature, sunshine and sun path, relative humidity, wind direction and speed-; topography –contour lines, nearby water surfaces, plateaus and mountains-; site limitations and considerations –main and service entrances, infrastructure connections, outdoor spaces-; urban design –surrounding urban fabric, road network, surrounding building materials and structural systems-; structural and technical issues –structural systems, building materials, mechanical and electrical requirements-.

The architecture curriculums in the theoretical level, as well as architectural design offices in the professional level, focus during the initial design phases and conceptual designing on these foregoing factors. In the practical and professional level, the environmental factors analysis used to be conducted at the final stages or after the design is completed, specifically in cases of small and medium size projects in which the designer is a small team of architects only with no other discipline specialists.

The Proposed Method

Architects should apply the analysis of different disciplines in order to achieve optimal design solutions in terms of electrical, mechanical, solar, thermal, energy, acoustic analysis. The early analysis in design stages helps reduce costs and increase occupant satisfaction.

The proposed method is to include the environmental factors analysis among the requirements of the design process, consequentially in the theoretical level the design evaluation inside the design studios of architecture schools should highlight this importance, and in the practical and professional level the different kinds of environmental analysis have to be performed to achieve the optimal solution before betaking to and relying on different assistances such as expensive insulation materials.

The proposed method employs BIM, building information modelling, as a single platform of the design to perform the environmental factors analysis, by using Autodesk programs, Revit as a BIM program and Green Building Studio as energy analysis and simulation program, versions 2014 and 2015. Having one platform would help architects (Kim et.al., 2015, Schlueter and Thesseling, 2009) and architecture students (Sebastian and van Berlo, 2010) effectively conduct the design and the environmental analysis without the assistance of other specialists or the knowledge of other energy analysis programs such as eQuest.

The research method focuses on the change that may need to be made in the design before starting in the working drawings and design details. Therefore, the researcher asked the students to model their previous conceptual designs using BIM, Revit program, and to conduct the environmental analysis on their designs. The objective is to discover whether these designs can be slightly modified and changed to improve the design performance or not.

The main reason of applying only slight change or modification is to keep the design satisfying other architectural design requirements that were discussed in the previous part of the research paper. This condition was highlighted to the students. Any change should not contradict with any other architectural requirement such as, function, structural, technical, etc.

Research Methodology

The research paper attempts at applying and investigating the proposed method in different areas of sustainability in design cases, by measuring environmental factors such as: solar analysis, thermal analysis, shadows and reflections, ventilation and air flow.

The research paper employs qualitative analysis of architectural design case studies to investigate the effectiveness of the proposed method. The researcher's students applied the environmental analysis in different design case studies. Each student was asked to apply one area of environmental analysis on a design, and to record the change in the design that may improve the design performance.

The research paper presents the qualitative analysis of each design case, and what were the benefits, if any, that were added to the design.

Architectural Design Cases and Qualitative Analysis

The researcher recorded the steps used in each design case according to the emphases area of environmental analysis. Any change in the architectural design was discussed with the students in order to ensure the consistency of modification level.

There were some architectural considerations, in few design cases, which are not used in this research paper because the change was architecturally significant. However, the proposed method was effective even in those cases. Clear

 DOI: 10.1051/e3sconf/20172302001

improvements were recorded after applying the environmental analysis in those design cases.

No comparison between design case studies was used or conducted due to having a different environmental area and application in each design case. Nevertheless, the improvement in each design case was evident.

Site Selection

The environmental analysis in this case study was conducted to compare between different proposed sites that satisfy the project requirements. Figure 1 displays the results of the best site in terms of electricity consumption while Figure 2 displays the environmental analysis of different sites.

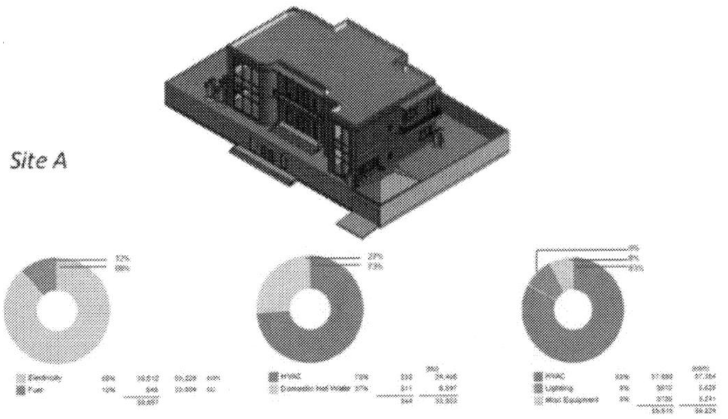

Figure 1 Environmental analysis of the selected site

DOI: 10.1051/e3sconf/20172302001

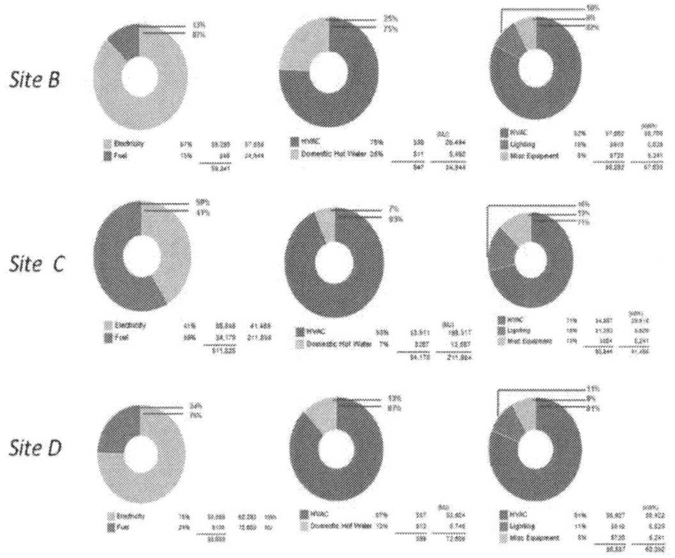

Figure 2 Environmental analysis of the proposed sites of a design project

Orientation Analysis

In this design case study, the question was how the angle between building masses would improve the building performance. The student conducted environmental analysis of some angles -30, 45, 60, 80 and 90 degrees- that would not apply any change in the conceptual design. According the analysis, the best angle between the building masses was 30 degrees to achieve the required wind speed, Figure 3.

 DOI: 10.1051/e3sconf/20172302001

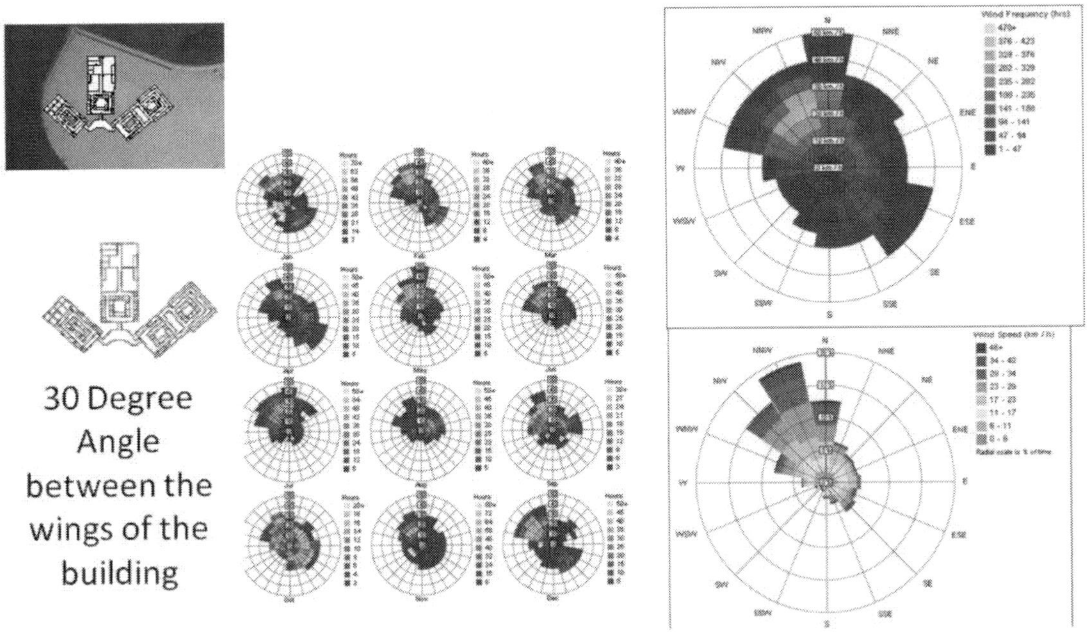

Figure 3, An example of BIM analysis to decide the best angle between the building masses

Another case study of orientation analysis was to specify the best orientation of a courtyard inside the design to achieve the optimal cooling solution, Figure 4 and 5.

Figure 4, Simulation of environmental analysis in the plan view

DOI: 10.1051/e3sconf/20172302001

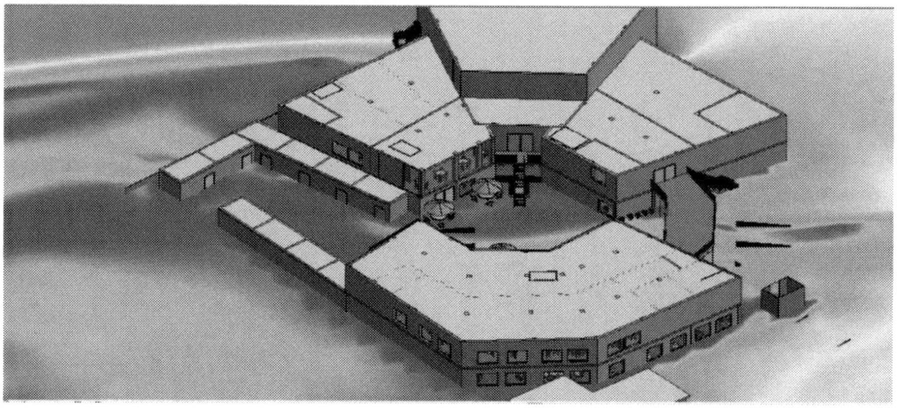

Figure 5, Simulation of environmental analysis to show the best cooling result

6.3. Energy Saving Analysis

The environmental analysis in the area of energy saving was conducted to get data of thermal, temperature and wind. Data analysis simulation on the building envelope helped the student decide simple changes in the architectural design, such as projecting the first floor in order to shade the main entrance, Figure 6 and 7.

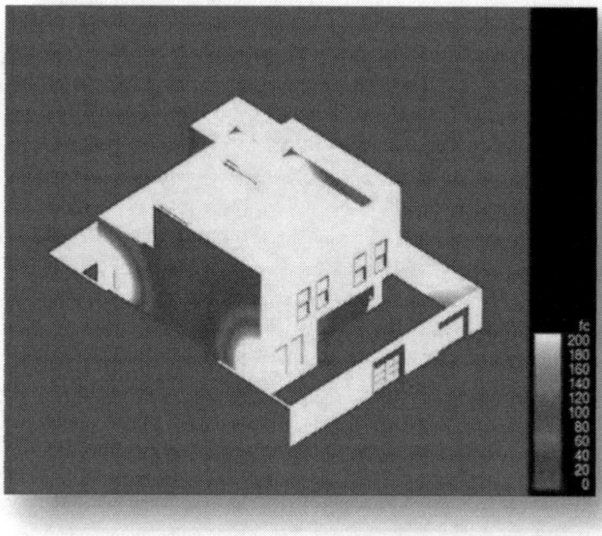

Figure 6, Data simulation on the building envelope.

DOI: 10.1051/e3sconf/20172302001

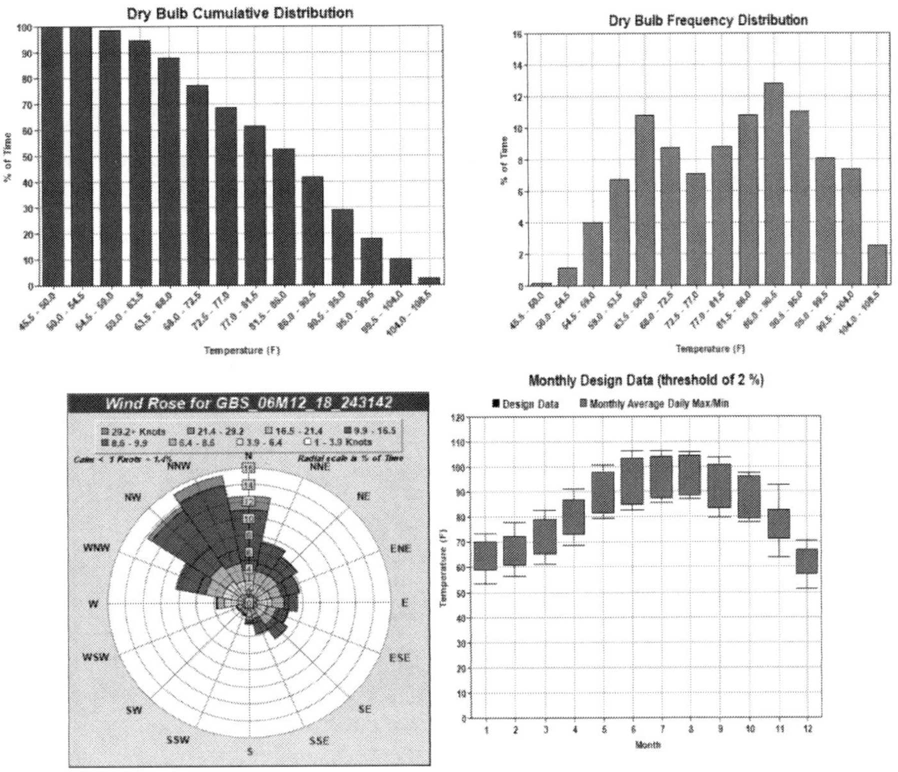

Figure 7, Energy data results of the modified solution

U-Value Calculations

Since the U-value is a measurement of heat flow, the lower the U- value, the more slowly does the material transfer heat in and out of a building. As per Bahrain constriction law, the overall thermal transmittance value -U-value- for the walls should not be more than 0.75 W/m².°C (Mohanadi, 2006).

In this case study, a plug-in for U value calculations was used inside the Revit program to help the student save time and effort. The plug-in automatically calculates U-Value -Overall heat transfer coefficient- for walls, floors, roofs, ceilings and foundation slabs.

Based on the result, the student used a double layer wall with an air cavity in the middle which reduced the U-Value to the target value 0.75 W/m².°C, Figure 8.

DOI: 10.1051/e3sconf/20172302001

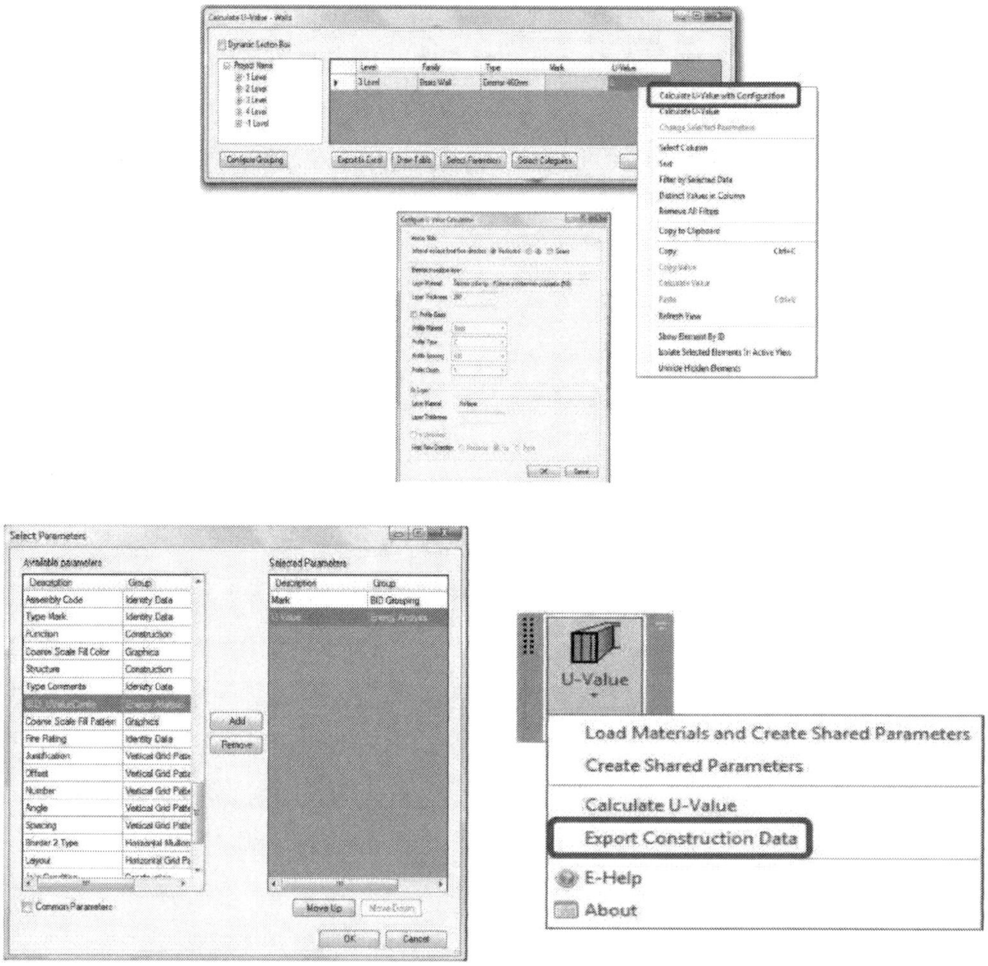

Figure 8, 3 screen shots of the Plug-In for U-Value calculations in the student project

In all design cases, every student presented a slight change in the conceptual design to clearly improve the existing conceptual design. This evidence not only proves the effectiveness of the proposed method but also highlights that the lack of awareness and the importance of environmental analysis.

Conclusion and Recommendations

The analysis recorded by the research paper proved the effectiveness of the proposed method. In all architectural design cases, the building performance was improved by

 DOI: 10.1051/e3sconf/20172302001

applying a slight architectural change that does not affect the main project requirements and objectives.

The different environmental analysis should be conducted to each design even at small size projects. More awareness should be achieved and spread among architects and stakeholders to highlight the importance of environmental analysis in the conceptual design phases.

Architecture curriculums should concentrate on the environmental analysis in design studios with the same value and scale as other design requirements such as structure, site, urban design, etc.

References

1. Kim, J.B., Jeong, W., Clayton, M.J., Haberl, J.S. and Yan, W.: Developing a physical BIM library for building thermal energy simulation, Automation in construction, 50, 16-28 (2015).
2. Lawson, B.: How designers think: the design process demystified, Elsevier/Architectural, (2005).
3. Schlueter, A. and Thesseling, F.: Building information model based energy/exergy performance assessment in early design stages, Automation in construction, 18(2), 153-163 (2009).
4. Sebastian, R. and van Berlo, L.: Tool for benchmarking BIM performance of design, engineering and construction firms in the Netherlands. Architectural Engineering and Design Management, 6(4), 254-263 (2010).
5. Welle, B., Haymaker, J. and Rogers, Z.: December. ThermalOpt: A methodology for automated BIM-based multidisciplinary thermal simulation for use in optimization environments, Building Simulation (4) 4, 293-313 (2011).
6. Mohanadi, K. k. A.: Thermal insulation in buildings (1st ed.). Kingdom of Bahrain: Ministry of electricity & water (2006).
7. Abdelhameed, W.: Visual design thinking in the design process as impacted by digital media. Architecture in the Network Society, the 22nd eCAADe Conference Proceedings, 90-94 (2004).
8. Barison, M.B. and Santos, E.T.: BIM teaching strategies: an overview of the current approaches, ICCCBE International Conference on Computing in Civil and Building Engineering (2010).

Solar radiation and street temperature as function of street orientation. An analysis of the status quo and simulation of future scenarios towards sustainability in Bahrain

Dr. Joao Pinelo Silva, Assistant Professor Architecture

College of Engineering at University of Bahrain

Abstract

This paper discusses the contribution of street orientation towards the development of a comfortable microclimate for pedestrians in Bahrain. Increasing walkability is a global agenda to address issues such as a) transportation, b) energy consumption, c) health, and d) air pollution, all of which are topics of the sustainability agenda. Thermal comfort is one of the pre-requisites for walkability. In warm climates, this is a challenging goal.

Street design is paramount for pedestrian comfort in warm climates. The roles of street orientation and aspect ratio are of particular importance as they determine the intake of solar radiation into the urban canyon. We investigate the state of affairs in Bahrain, by measuring the frequency with which the street orientations E-W, N-S, NE-SW, and NW-SE, currently occur. Research suggests that the street orientation E-W presents the lesser performance for mitigating the effects of heat gain. The ideal grid orientation would, therefore, be N-S, and NE-SW - NW-SE, avoiding street segments with E-W orientation.

A countrywide analysis shows that E-W orientation accounts for the highest overall street length with 37%. The second most frequent orientation is N-S (29%), the best performer. NW-SE and NE-SW both have frequencies of only 17%. Preference for a street grid with N-S, NW-SE, and NE-SW orientation would improve the thermal performance of streets and provide a continuous network of a comfortable pedestrian environment. We simulate two future scenarios based on avoiding new E-W streets, or not. We measure their potential reduction in thermal gain and conclude that a simple policy could reduce solar exposition in 40%.

Keywords: Walkability, Pedestrian, Street orientation, Temperature, Street design, Street canyon

DOI: 10.1051/e3sconf/20172302002

Introduction

The global trend towards the development of a safe pedestrian infrastructure has been pursued for decades [1][2] as a way to encourage walkability, as it promotes street livability [3], local economic vitality [4], and social interaction [5]. More recently, the pursue to increase pedestrianism has been driven by significant public health concerns due to soaring levels of inactivity across the globe [6][7]. In response to global goals to reduce inactivity [8], urban planning policies have focused on increasing walking for transportation [9][10][11][12], which is leading to a reinvigorated pursue of a connected and attractive network of public space.

Progress in urban climatology and biometeorology in the last decade have created conditions for the development of studies of behavior in the urban context, which shown that outdoor activities are affected by the microclimate [13][14]. In regions where the weather can severely influence the use of outdoor environments, it is essential to mitigate its effects. This mitigation can be achieved by the manipulation of the microclimate at street level.

Due to their direct exposure to the weather, the manipulation of outdoor microclimates is more limited in scope and depth than that of indoor environments. For this reason, the creation of outdoor environments is challenging, and achieving the maximum mitigation of the climate implies joint planning and design towards that particular goal. To plan and design for particular microclimatic goals, planners and designers need to have baseline information on the effect of design solutions to the microclimate. The main design component with influence on microclimate is the solar orientation of streets. This work measures the country-wise overall practice up to date, establishing a baseline that could serve as a reference point for guiding future development.

Outdoor thermal comfort

Persistent high temperatures due to climate might seem impossible to overcome by passive means. However, the manipulation of the microclimate to increase the thermal comfort outdoors does not imply reducing the summer temperatures to the level of winter ones. Thermal comfort does not depend entirely on the physics of the environment, such as temperature or humidity, but on psychological aspects as well

 DOI: 10.1051/e3sconf/20172302002

[15]. Preliminary research on outdoor thermal comfort suggested a significant discrepancy between an objective description of the thermal comfort (Predicted Mean Vote (PMV) [16], based on environmental physics) and the subjective description (Actual Sensation Vote) [17]. The discrepancy lead the authors to conclude that the evaluation of thermal comfort must not be limited to physical indicators but to include physiological and psychological ones [15] relative feeling of comfort is, therefore, to a certain extent, dependent on the larger thermal context. Studies show that even declared indoor comfort depends on the contextual outdoor temperatures. Rupp and colleagues found that participants reported equivalent levels of comfort when exposed to different indoor temperatures, responding to variations in the outdoor temperature, with higher levels of outdoor temperature affording the participants comfort indoors at higher temperatures [18]. The findings strongly suggest a sense of adaptation on the part of the user, which can bridge some magnitude of the shortcomings of passive (design-based) microclimate manipulations.

The literature on urban climatology and biometeorology does not prescribe high nor low thresholds for parameters such as temperature, as cut off points for the use of outdoor environments. Physiological differences between groups seem to have a bearing on temperature preferences and sensitivity [19][20][21][22]. However, these studies have focused on indoor environments and did not include very high temperatures. Nonetheless, it seems likely that some groups are more vulnerable to high temperatures than others. Knowledge of such details would be informative to the design of outdoor environments. While research carried out locally might eventually reveal more precise guidelines for exactly which temperatures outdoor pedestrian areas should aim to deliver, it is an undeniable fact that in warm climates, an influence at the level of the microclimate is desirable.

The urban canyon – street

Climate physics are closely related to latitude, due to the effect that this has on the exposition to solar radiation. The impact of the climate on the microclimate is therefore studied at the regional level. The dulling of properties of the climate, such as temperature, can be achieved through the design of the built environment, namely through the manipulation of direct sun exposure. The systematic study of the impact of design features on the microclimate provides planners and designers with valuable information to make goal-oriented design decisions.

The streets make up for most of the urban public space used by pedestrians. Two properties of street design have been shown to have significant impact on the microclimate: a) street orientation, the relation of the street to the sun path, commonly represented by the cardinal directions; b) aspect-ratio, the proportion between the height of the buildings and the distance between them (width of the street). Such characteristics are at the genesis of any urban design, and, to a certain extent, depend on urban planning, and consequently, make a useful starting point for the investigation of local conditions. Street orientation is of particular interest because it is hardest to change after it has been built, while the ratio can be altered through the redesign of individual buildings.

Due to the geometrical and microclimatic similarities with canyons, urban streets are commonly described as urban canyons, canyons, or street canyons, after [23]. These are important spatial units for the analysis of the energy input, partitioning and output of a structure that is repeated throughout a city [23]. Since the exposure influences the amount of energy input, the study of typical exposures is necessary for a comprehensive documentation of the thermal behavior of each configuration. In the absence of a local study, we use as reference one study completed in Ghadir, Tunisia, at a latitude of 32.40N, 3.80E [24] approximately six degrees of latitude North of Bahrain (26.13°N, 50.35°E). Shading as a design strategy to mitigate extreme heat has been confirmed by [25] and [26].

While the solar gain of one street contributes to its own overheating, the citywide sum of the thermal gain of individual street canyons contributes to raising the heat of the area, contributing to the urban heat island effect, as described by [27]. The effect results in the city having higher temperatures than surrounding non-urbanized areas. A perverse consequence of this effect is reached when the temperature is raised to a level that leads to the use of active cooling methods such as air conditioners. The motors of such equipment generate heat to produce cool. While the cold air is fed to indoor environments, the hot air is dissipated outdoors, thus contributing to a further increase of the temperature within the canyons.

DOI: 10.1051/e3sconf/20172302002

The Orientation of Street Canyons

Ali-Toudert and Mayer [24] show through numerical analysis of digital modeling of the microclimate of street canyons that the orientation of the canyon significantly shapes thermal gain, thus influencing air temperature at 1.2meters from the ground (the pedestrian domain). The model is based on a typical summer day (1st August) in Ghardia, Algeria, approximately 6° North of Bahrain. The climate is hot and dry (Saharan). The simulation duration is from 7:00 to 20:00. The model had a spatial resolution of one meter horizontally, and two meters vertically. The results report that the highest differences between orientation are felt in the afternoon (around 15:00), and are only significant between 9:00 and 19:00. As the aspect ratio increases, East-West (E-W) streets tend to be warmer than North-South (N-S) streets, due to their longer exposure to solar radiation. Overall, the results show that the most favorable orientation regarding thermal comfort is N-S, and the least comfortable E-W, for their longer periods of exposure to solar radiation. The intermediate orientations Northeast-Southwest (NE-SW), and Northwest-Southeast (NW-SE), present equivalent and intermediate levels of comfort, but are noticeably different from E-W streets [24] by offering lower temperatures and smaller duration of peak exposure.

E-W streets with low aspect ratio present PET patterns with extremely high values for over eight hours per day. Furthermore, the increase of height/width ratio to mitigate heat gain in E-W orientations is extremely limited [24, p.102]. Comparatively, N-S streets with equivalent aspect ratios present lower temperatures in general and a smaller duration of peak temperatures. On streets with an aspect ratio ≥ 2, the peak PET temperature is reduced and only occurs for 2 hours, around noon. Streets with intermediate orientations (NE-SW, NW-SE), present (in equivalent aspect ratios), a compromise between the two main orientations (N-S and E-W), with lower peak temperatures and shorter length of about 4 hours.

Overall, the analysis of the orientation makes sense in isolation only at an aggregate level, as a way to measure the potential to increase comfort at street level in the country's portfolio. The real solar gain highly depends on width to height ratio. At very low ratios (approx. 0.5) the differences in solar gain between orientations are minimal. However, the as the ratio increases, the reduction in solar gain increases significantly in N-S and NE-SW and NW-SE, while it is barely noticeable on E-W streets. Current orientation of segments with an aspect ratio < 2 is a measurement of

DOI: 10.1051/e3sconf/20172302002

the dormant potential to more comfortable streets. For segments with aspect ratio ≥ 2, the orientation reflects the current intake of solar radiation.

According to the results discussed above, the ideal orientation would be to use would be N-S, which is roughly the bearing of the island of Bahrain, and is likely to be naturally frequent. Nonetheless, grids tend to be orthogonal, and therefore, E-W streets, which are the poorest thermal performers, would complement N-S ones. The optimal solution seems likely to be a mix of N-S streets with a grid expanding NE-SW and NW-SE. Furthermore, a crucial feature of a pedestrian network is continuity. A grid with N-S and E-W elements would have its continuity challenged by the lack of thermal comfort in E-W streets, even if these had high aspect ratios. Figure 1 illustrates the four street orientations discussed and the binning factor used to categorize the current network of Bahrain, discussed below.

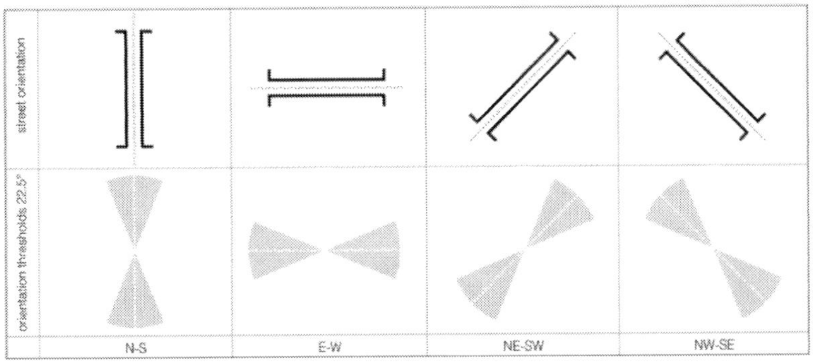

Figure 1 – Orientation of street canyons and the binning thresholds used for the categorization of the segments in the case studies of Bahrain and Manhattan N.Y.

Site Climate

DOI: 10.1051/e3sconf/20172302002

Bahrain, the main island of the archipelago Kingdom of Bahrain, has an area of 572 square kilometers. The island measures 48 kilometers from north to south, and 16 kilometers east to west. The highest point is Jabal al Dukham at 134 meters above the mean sea level. The largest city is the capital Manama, which originally developed around the seaport, lays just above sea level. The climate presents pleasant temperatures during the largest part of the year, with the minimum average low temperature of 14 °C Celsius taking place in January. However, from May through September, the temperatures are higher, and in the summer temperatures in the afternoon frequently reaching the high forties. The average temperature for the months of May through September is above 30°C Celsius (86°F Fahrenheit), with the average of the minimum temperature during June through September being over 30°C (86°F Fahrenheit), as illustrated in Table 1.

Table 1. Average monthly temperature, relative humidity and precipitation in Bahrain

	Jan.	Feb.	Mar.	Apr.	May	Jun.	Jul.	Aug.	Sep.	Oct.	Nov.	Dec.
Average High Temperature (°C)	20.3	22.4	26.1	30.7	36.9	39.0	39.9	39.8	37.9	33.7	27.7	22.4
Average Temperature (°C)	17.2	18.5	21.7	26.2	32.0	34.6	35.6	35.5	33.7	30.0	24.6	19.3
Average Low Temperature (°C)	14.4	15.5	18.5	22.8	28.3	31.1	32.2	32.2	30.4	26.9	21.9	16.6
Maximum Relative Humidity (%)	88	87	84	81	78	77	80	82	85	86	85	87
Minimum Relative Humidity (%)	57	55	47	42	39	39	40	44	44	47	52	56
Precipitation (mm)	14.6	1.5	1.9	5.1	0.4	0.0	0.0	0.0	0.0	0.0	10.2	14.5

Source:

The relative humidity is high throughout the year. The mean daily maximum relative humidity reaches the annual highest value of 88% in January, and the minimum of 77% in June. The mean daily minimum relative humidity reaches a maximum of 57% in January and a minimum of 39% in May and June. The precipitation is irregular and low, with a maximum of approximately 15mm in December and January, and 0mm from June through October (Table 1).

Methodology

We use the segment map of Bahrain, which is similar to a road-center-line map, except for the fact that it includes pedestrian areas and not only vehicular ones, and it is made exclusively of straight segments, as opposed to including arcs [28]. Therefore such map presents nodes at every change in direction, making it a suitable candidate for the measurement of the orientation of each component of the street

DOI: 10.1051/e3sconf/20172302002

network. For comparison, we use the segment of Manhattan (New York) to demonstrate an alternative balance between the main four cardinal orientations (Figure 2). The maps were drawn on a Geographic Information Systems (GIS) software package, and the azimuth of each street segment measured.

The data was then exported to a statistical analysis package [29] where the analysis is processed and charts created [30]. We quantify each orientation category in two ways: a) the total number of street segments, and b) the total sum of street

length. Remove lines with length less than 30 m, as they are not likely to represent formed street canyons.

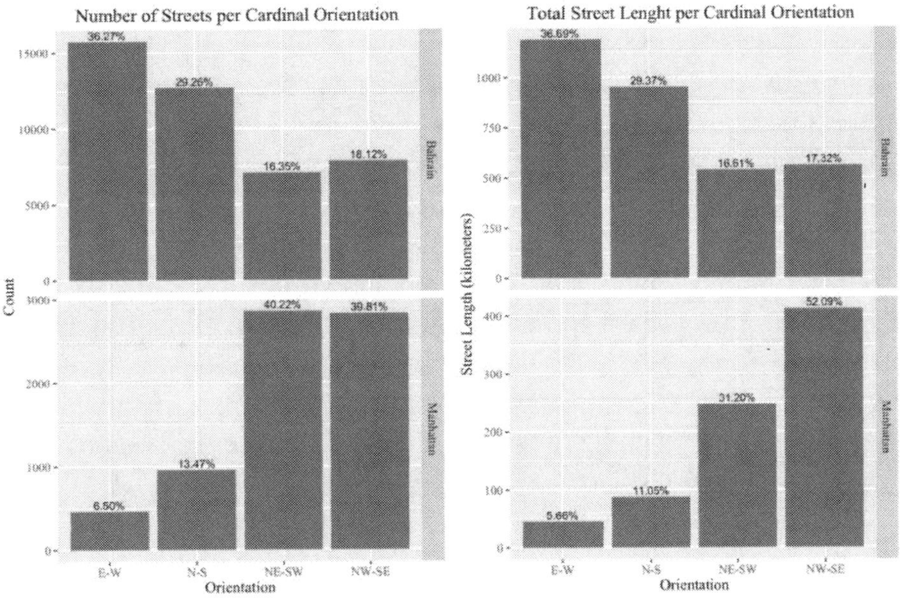

Figure 2 – The relative proportions of each orientation towards the number of streets and total street length in Bahrain and Manhattan present opposite trends.

Based on the spatial and temporal variation of the physiologically equivalent temperature PET at 1.2 meters above ground, for the aspect ratio (height/width) 2,

 DOI: 10.1051/e3sconf/20172302002

as calculated per [24], we calculate the thermal coefficient of each street orientation (TCSO). The TCSO is calculated by multiplying the time interval of duration of PET over 42°C (107.6 Fahrenheit) by the maximum temperature reached. The units of the TCSO are hours per degrees Celsius. The TCSO for each orientation are shown in Table 2.

Table 2. Thermal Coefficient of Street Orientation based on Ali-Toudert and Mayer (2006)

Orientation	Interval PET > 42°C (time)	Duration (hours)	Max. PET (°C)	TCSO
E-W	09:00 to 18:00	9.00	68	612
N-S	11:00 to 15:00	4.00	56	224
NE-SW	11:30 to 16:00	4.50	66	297
NW-SE	10:15 to 16:30	6.25	62	388

As expected, the TCSO is significantly larger for the E-W orientation. We calculate the current 'cost' of each orientation by multiplying the total street length by the corresponding TCSO. We then simulate two alternative scenarios for the future, when the network is doubled. Figure 3 illustrates the potential impact on the growth of TCSO for the same total street length. Scenario 1 (Future non Regulated) describes a situation that can be expected if no regulation addresses the choice of street orientation. Scenario 2 describes a future where a planning policy determines that no streets should be made with orientation E-W. In this scenario, the other types grow by the same proportion (2.33333). The potential positive effect is noticeable by the lesser increase of TCSO (approximately less 40%) in for the regulated future scenario. The increase in TCSO for double the network length would represent an increase of 200%, to 2637072 Celsius per Hour per Kilometre. The same increase in the length of the network, guided by a policy that would prevent the creation of E-W streets, would increase the TSCO to 2106372 (160%)(see Figure 3).

Results, Discussion and Conclusion

The analysis is summarised in Figure 2, which shows that the prevailing orientation of streets in Bahrain is E-W, with 36% of the total number of street segments, and 37% of the total street length. The second prevailing direction is N-S, with 29% of both the number of street segments of total street length. The third prevailing direction is NE-SW with 16% street segments and 17% of total street length. The least frequent direction is NW-SE with 18% street segments and 17% total street

 DOI: 10.1051/e3sconf/20172302002

length. Cumulatively, the less desired orientations (E-W and N-S) is over represented with 66% of the total number of streets segments and of the total street length; while the desired orientations are under represented with 34%.

In regions with harsh climatic conditions, the manipulation of the microclimate is a crucial element towards creating favourable conditions for the use of outdoor spaces, complementing an appropriate land use mix and infrastructure such as sidewalks and amenities. The cardinal orientation of street canyons is of paramount importance because of its passive role on solar intake, and also its perpetual character. Less permanent features being, for example, aspect ratio, building materials or vegetation.

The results show a clear tendency for streets to follow N-S and E-W orientation, rather than the desired NE-SW and NW-SE. The situation implies that the current situation in Bahrain is neither neutral (equally favouring all orientations), nor towards the desired orientation, but the opposite. This is likely to be due to the oblong shape, and orientation the island itself, the longest axis being a few degrees off north. The tendency to follow the E-W and N-S directions when designing infrastructure aims at shortening the distance between the geographic extremes. However, such concern, though legitimate when considering the largest distances, could be avoided at local scale, in the design of the complementary network, such as within cities and neighbourhoods. Regulating street orientation through policy could significantly decrease the solar gain of street canyons. The aggregate effect of such policy would be a reduction of solar intake of approximately 40%, as illustrated in Figure 3.

DOI: 10.1051/e3sconf/20172302002

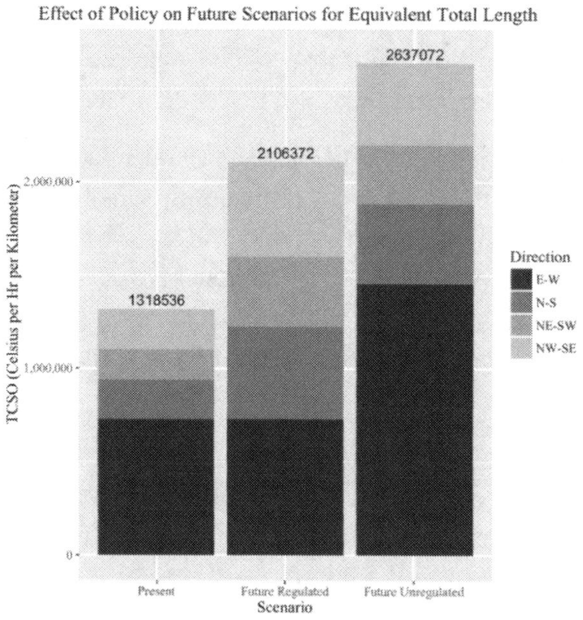

Figure 3 - Illustration of current exposure to current high temperatures and two potential future scenarios; with regulated and unregulated street orientation.

Inverting the current balance between orientations would significantly increase the baseline potential for reaching thermal comfort at the street level, thus creating a basic condition for the development of pedestrianism. Such conditions would impact the reliance on motor vehicles for short distances and generate a local distribution network for public transport. As a consequence, motorized traffic should diminish, bringing down air pollution. The population would have more opportunities to walk for transportation, thus increasing levels of physical activity, and therefore decreasing public health problems.

Reaching the maximum potential for thermal street comfort would involve concerted efforts, starting with adoption the desired orientations (NE-SW and NW-SE) as default, followed by other design principles. Further to street design, a strategy for the spatial distribution of land use, could diminish the dependence on longer trips. Furthermore, a marketing campaign to sensitize the population to the benefits of walking would fix an eventual stigma in favour of the use of private transportation.

DOI: 10.1051/e3sconf/20172302002

A limitation of the application of a favourable street orientation is that its effects are only felt on aspects ratios larger than 0.5 [24], [31]. Therefore, they do not constitute an advantage on residential neighbourhoods with lower aspect ratios, as is the case with houses, unless the buildings are placed right on the street, with small set-backs, no front gardens, and without spacing in-between properties, resembling multi-story housing.

The suggestion that air temperature might be more sensitive to the aspect ratio than to street orientation [24,p.99], suggests that re-profiling streets towards increasing their aspect ratio (height/width ≥2) could lead to a thermal comfort increase on existing streets. However, the efficiency of aspect ratio on streets with E-W orientation is very limited [24,p.102].

Ali-Toudert and Mayer [24] also conclude that, based on their results, air temperature might be more sensitive to the aspect ratio than to street orientation [24,p.99]. This finding suggests that it might be worth studying the potential impact of a policy that regulates the height/width ratio of new streets.

Study Limitations

Due to the lack of better data, the study assumes that the thermal performance of the canyon orientation/ratio in Bahrain (26.13°N, 50.35°E) is similar to that on Ghardaia, Algeria (32.40°N, 3.80°E), which is the, geographically closer, numerical study made [24]. The difference consisting on approximately six degrees of latitude. A local detailed numerical analysis of the thermal behaviour of street canyons in Bahrain as a function of orientation and aspect ratio, considering local climatic conditions such as temperature, wind and humidity, would make it possible to clearly estimate the impact of street orientation, and its cumulative effect.

Funding

Deanship for Scientific Research, University of Bahrain, Project 12/2014.

References

[1] APPLEYARD, D., GERSON, M. S., LINTELL, M., ÊTATS, U. & FEDERAL HIGHWAY, A. 1976. Liveable urban streets managing auto traffic in neighborhoods : final report : prepared for Federal Highway Administration, Department of Transportation, Washington, U.S. Dept. of Transportation, Federal Highway Administration : for sale by the Supt. of Docs., U.S. Govt. Print. Off.

[2] GEHL, J. & GEMZØE, L. 1999. Public spaces - public life : Copenhagen 1996, Copenhagen, Danish Architectural Press.

[3] GEHL, J. 2006. Life between the buildings : using public space, Copenhagen, Danish Architectural Press.

[4] MARSHALL, S. 2005. Streets and patterns : the structure of urban geometry, London : Spon.

[5] HILLIER, B. & HANSON, J. 1984. The social logic of space, Cambridge : Cambridge University Press.

[6] JEON, C. Y., LOKKEN, R. P., HU, F. B. & VAN DAM, R. M. 2007. Physical Activity of Moderate Intensity and Risk of Type 2 Diabetes: A systematic review. Diabetes Care, 30, 744-752.

[7] LEE, I. M., SHIROMA, E. J., LOBELO, F., PUSKA, P., BLAIR, S. N. & KATZMARZYK, P. T. 2012. Impact of Physical Inactivity on the World's Major Non-Communicable Diseases. Lancet, 380, 219-229.

[8] WORLD HEALTH ORGANIZATION, 2013, Global Action Plan for the Prevention and control of Noncommunicable Diseases 2013-2020. WHO Press, Geneva.

[9] TRANSPORT FOR LONDON, L. 2005. Improving Walkability. London, England.

[10] LAND TRANSPORT NEW ZEALAND, Z. 2007. Pedestrian Planning and Design Guide. Wellington: New Zealand.

[11] EWING, R. & CERVERO, R. 2010. Travel and the Built Environment. Journal of the American Planning Association, 76, 265-294.

[12] RYUS, P. 2014. Guidebook on pedestrian and bicycle volume data collection, Washington, DC, Transportation Research Board of the National Academies.

[13] ZACHARIAS, J., STATHOPOULOS, T. & WU, H. 2004. Spatial Behavior in San Francisco's Plazas: The Effects of Microclimate, Other People, and Environmental Design. Environment and Behavior, 36, 638-658.

DOI: 10.1051/e3sconf/20172302002

[14] ALJAWABRA, F. & NIKOLOPOULOU, M. 2010. Influence of hot arid climate on the use of outdoor urban spaces and thermal comfort: Do cultural and social backgrounds matter? Intelligent Buildings International, 2, 198-217.

[15] NIKOLOPOULOU, M. & STEEMERS, K. 2003. Thermal comfort and psychological adaptation as a guide for designing urban spaces. Energy and Buildings, 35, 95-101.

[16] FANGER, P. O. 1982. Thermal comfort : analysis and applications in environmental engineering, Malabar, Fla., R.E. Krieger Pub. Co.

[17] NIKOLOPOULOU, M. A. B., NICK AND STEEMERS, KOEN 2001. Thermal comfort in outdoor urban spaces: understanding the human parameter. Solar energy, 70, 227-235.

[18] RUPP, R. F., VÁSQUEZ, N. G. & LAMBERTS, R. 2015. A review of human thermal comfort in the built environment. Energy and Buildings, 105, 178-205.

[19] WANG, Z. 2006. A field study of the thermal comfort in residential buildings in Harbin. Building and Environment, 41, 1034-1039.

[20] KARJALAINEN, S. 2007. Gender differences in thermal comfort and use of thermostats in everyday thermal environments. Building and Environment, 42, 1594-1603.

[21] KARJALAINEN, S. 2012. Thermal comfort and gender: a literature review. Indoor Air, 22, 96-109.

[22] TUNG, C.-H., CHEN, C.-P., TSAI, K.-T., KÁNTOR, N., HWANG, R.-L., MATZARAKIS, A. & LIN, T.-P. 2014. Outdoor thermal comfort characteristics in the hot and humid region from a gender perspective. International journal of biometeorology, 58, 1927-1939 %@ 0020-7128.

[23] NUNEZ, M. & OKE, T. R. 1977. The Energy Balance of an Urban Canyon. Journal of Applied Meteorology, 16, 11-19.

[24] ALI-TOUDERT, F. & MAYER, H. 2006. Numerical study on the effects of aspect ratio and orientation of an urban street canyon on outdoor thermal comfort in hot and dry climate. Building and Environment, 41, 94-108.

[25] MAYER, H. & HÖPPE, P. 1987. Thermal comfort of man in different urban environments. Theoretical and Applied Climatology, 38, 43-49.

[26] MAYER, M. 1993. Urban bioclimatology. Experientia, 49, 957.

[27] HOWARD, L. F. R. S. 1818. The climate of London, deduced from Meteorological observations, made at different places in the neighbourhood of the metropolis, London.

[28] TURNER, A. 2007. From axial to road-centre lines: a new representation for space syntax and a new model of route choice for transport network analysis. Environment and Planning B-Planning & Design, 34, 539-555.

[29] R Core Team (2016). R: A language and environment for statistical computing. R Foundation for Statistical Computing, Vienna, Austria. URL https://www.R-project.org/

DOI: 10.1051/e3sconf/20172302002

[30] H. WICKAM. ggplot2: Elegant Graphics for Data Analysis. Springer-Verlag New York, 2009.

[31] KARIMINIA, S., AHMAD, S. S. & SABERI, A. 2015. Microclimatic Conditions of an Urban Square: Role of Built Environment and Geometry. Procedia - Social and Behavioral Sciences, 170, 718-727.

E3S Web of Conferences **23**, 02003 (2017) DOI: 10.1051/e3sconf/20172302003
World Renewable Energy Congress-17

Human Behavior and Environmental Sustainability: promoting a pro-environmental behavior by harnessing the social, psychological and physical influences of the built environment

Shaden Abusafieh arch_shaden@yahoo.com *Amman, Jordan*
Maiss Razem maissrazem@hotmail.com *Amman, Jordan*

Al-Ahliyya Amman University
Amman-Jordan- Zip-Code (19328)
Phone +962 5 3500211 *Mobile +962 0787641111* *Fax +962 6 5333315*

Recently, technological advancements in the sustainable design field have allowed us to reduce the ecological impact of the built environment, to reduce consumption of non-renewable resources, to create healthy environments and in some cases may even rehabilitate the ecosystem. Nevertheless, several studies have been carried out showing that sustainable technology does not automatically lead to environmentally friendly behaviors in its users.

Various environmental problems threaten environmental sustainability and many of these problems are rooted in human behavior. Unfortunately, there is a lack in studies which take into consideration the human behavior influences within a sustainable built environment. We believe that the built environment should be used to support human goals and requirements, but at the same time we should consider it as a context in which human values and behaviors are cultivated.

This research aimed to help in promoting environmental sustainability by using architectural design in changing relevant human behavior toward an environmentally friendly behavior. In order to achieve this, the research adopted Environment-centered Approach to gain more acute perspective into the relationship between the physical environment and human behavior, focusing on social, psychological and physical influences of the built environment.

It appears that environmental psychology's merits have high potential in changing behavior within the built environment. The research provides a systematic approach for selecting, assessing, evaluating the behaviors to be changed and the factors that determine them.

DOI: 10.1051/e3sconf/20172302003

Furthermore, this approach helps in choosing the best interventions that could be applied in built environment to encourage such a sustainable behavior.

This study tried to construct an agenda for further researches to find particular architectural design elements and strategies that we can harness to develop a pro-environment human behavior.

Introduction

Environmental psychology is the study of transactions between individuals and their physical settings (Gifford, 2007). It is the branch of Psychology concerned with providing a systematic account of the relationship between a person and the environment (Russell & Snodgrass, 1987).

For many decades, Environmental Psychologists have tried to study the relationship between environment settings and human behavior as well as how they affect one another. Why are some spaces comfortable and others not? How can we promote our environments to reduce stress, create more efficiency spaces and minimize losses in emergency cases? These are some concerns that environmental psychologists and designers as well have been working with for many decades. Recently, many researches in this field start considering the psychological aspect of sustainability trying to answer the question: Can the built environment enhance and promote a sustainable behavior?

Environmental psychologists rely on many theoretical approaches. Most psychologists examine the effects of environmental stimuli on human perception and responses in one way or another. Ittelson addressed in his note the transactional theory framework "Man is never encountered independent of the situation through which he acts, nor is the environment ever encountered independent of the encountering individual." (Ittelson, 1973). In order to predict a human behavior, transactional perspective to human-environmental relations requires that we study equally people, their perceptions, intentions, purposes and the physical settings in which they transact.

DOI: 10.1051/e3sconf/20172302003

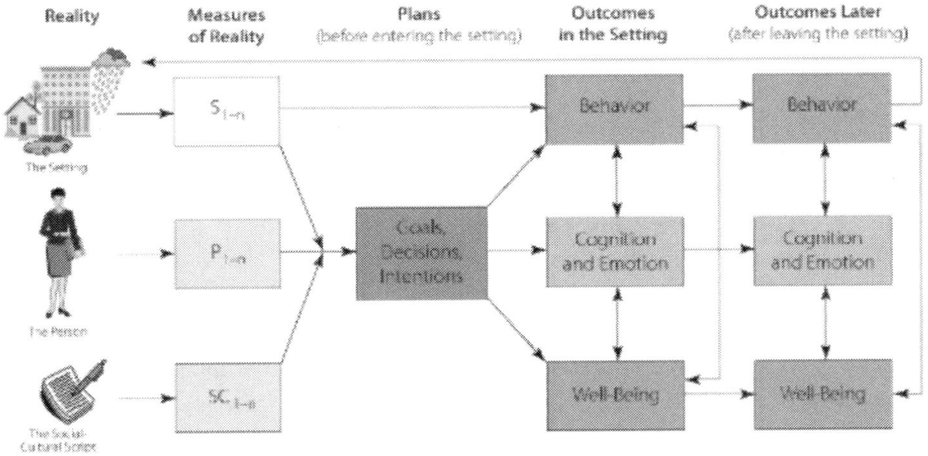

Figure1. An overview of environmental psychology's scope (Gifford, 2007)

Theoretical Bases

Over the past three decades, researchers who focus on person-environment-behavior relations conclude that they cannot simply assess people or the environment (Altman, 1975; Churchman, 2002). In classical and environmental centered theories, people are assumed to make rational decisions while performing in their built environments. We all know that people are not always rational about decision making. "Humans rarely choose things in absolute terms. We don't have an internal value meter that tells us how much things are worth" (Dan Ariely, 2010). So, we cannot assume that for example a sustainable well designed built environment will drive the users to have sustainable behaviors as well.

In this research we will adopt the transactional theory which applies equal consideration to human behavior and the settings they transact with on their daily life. Transaction more clearly exemplifies the nature of person-environment-behavior by considering people and environment inclusive and one defines the other, rather than dealing with them as separate entities. The goal of this research is to provide a theoretical framework that allows designers to understand the human-environment dynamics helping them in decision making in constructing a better environment that help in promoting pro-environmental behaviors for users.

DOI: 10.1051/e3sconf/20172302003

Changing the habit by changing the setting

There are various environmental problems that threaten sustainability. Many of these problems such as air pollution, water shortages, environmental noise, and littering are rooted in human behavior (Gardner & Stern, 2002). That means in order to attempt environmental quality improvement we should change relevant behaviour so as to reduce its environmental impacts. For example, sustainable built environment apply physical and technical innovations to gain energy efficiency. But at the same time, individuals need to accept and understand them, buy them, and use them in proper ways. Fortunately, there's a belief that buildings can change behavior. What makes people change their behavior through interaction with their environment which they perceive? To answer this question we will follow Geller (2002) in his method of choosing the right behavior that we should change to pro-environmental behavior that harms the environment as little as possible or even benefits the environment. Behavior change for better is more effective when one (1) selects the appropriate behaviors to be changed in order to promote environmental quality and human quality of life, (2) examines which factors cause those behaviors, (3) applies the best interventions in built environment to change relevant behaviors, (4) Evaluates effectiveness of the interventions (Geller, 2002; Steg & Vlek, 2009). In many countries, about half of total household energy use is direct energy use (use of electricity, natural gas and other fossil fuels); the other half is indirect energy use (the energy used in the production, transportation and disposal of goods and services). This research will focus on analyzing a case study of a house in Amman-Jordan along with examining some factors influencing household direct energy use, attempting to promote household energy conservation.

Factors that influence decisions that encourage pro-environmental action

Many theories have studied which factors promote or inhibit pro-environmental behavior. We have to reveal these factors in order to decide which should be chosen to encourage sustainable behaviors. The Theory of Planned Behavior assumes that individuals choose alternatives with highest benefits against lowest costs (TPB; Ajzen, 1991). This approach proposes that attitudes, social norms and perceived behavioral control, together shape an individual's behavior. The TPB was successful in explaining various types of environmental behavior such as household recycling, waste composting, water use and general pro-environmental behavior (Heath & Gifford, 2002). The norm-activation model (NAM; Schwartz, 1977) and the value-belief-norm theory (VBN theory; Stern, 2000) assume that

 DOI: 10.1051/e3sconf/20172302003

people act pro-environmentally when they feel a moral obligation to do so. In order to activate sustainable personal norms for behavior, they suggest different variables from values and general environmental concern to which people are aware of negative consequences of certain actions and their ability and responsibility to avoid these negative consequences. The NAM and VBN theories have been successful in explaining low-cost environmental behavior and good intentions such as willingness to change behavior.

Recently, behavioral economics have become a credible field that integrates behavioral psychology and the environment in order to improve models of decision-making. There are many great readings in this field such as: *Predictably Irrational, Nudge*, and *Thinking, Fast and Slow.* Deriving from behavioral economics, we can follow two routes to achieve changes. First, by changing minds we can directly change behaviors and habits. Second, we can change the decision making by changing the architectural context. Although we can evaluate our options that change our behaviors, our subconscious usually guides our choices. Our decision making do not always relay on objective rational choices, they rather usually come from emotions and irrational thinking (Dan Ariely, 2010). Based on this fact, how can we promote people decision making when it comes to their impact on the environment? Richard Thaler, the author of *Nudge,* argues "the solution is to apply the single most useful bit of psychology one can ever learn: If you want to encourage people to do something, make it easy - or even better, automatic". Nudge is a collective term for different policy tools that can be used in order to influence individuals' behavior. Table 1 summaries different policy tools and on how they influence the choice of individuals (House of Lords, 2011).

Interventions to promote pro-environmental behavior

Pro-environmental energy behaviors have been studied from different theoretical perspectives. And many motivational factors proved to be related to specific energy behaviors. In this research we will focus on three types of motivations that are considered behavioral economic concepts influence "via nudges": *Salience (*as a physical influence*), Social norms* (as a social influence) and *Defaults* (as a psychological influence), and how they impact the direct energy use. Nudges are ways of changing the context in which we make decisions to potentially reach different outcomes, while not limiting consumer choice (Richard Thaler, 2009).

We can use these concepts to create interventions that promote pro-environmental behavior (energy conservation as a case study).

Table 1. Policy tools to influence individual behavior (House of Lords, 2011)

Regulation of the individual	**Fiscal measures directed at the individual**		**Non-regulatory and non-fiscal measures with relation to the individual**				
Eliminate and restrict choice	*Guide and enable choice*						
	Incentives and information			Nudging			
Laws and regulations	Fiscal incentives	Non-fiscal incentives	Provision of information	Simplification and framing of information	Changes to physical environment	Changes to the default policy	Use of social norms

Salience (physical influence)

Salience is an external process for individuals. The information that stands out and seems relevant has more effect on our thinking and actions (Dolan et al., 2010). Salience guides the judgments that are based on external cues, in other words it simplifies our decision making. According to behavior economics, most of people choices are not the result of careful deliberation. Our thinking is subject to insufficient knowledge, uncertainty feedback, and the context in which we make decisions. For instance, when people assess the energy use of household appliances, they tend to rely on a simple heuristic: The larger the appliance, the more energy it is believed to use and which is not always accurate (Schuitema and Steg, 2005).

We are affected by legible available information, feedback, influences of the context, and salient information in the environment. For example, meter readings reflect how much electricity has been used by a particular household. However, meter readings do not reveal which specific behaviors contributed most to total energy use (heater, oven, refrigerator, ex…). This is consider a problematic for people, that they do not know which and whose behaviors significantly affect energy use, and that they cannot receive specific feedback on the results of their behavioral changes (Gatersleben et al., 2002). Also, we can manipulate salience by

DOI: 10.1051/e3sconf/20172302003

rearranging the physical environment; for instance, "I want to climb the stair to fitness" sign at Bugis MRT station in Singapore prompts people to take the stairs rather than the escalator. An essential finding in behavioral science is to make it easy for people to choose the preferred option rather than to force them. So, a light-filled staircase can attract people to use it more since people are naturally drawn to light.

Figure 2. Household reminder of the bigger picture when it comes to flipping a switch.
Reference:http://blog.sustainability.colostate.edu/?q=node/59http://blog.sustainability.colostate.edu

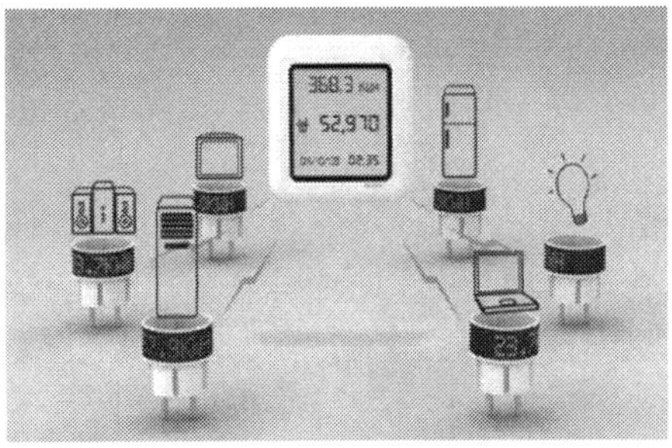

Figure 3. The Diet Watt monitor shows the electrical usage readings in real.
Reference: http://www.ubergizmo.com/2010/01/diet-watt-to-help-you-monitor-power-usage/

Figure 4: Yellow stairs in the North Carolina University Library placed along the main circulation lines. Reference: https://sourceable.net/prompting-sustainable-behaviour/

Figure 5: Bugis MRT station, Singapore. Reference: http://inudgeyou.com/archives/780

Social norms (social influence)

Environmental concerns and social norms are important in promoting energy conservation behavior, because they provide the most solid basis for it (Lindenberg and Steg, 2007). When energy pro-environmental behavior results from normative concerns, it will be sturdier against changes. Furthermore, some strategies are based on the assumption that positive or negative consequences will influence behavioral choices, so pro-environmental behavior will become a more attractive alternative when positive consequences are attached to it.

Social norms are a strong force that influences human behavior. There are two ways in which social norms affect individuals, as injunctive norms and as descriptive norms. The injunctive norms act as a moral implication (what should be done and what should not to be done). The descriptive norms refer to the simple observation of how most others behave, which will help an individual to choose the best action when he or she is in an uncertain situation (Cialdini et al., 1990). A relevant study found that hotel guests re-used towels more frequently when they were provided with information that emphasized descriptive social norms in favor of re-use ('did you know 75% of our guests help save the environment by reusing their towels') compared to when they were given environmental information only ('help save the environment by re-using your towels') (Goldstein, Cialdini, & Griskevicius, 2008). Another research finds that when the company OPower shows how much energy a household is using compared to similar neighbors, a significant impact on changing household energy use was noticed. The households reduced energy use in the short long run (Allcott, 2011).

Defaults (psychological influence)

According to psychologists, people tend in their behavior to go along with the default options that make life easier. So, can we make the default setting in a way that users will choose the behavior we would prefer to perform? Thaler said: "By paternalism, all we mean is caring about peoples' outcomes. We want to devise policies that will make people better off, not worse off, as judged by them". He added: "we can help people make choices that they themselves think are better" (Thaler, 2008). This goal can be achieved with "choice architecture" which refer to the influence that architects can do so people may change their decision making.

 DOI: 10.1051/e3sconf/20172302003

Default is one of the key features of "architecture choice". Whatever you choose as a default, it will have a better chance to be selected. We need to nudge people toward the right choices hoping to create better and easier lives. For example, Placing staircases directly in obvious spaces and hiding elevators to the side as well as designing interesting and cheerful landings, will highlight the staircases as a default choice. Accordingly, this will encourage a pro-environmental behavior and people will tend to use the stairs more than elevators.

Effectiveness Evaluation of the interventions

In this research, we aimed to evaluate interventions based on the three nudges concepts: Salience, Default and Social norms by analyzing a case study of a greenhouse located in Amman, Jordan. We tried to reveal the effectiveness of a combination of design interventions on energy conservation behavior. Effectiveness measurements should focus on changing of energy use behaviors and changing in the people's quality of life, as well as people's commitment and their maintain of these changes.

Al-Kamaliya Residence - Environmental Aspects

Al-Kamaliya residence chosen for this study is considered a green home for its environmentally friendly features. It won the Green Apple Award. The owner of the residence is Dr. Ayoub Abu Dayyeh who also designed and constructed the house, as he is an energy consultant and an active environmentalist. The residence was built in 2009 and is occupied by himself, his wife and his daughter intermittently; three days a week, while they stay at another home closer to the city center the rest of the week. An informal interview was conducted between the authors and Dr. Aby Dayyeh and his wife. The questions during the interview addressed the level of contentment and consequent behavioral change between the occupants and their green home, as they revealed how the design of the house influenced the occupants with varying satisfaction levels from Dr. Abu Dayyeh to his wife.

The house has integrated Solar thermal, P.V solar cells, geothermal energy, rain water harvesting, unique thermal design and passive architectural design to produce a zero carbon emission house at a feasible economical cost. The construction took two years and finished in 2009. During that period all material waste produced on site was reused in an eco-friendly modes within the site and all trees displaced by the construction were successfully replanted. A dry garden was

DOI: 10.1051/e3sconf/20172302003

established too that needs no irrigation and a wild garden was confined to allow for indigenous plants and wild life to prosper. The house is not heated by fossil fuels nor does it use HVAC. Indoor temperature ranges between 21 to 19 degrees in summer and winter. In January of 2015, the family only needed one electric direct-heat-source to be operated during the winter time for one day, when the indoor temperature fell below 18 degrees Celsius.

Due to the cold winds of Amman, the western façade was completely closed off as seen in figure.6, while the southern elevation has the largest percentage of windows that allow winter solar rays to heat the indoor space as they block the hot summer radiation to enter with the help of sun-breakers which is illustrated in figure.7. Fewer areas of windows exist in the east and north facades as shown by figure.8. The southern living room has two fixed large double glazed windows, while upper floor spaces have operable and fixed windows. It is uncommon in Jordanian homes to incorporate fixed windows, and the impact of the placement of such windows at Al-Kamaliya house was highly noticeable when we held the interview inside. It was a windy day accompanied by noisy sounds that we could hardly hear each other outside, but as we entered the house all the cold, air movement and sound were cut off; a truly thermally and sound insulated home with veritable air sealing. Because Al-Kamaliya house has no mechanical air circulation, the operable windows induce air flow when prompted by the beep of the CO_2 level reader device if concentrations exceeded 1000ppm.

Figure6. Western elevation, Al-Kamaliya house *Reference: Researchers*

Figure7. Southern elevation, Al-Kamaliya house *Reference: Researchers*

Figure8. Eastern elevation, Al-Kamaliya house *Reference: Researchers*

As the house clearly exhibits careful design and implementation of green technologies, the study aimed to gain more acute perspective into the relationship between the physical environment and human behavior, focusing on social, psychological and physical influences of the built environment. In an informally structured interview, Dr. Abu Dayyeh was asked whether Al-Kamaliya house instigated pro-environmental behaviors in this house and his other home in central Amman for himself and his family.

Physical Influences "Salience"

Pro-environmental behaviors should be convenient, directly beneficial, and easy to implement in order for them to be effective and sustainable in the long-term. However, the main drawback with environmental enthusiasts' mission is resorting to spreading awareness as the only and most effective means to solve imminent problems. From the various studies conducted by social environmental psychologists and behavioral economists to produce influential outcomes in behavioral change, the most effective approach is to change the social context and driving norms.

As seen in figure.9, signage and environmental paintings were hung on the house's walls which represent a medium of educating the home occupants and constant messaging of the unique nature of the house that activates conscious awareness to maintain their pro-environmental behaviors. These types of Informational strategies are influential when pro-environmental behavior is relatively convenient and economical (Steg and Vlek, 2009).

Another Salience strategy that instigated pro-environmental behavior, and that led to turning off lights when not used and so reducing electricity consumption, is the installation of the electricity consumption reading meter (OWL meter) figure.10. Mrs. Yasmeen after a while started closely following the home's consumption and the active reading on the meter as a daily behavior. The family is attentively engaged with the meter's feedback and their energy saving behavior becomes more instilled. This salience strategy values information feedback especially when it is visually represented using recent technologies in the design of homes, geared towards changing behaviors. It shows that the more specific, personalized, and beneficial the normative information the better desired behavioral change is achieved.

DOI: 10.1051/e3sconf/20172302003

Figure9. Environmental paintings, *Al-Kamaliya house*
Reference: Researchers

Figure10. OWL meter, Al-Kamaliya house
Reference: Researchers

Social Influences "Social norms"

The more people engage in a certain behavior the more the individual follows suit with the group's behavior by following social norms inducted by either unwritten cultural rules, or by policies. Within the confines of the home, the occupants themselves serve as a social group for its individual members to emulate. Such is the role of the informed homeowner whose aim is to reduce the energy bills and is thus incentivized to behave sustainably. The OWL meter, for example, introduced a new social norm in the household that engaged every occupant to monitor and reduce electricity consumption.

A larger social influence of Al-Kamaliya house was when it won the Green Apple Award and was publicized in the media channels. The transfer of the best practice attracted clients who wanted to build a green home. Here, a potent social norm of building green is at play, where telling people about the actual achievements of Al-Kamaliya's environmental benefits and economic feasibility has attracted interested individuals. Depending on measured green homes' viability and regardless of regulations that might drive a pro-environmental behavior, the impact of the social norm is likely to be more profound.

DOI: 10.1051/e3sconf/20172302003

Psychological Influences "Default concept"

Many windows are designed as fixed to maintain thermal comfort. According to Dr. Abu Dayyeh, the need to open windows all the time and maintain air flow is inefficient when a controlled indoor thermal comfort is to be achieved, thus he decided to minimize operable windows The Fixed windows was a default feature inside the house figure.11. In their other home in Amman, which consists of operable windows, the behavior of Mrs. Yasmeen and her daughter seemed to have altered a bit in a reduced urge to open the windows. This behavior change must have been instigated by Al-Kamaliya house. That they realized added benefits of keeping the indoor thermal comfort within range, classically exhibiting the case of how a changed context can produce new habits. It is worth mentioning here that 'opening windows for ventilation' even at winter time to introduce fresh Oxygen , sometimes longer than needed, seems to be a behavioral pattern among ladies of Jordanian households, becoming a social cultural norm.

Figure11. Fixed windows as a default feature *Reference: Researchers*

Conclusion

Social Psychologist Daryl Bem once observed that an attitude requires a "fact, a feeling, and a following". It is paramount to comprehend and incorporate the full spectrum of the various users' attitudes and behaviors when designing a green building, more so homes. Human behaviors are as complex factors in design as the technical accomplishment of environmentally friendly features, granting the integrative design process more potency when pro-environmental behaviors are to be nurtured sustainably from factual, emotional and social aspects.

Nudges coming from different concepts such as salience, social norms and default have an important role to play in promoting energy conservation via behavioral changes. Environmental interventions are generally more effective when they are systematically planned, implemented and evaluated in order to choose the best behavior to be changed. This can be achieve by going through the four steps addressed in this research:

To achieve and to maintain environmental sustainability, individuals can contribute significantly by promoting sustainable behaviors that can reduce household energy use. The challenge for Architects is to understand the individual, factors that influence their behaviors and the processes of choosing architectural choices that create pro-environmental behaviors.

Further research may be conducted to further improve intervention studies by considering energy conservation from an interdisciplinary perspective. For instance, according to the ventilation issue, a question arises here whether opening windows for ventilation is an exaggerated and unwarranted behavior in our culture. Such inquiry indeed heralds further research to include a better informed holistic understanding instead of following a reductionist reasoning. What activates the behavior of ventilation; social norms, habits, individual attitude, or physiological needs? Who is most motivated by which behavior driver in a typical household? Moreover, can aesthetic taste hinder sustainable behaviors if the market does not afford environmentally aesthetic choices? Therefore, more studies should focus more on attitudes, emotions as well as cognitive so we can understand humans and control their behaviors in different context. To elucidate the power of affect in driving behaviors, T. Heberlein concludes in his book 'Navigating Environmental Attitudes' that: *"We often forget that attitudes have two parts, the cognitive and the emotional, and too often research focuses on the cognitive. The real power of attitudes is the emotional dimension*."

 DOI: 10.1051/e3sconf/20172302003

References

Altman, I. (1975). The environment and social behavior: Privacy, personal space, territoriality, and crowding. Monterey, CA: Brooks/Cole.

Altman, I., & Rogoff, B. (1987). World views in psychology and environmental psychology: Trait, interactional, organismic and transactional perspectives. In I. Altman & D. Stokols (Eds.), Handbook of environmental psychology. pp. 245–281, New York: Wiley.

Ariely, Dan. (2010). Predictably Irrational: The Hidden Forces That Shape Our Decisions. Harper Collins. A well written introduction to the implicit influences on our behaviour by one of the leading behavioural economists.

Bechtel, R. B., & Churchman, A. (Eds.). (2002). Handbook of environmental psychology. New York: Wiley.

Bem, D. J. (1972). Self-Perception Theory. In L. Berkowitz (Ed.), Advances in Experimental Social Psychology, Vol. 6, pp. 1-62. New York: Academic Press.

Cialdini, R. B, Reno, R. R., & Kallgren, C. A. (1990). A focus theory of normative conduct: Recycling the concept of norms to reduce littering in public places. Journal of Personality and Social Psychology, 58, pp. 1015-1026

Dolan, P., Hallsworth, M., Halpern, D., King, D. and Vlaev, I. (2010). MINDSPACE: Influencing Behavior Through Public Policy. Cabinet Office: London.

Gardner, G. T., & Stern, P. C. (2002). Environmental problems and human behavior (2nd edition). Boston, MA: Pearson Custom Publishing.

Gatersleben, B., Steg, L., Vlek, C. 2002. Measurement and determinants of environmentally significant consumer behaviour. Environment and Behavior, 34 (3), pp. 335–362.

Geller, E. S. (2002). The challenge of increasing proenvironmental behavior. In R. B. Bechtel & A. Churchman, Handbook of environmental psychology . pp. 525–540, New York: Wiley.

Gifford, R. (2007). Environmental psychology: Principles and practice (4th Ed.). Colville, WA: Optimal Books.

Goldstein, N. J., Cialdini, R. B., & Griskevicius, V. (2008). A room with a viewpoint: Using social norms to motivate environmental conservation in hotels. Journal of Consumer Research, 35, pp. 472-482.

Heath, Y., & Gifford, R. (2002). Extending the theory of planned behaviour: Predicting the use of public transportation. Journal of Applied Social Psychology, 32, pp. 2154–2185.

House of Lords (2011), Behaviour Change, The House of Lords, pp. 111.

Hunt Allcott Social norms and energy conservation Journal of Public Economics, 2011, vol. 95, issue 9-10, pages 1082-1095

Linda Steg, Charles Vlek, "Encouraging pro-environmental behavior: An integrative review and research agenda", Journal of Environmental Psychology · September 2009

Lindenberg, S., & Steg, L. (2007). Normative, gain and hedonic goal-frames guiding environmental behavior. Journal of Social Issues, 63 (1), pp. 117-137.

MarktOnderzoekAssociatie (Developments in marketing research. Yearbook 2005), pp. 165-180. Haarlem (NL): De Vrieseborch.

Richard H. Thaler, Prof. Cass R. Sunstein-Nudge_ Improving Decisions About Health, Wealth, and Happiness-Yale University Press (2008)

Russell, J. and Snodgrass, J. (1987) "Emotion and Environment" in D Stokols and I. Altman (eds) Handbook of environmental psychology. Ch. 8, pp245-281, New York: Wiley.

Schuitema, G., & Steg, L. (2005). Perception of energy use of domestic appliances, 10, pp. 221–279. New York: Academic Press.

Steg, L., & Vlek, C. (2009). Encouraging pro-environmental behavior: An integrative review and research agenda. Journal of Environmental Psychology, 29, pp. 309–317.

Stern, P. C. (2000). Toward a coherent theory of environmentally significant behavior. Journal of Social Issues, 56(3), pp. 407–424.

Thaler, R.H., Sunstein, C.R. (2009) Nudge, improving decisions about health, wealth and happiness. Penguin Books: London

Thomas A. Heberlein. Navigating Environmental Attitudes Oxford University Press; 1 edition. September 24, 2012, Page 52

Water Collection from Air Humidity in Bahrain

Nidal A. Dahman. Khalil J. Al Juboori, Eman A. BuKamal, Fatima M. Ali, Khadija K., AlSharooqi, and Shaima A. Al-Banna.

Civil Engineering Department, University of Bahrain, Isa Town, Kingdom of Bahrain

ndahman@uob.edu.bh, kaljuboori@uob.edu.bh , emanz1994@gmail.com , Fatima.mousa8@gmail.com , Khadija.alshurooqi2@gmail.com , Shaima.albanna.11@gmail.com

Abstract

The Kingdom of Bahrain falls geographically in one of the driest regions in the world. Conventional fresh surface water bodies, such as rivers and lakes, are nonexistent and for water consumption, Bahrain prominently relies on the desalination of sea water. This paper presents an ongoing project that is being pursued by a group of student and their advising professors to investigate the viability of extracting water from air humidity. Dehumidifiers have been utilized as water extraction devices. Those devices have been distributed on six areas that were selected based on a rigorous geospatial modeling of historical meteorological data. The areas fall in residential and industrial neighborhoods that are located in the main island and the island of Muharraq. Water samples have been collected three times every week since May of 2016 and the collection process will continue until May of 2017. The collected water samples have been analyzed against numerous variables individually and in combinations including: amount of water collected per hour versus geographical location, amount of water collected per hour versus meteorological factors, suitability of collected water for potable human consumption, detection of air pollution in the areas of collection and the economy of this method of water collection in comparison to other nonconventional methods. An overview of the completed analysis results is presented in this paper.

Keywords: Bahrain, GIS, Renewable Projects and Applications, Sustainable Water Supply, Multi criteria modeling.

Introduction

It is a common knowledge that planet Earth's atmosphere is composed mainly of gases such as nitrogen, oxygen and carbon dioxide. The lower altitude of the atmosphere, called the troposphere, also contains quantities of water in its gaseous form of vapor. Quantities of water vapor in the troposphere near the earth surface span from a negligible 0.01% to about 4.25% of air volume [2]. Therefore, on average, about 2.1% of the air we breathe is water. The geography, climate and closeness to significant water bodies determine how moist or dry the atmosphere of an area is. By mass and volume, atmospheric moisture ranges from zero grams per cubic meter in dry air to more than 50 grams per cubic meter, when the air is saturated with water vapor at 40°C (Figure 1) [16, 17]. Considering that an average of 2% of the troposphere is made of water vapor, a simple calculation of the troposphere volume multiplied by 0.02 reveals that earth's troposphere contains close to 1.42×10^7 cubic kilometers of water, in its gaseous vapor form. This astounding amount of pure water can be harvested for various uses of humanity via passive and active means.

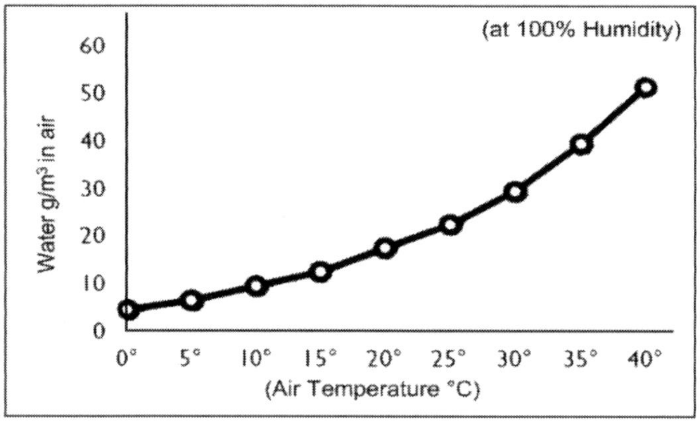

Figure 1: Saturation Atmospheric Moisture Content at Different Temperatures. [After Ref. 2]

The kingdom of Bahrain is an archipelago in the Arabian Gulf. It geographically falls in UTM 39R, at a latitude and longitude of 26° 1′ 39″N and 50° 33′ 0″E respectively. These geographical coordinates place Bahrain in a very arid region, where the daily average high temperature is in excess of 26°C and the average yearly precipitation is less than 25mm. Moreover, being an island-nation makes the weather humid all year long with an average low relative humidity of 47% and an average high relative humidity of 83%. Figure 2 presents the monthly averages of temperature, humidity and dew point for the Kingdom of Bahrain the last 5 years. This figure was produced from a data that was obtained from Meteorological Directorate Climate & Observation section of the Bahrain Ministry of Transportation and Telecommunications [13].

 DOI: 10.1051/e3sconf/20172303001

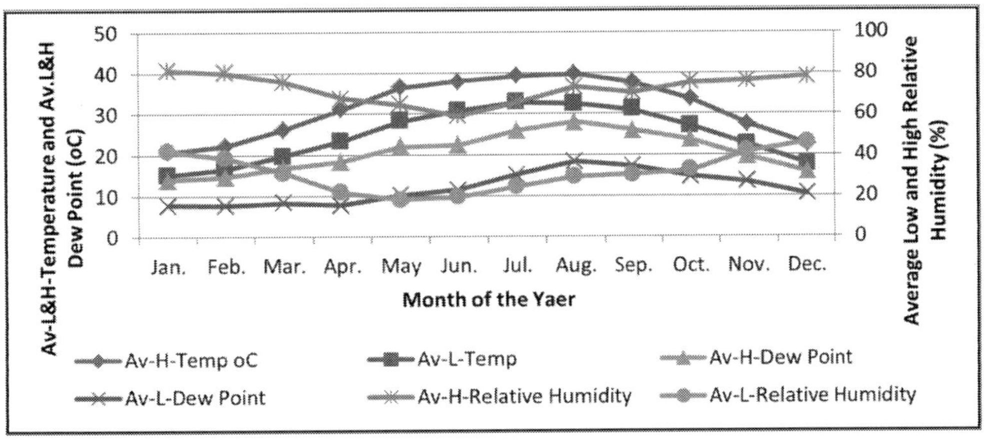

Figure 2: Average Weather Elements of Bahrain based on the last 5-year data.

This paper presents an ongoing project that is pursued at the civil engineering department of the University of Bahrain to investigate the viability of extracting water from air humidity and outlines the results that have been achieved thus far. The project involves many facets and goals, including the quantities of water that can be collected versus geographical locations and meteorological factors, suitability of the collected water for human consumption, detection of air pollution in the collection areas and the economy of this method in comparison with desalination of sea water.

The next section of this paper presents a brief background from meteorology and engineering. Thereafter, a quick literature review is followed by research methodology, analysis of results and finally, conclusions and recommendations for further studies.

Background

The water vapor content of air in a certain location is referred to as humidity and there are numerous ways to measure it, including among others: absolute humidity, specific humidity, mixing ratio, relative humidity and dew point. [2]

Absolute humidity (AH) is the mass of water vapor existing in a unit volume of air and is expressed in grams per cubic meter. *Specific humidity* (SH) is the ratio of water vapor mass to the total mass of an air parcel (dry air plus water vapor). An alternative term is the Mixing Ratio (MR), which indicates the ratio of the mass of water vapor in a given volume of air to the mass of dry air in the same volume.

Relative humidity (RH) is the ratio of the current moisture content of a unit volume of air to its saturation moisture content at the same temperature and pressure and is expressed as a percentage. RH can't exceed 100% and it is the measure used by meteorologists in their forecasts of rain, dew or fog. *Dew point* (DP) is the temperature at which RH reaches 100%. At that point, if more vapor is induced, air humidity is relinquished as liquid water. The amount of humidity that air can tolerate depends on temperature as warmer air can bear more vapor than cold air. [2]. For the sake of this research, AH is more straightforward to apply, but all meteorological data obtained from authorities is expressed in RH terms. AH can be computed from RH at a given temperature using the following famous Equation (1). This equation was used to produce the curve in Figure 1 and it is based on the physics of ideal gases [11].

$$AH\left(\frac{gm}{m^3}\right) = \frac{6.112 \times e^{\left[\frac{17.67\times T}{T+243.5}\right]} \times RH \times 2.1674}{273.15+T} \quad (1)$$

1- Previous Works Investigation

Atmospheric water can be harvested by passive methods or active methods. Passive methods are suitable for atmospheric humidity that has already condensed to liquid rain, dew or fog. They just collect naturally-condensed water droplets from wet air and stream them into collection tanks. For many centuries people of arid and semi-arid regions have collected winter rain flowing down from building roofs and stored it in tanks and land pits for later human, animal and agricultural consumption [9]. A more recent passive method of atmospheric moisture collection utilizes screens and meshes to hunt dew mist and fog droplets and deposit them into gutters that stream them into collection tanks [3, 12, 15, 24]. A third method of passive atmospheric water harvesting involves chemical treatment of cotton fabrics to enable them to absorb considerable amounts of water from misty air that can be more than three times of their own weights [27].

DOI: 10.1051/e3sconf/20172303001

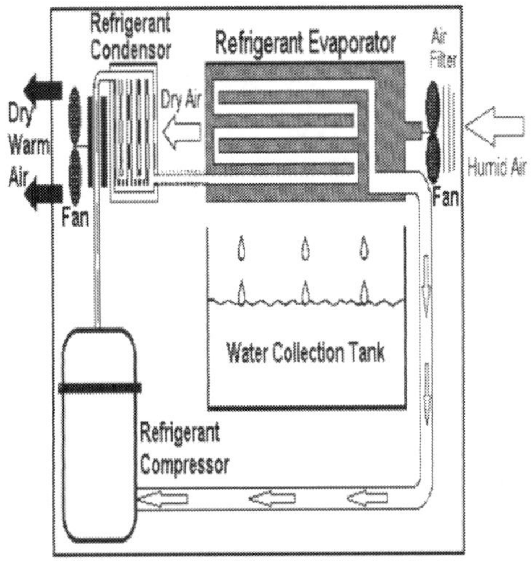

Figure 3: The Architecture of an Active Atmospheric Water Harvesting Device.

Active harvesting of atmospheric humidity requires special dehumidifiers that condense atmospheric moisture to produce liquid water. Electro-mechanical/refrigerative dehumidifiers come with refrigerative coils, where moist atmospheric air is forced to pass through the cold coils. This causes the condensation of gaseous vapor into liquid water. This water is then streamed into collection tanks as shown in Figure 3 [7, 10, 16]. In the last few years these active methods of harvesting atmospheric moisture have been researched extensively. Numerous research papers have been published [1, 4, 5, 19] and the research has traversed in many directions including: new harvesting methods and equipment [6, 8, 14, 21, 22], improving the efficiency of existing hardware [18, 20, 25] and examining the viability of these methods to geographic locales [17, 24].

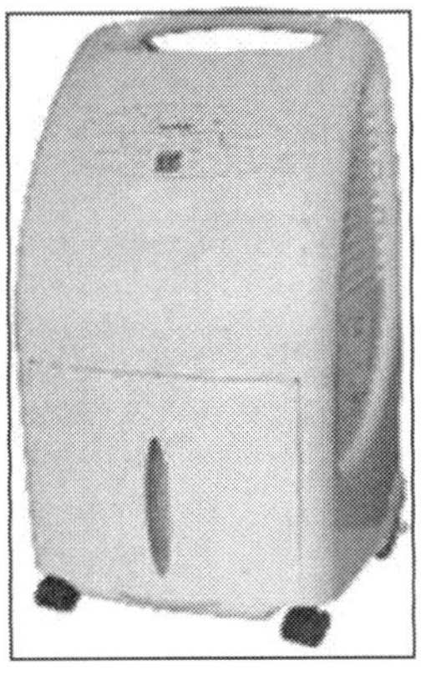

Figure 4: Over-the-Shelf Dehumidifier.

 DOI: 10.1051/e3sconf/20172303001

2- Methodology

For the purpose of this project, six of over-the-shelf dehumidifiers, shown in Figure 4, were utilized as air-humidity water-harvesting devices. According to the manufacturer's sticker on the rear side, the device has a rated voltage of 220-240 Volts of alternative current at 50 Hz frequency, an input power of 330 Watts, and a rated current of 1.5 Amperes. According to the manufacturer, the device is capable of extracting up to 20 liters of water per day at 30°C and 80% RH by processing 180 m^3 of air per hour with a suction pressure of 0.5 MPa and discharge fan pressure of 1.5 MPa. The devices were acquired from a local department store and installed in six cities of the kingdom of Bahrain as shown in Figure 5. A merger of Geographic Information Systems (GIS) and meteorological data tables assisted the selection of the locations of the water collection devices [13]. The six devices were placed in six geographical locations that cover diverse characteristics, such as closeness to shores land-use classification (residential versus industrial). Those locations are: Hidd, Manama, Barbar, Salmabad, Sanad and Hamad Town. The first city is in the island of Muharraq, while the later five are in located the main island. The experiment of water collection, using these devices, started on May of 2016 and will continue until May of 2017. The devices are operated by third to 6th coauthors of this paper for one hour every time, three times a week with three different starting and ending collection times that cover early morning (6:00 to 7:00 AM), mid-day (1:00 to 2:00 PM) and evening (8:00 to 9:00PM). By the end of each operation session, the amount of harvested raw water is measured and logged into a spreadsheet along with other concurrent meteorological data, such as weather status, relative humidity, temperature, wind direction and due point.

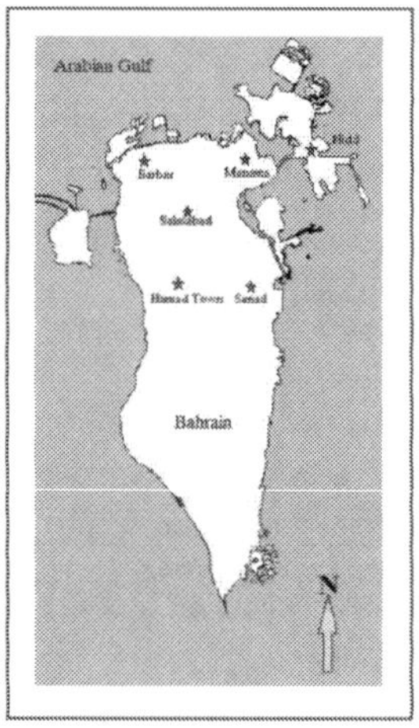

Figure 5: Locations of the water collection devices in the Kingdom of Bahrain.

DOI: 10.1051/e3sconf/20172303001

By design, harvested water samples do not undergo any filtration and are kept raw. In theory, the water harvested from atmospheric humidity should be pure and distilled, but without any filtration, it is fair to expect it to contain many air-borne ingredients such as, dust particles, pollen, mold spores and bacteria. Identifying those ingredients can provide researchers with a plethora of environmental air quality indicators about the air pollutants in the water collection sites.

To calculate the efficiency of the harvesting device, the following general expression of efficiency can be adopted;

$$\eta = \frac{Process\ Output}{Process\ Input} X100\% \tag{2}$$

Equation (2) can be tailored for the utilized harvesting device as follows:

$$\eta = \frac{Volume\ of\ Water\ Harvested\ per\ Unit\ Time}{(Total\ Volume\ of\ Air\ Processed\ per\ Unit\ Time)\ X\ (AH)} X100\% \tag{3}$$

Then, the per-unit-time capacity *(C)* of the harvesting device in liters can be expressed as:

$$C = Airflow\ (m^3/hr)\ X\ AH\ (gm/m^3)\ X\ 0.001\ (L/gm)\ X\ \eta \tag{4}$$

3- Results and Analysis

For the sake of this paper, a representative sample of the harvested waters from each of the six collection sites was tested in the civil engineering instructional laboratories for acidity/basicity (pH value), electric conductivity, total dissolved solids, and the dissolved ions of nitrates that are detectable by the accessible spectrometer. In addition, a bacteria count test was performed on freshly-collected samples, and to the authors' surprise, the number of colonies was zero in 100 ml. The results are presented in the following Table 1.

DOI: 10.1051/e3sconf/20172303001

Table 1: Preliminary Chemical Analysis Results on Some Harvested Water Samples.

Sample Origin	pH	Conductivity (μS/m)	Nitrate (mg/l)	TSS (mg/l)
BarBar	7.1	83	35.8	2 x 10^{-4}
Manama	7.0	50	31.2	5.4 x 10^{-4}
Hidd	7.0	46	35.8	0.00
Salmabad	6.9	44	35.8	4.1 x 10^{-4}
Sanad	7.0	53	11.1	4.1 x 10^{-4}
Hamad Town	7.0	25	5.9	0.00

The preliminary results shown in Table 1 look very promising and are in agreement of what has been reported in the health literature [26]. Nonetheless, to overcome the limitations of the instructional labs, a set of water samples from each collection site has been sent to the official water labs of the Electricity and Water Authority (EWA) and the official labs of the Ministry of Health for comprehensive chemical, bacteriology tests and pathogen analysis. As of the time of writing this version of this paper, results of those official tests have not been received yet.

In the meantime a look can be taken at the harvested water data. Figure 6 shows the hourly average quantities of harvested waters in each of the six locations. The figure shows that there is more potential to collect more water in Barbar and Hidd because of their closeness to the see in comparison to Hamad Town and Salmabad that are far away from the shores.

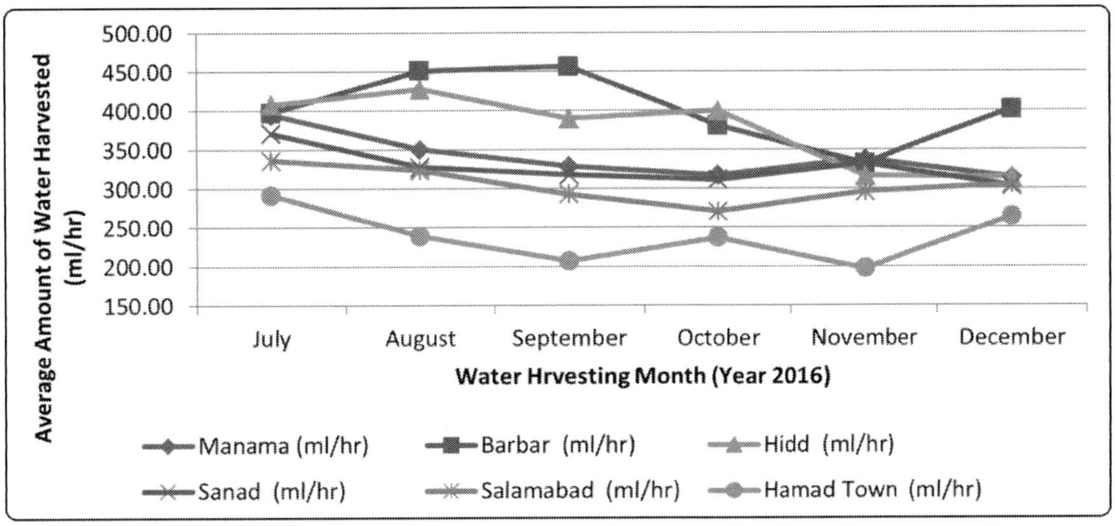

Figure 6: Hourly average quantities of harvested waters in each of the six locations

It was found that for the first six months of operation, the used harvesting devices could produce an average of 332 milliliters of water per hour. Based on this 332 ml, and based on the manufacturer's data that the device processes 180m^3/hr, and with an average daily absolute humidity (AH) of 36.03 gm/m^3 that was calculated using Equation (1) from meteorological data provided by Bahrain's Meteorological Authority, the efficiency of used device is:

$$\eta = \frac{332.0}{180\ X\ 36.03}\ X100\% = 5.12\% \tag{5}$$

The pressing question is how economical is this method? And how much water can be harvested per unit consumption of energy? And how much is the cost per unit volume of harvested waters?

The manufacturer's sticker on the back of the dehumidifier states that the device draws 300 watts of electric power. According to the electricity bill of the first author, the government of Bahrain sells electricity to citizens and residents for 0.003 Bahrain Dinar (BHD) per kilo-watt-hour (kWh). The exchange rate against the American Dollar is BHD 1.00 = \$2.65 makes the cost of electricity in Bahrain \$0.008/kWh.

➔300 watts × 1 hours of operation = 300 watt-hour per hour of operation

➔300 watt-hours ÷1000 = 0.3 kWh

➔0.3 kWh × \$0.008/kWh = \$0.0024 for every hour of operation.

 DOI: 10.1051/e3sconf/20172303001

Given the fact that the harvesting device has been yielding an average of 332.0 ml per hour, the device will cost $7.23 of electricity to produce 1.0 m3 of water. This is in expensive for a device that has only 5.12% efficiency. This cost can be reduced by a factor of 20 or even more, if a device with a harvesting efficiency of 80% or more is used, which puts it in a real competition with other nonconventional potable water production methods like desalination.

Finally, Data of the average quantities of water that could be harvested in each of the six collection locations over the first six months of the experiment, time of the year, average temperature, average relative humidity, average absolute humidity, and the average dew point are presented in Table 2.

Table 2: Harvested waters per month with associated meteorological factors

Month	**Manama (ml/hr)**	**Av. Temp. (°C)**	**Av. RH %**	**Av. AH gm/m³**	**Dew Point (°C)**
Jul.	394.88	36.20	50.70	31.11	23.20
Aug.t	350.27	36.69	52.69	34.58	24.46
Sep.	328.50	33.54	59.54	42.99	24.31
Oct.	317.20	29.25	52.75	27.64	17.25
Nov.	337.82	25.00	64.85	40.27	17.69
Dec.	314.36	21.08	70.58	43.09	15.91
Month	**Barbar (ml/hr)**	**Av. Temp. (°C)**	**Av. RH %**	**Av. AH gm/m³**	**Dew Point (°C)**
Jul.	397.31	36.18	49.90	29.96	23.10
Aug.t	451.46	36.77	52.46	34.29	24.46
Sep.	456.92	33.54	59.54	42.99	24.31
Oct.	380.55	29.25	52.75	27.64	17.25
Nov.	332.10	25.00	64.85	40.27	17.69
Dec.	401.64	21.08	70.58	43.09	15.91
Month	**Hidd (ml/hr)**	**Av. Temp (°C)**	**Av. RH %**	**Av. AH gm/m³**	**Dew Point (°C)**
Jul.	407.78	35.80	51.20	31.49	22.89
Aug.t	427.73	37.00	51.00	32.25	24.25
Sep.	390.00	33.62	59.54	43.09	24.38
Oct.	400.00	29.00	56.43	32.39	18.36
Nov.	315.82	25.08	61.67	35.28	23.30

Dec.	314.58	21.67	72.25	47.38	16.42
Month	**Sanad (ml/hr)**	**Av. Temp (°C)**	**Av. RH %**	**Av. AH gm/m³**	**Dew Point (°C)**
Jul.	371.11	35.82	52.18	32.98	23.73
Aug.t	327.92	36.64	53.71	36.19	24.79
Sep.	317.92	33.83	59.58	43.45	24.50
Oct.	311.36	29.15	55.31	30.96	17.77
Nov.	332.31	24.92	63.25	37.50	14.75
Dec.	303.62	20.82	70.18	41.86	14.70
Month	**Salamabad (ml/hr)**	**Av. Temp (°C)**	**Av. RH %**	**Av. AH gm/m³**	**Dew Point (°C)**
Jul.	336.25	35.82	52.18	32.98	23.73
Aug.t	323.55	36.64	53.71	36.19	24.79
Sep.	292.92	33.83	59.00	42.35	17.33
Oct.	271.00	29.42	53.00	28.12	17.33
Nov.	296.08	25.08	60.58	33.65	14.33
Dec.	305.67	20.82	53.71	20.56	14.70
Month	**HamadTown (ml/hr)**	**Av. Temp (°C)**	**Ave. RH %**	**Av. AH gm/m³**	**Dew Point (°C)**
Jul.	291.94	35.75	52.92	34.05	22.40
Aug.t	238.89	36.62	53.15	35.25	25.43
Sep.	207.00	33.54	59.54	42.99	24.08
Oct.	236.89	29.17	54.67	30.08	17.58
Nov.	197.29	25.23	63.54	38.45	15.46
Dec.	263.55	20.22	69.56	39.63	14.33

This data is presented in this paper and is offered to other researchers along with more data as the project progresses. After the completion of this project, the full 12-month data with all details will be made available to other researchers upon contacting the first author with their request.

4- Conclusions

The following are remarks that could be drawn from the aforementioned work:

- In this paper the authors presented an ongoing project that is being conducted at the civil engineering department of the University of Bahrain. At the time of writing this article, the research is in its midway and currently, some results have been attained but many others are in the making.

 DOI: 10.1051/e3sconf/20172303001

- Preliminary chemical analysis of the harvested indicated that the harvested water, which was kept raw without any filtration, was close to pure. But more tests are are needed to verify this finding. Actually, 6 samples from each collection site were sent to the official Bahrain government water quality and ministry of health labs for a comprehensive list of tests that covers all of the world health organization requirements for drinking water.

- The efficiency of the utilized dehumidifier is very low. This efficiency is expected, since the surface are of the condenser inside the used device is only 15cm×15cm. Furthermore, the utilized dehumidifier is rated for indoor use and in this experiment it was stretched to its limits by being operated in the outdoors with full exposure to the climatic elements, especially the dust in sandstorm days.

- Results show that locations that are close to the shores have higher potential to harvest more water from atmospheric humidity. Results have also shown that absolute humidity and not relative humidity is the factor with direct correlation to the amount of harvestable air humidity.

- Preliminary cost analysis of the harvested water based on a 5%-efficient harvesting device is very promising, yet with the advent of more data in the coming few weeks/months more conclusive findings on a more solid ground can be reached.

- It is the belief of the authors of this paper that harvesting water from air humidity, especially in a geographical setup like that of the kingdom of Bahrain, is very promising for many reasons. Among those are: its simple design, substantially, smaller size, area and operation cost than desalination plants, and savings in installation costs since water is drawn from air without the need for of elaborate piping system.

- Finally, this study can be considered as a door opener to more comprehensive studies about water harvesting of atmospheric humidity in the Kingdom of Bahrain in which, a full-scale production system is utilized.

DOI: 10.1051/e3sconf/20172303001

Acknowledgement:

The authors would like to express their sincere thanks for the all the nice people, who supported this budget-less project. Thanks are extended to H.E. Sh. Nawaf Al-Khalifa and his wonderful team at Electricity and Water Authority (EWA) for their support and help in performing water quality tests on EWA labs. Thanks are also extended to the Dean of Engineering, Dr. Fuad Al-Ansari, for his support of this project and to all of whom giving us any direct or indirect help.

References

1. Abualhamayel, H. and Gandhidasan, P. "A Method of Obtaining Fresh Water from the Humid Atmosphere," Elsevier Desalination Journal, Vol. 113, November 1997, pp. 51-63.
2. Ahrens D., "Essentials of Meteorology: An Invitation to the Atmosphere," 7th edition, Cengage Learning Pulishing, Boston, MA, USA, January 2014.
3. Alekseev, V., Rustamo, N., Ivanov, V. and Dubovskaya V. "Experimental Study of Overland Moisture Condensation," Doklady Earth Sciences, Vol. 393, No. 8, August 2003, pp. 1156–1159.
4. Alnaser, W. and Barakat, A. "Use of Condensed Water Vapor from the Atmosphere for Irrigation in Bahrain," Elsevier Applied Energy Journal, Vol. 65, No. 1, January 2000, pp. 3-18.
5. Alipour, V., Mahvi, A. and Rezaei, L. "Water Condensate Management of Atmosphere Humidity in Bandar Abbas, Iran," Proceedings of WSEAS 9th International Conference on Energy, Environment, Ecosystems and Sustainable Development (EEESD '13), Lemesos, Cyprus, March 2013, pp. 279-284.
6. Atta, R. "Solar Water Condensation Using Thermoelectric Coolers," International Journal of Water Resources and Arid Environments, vol. 1, No. 2, April 2011, pp. 142-145.
7. Alayli, Y., Hadji, N. and Leblond, J. "A New Process for the Extraction of Water from Air," Elsevier Desalination, Vol. 67, December 1987, pp. 227-229.
8. Bergmair, D. "Design of a System for Humidity Harvesting Using Water Vapor Selective Membranes," A Ph.D. Thesis submitted to the Mechanical Engineering Department of the Eindhoven University of Technology, Austria, 2015.
9. Beysens, D., and Milimouk, I. "The Case for Alternative Fresh Water Sources," Translated by T. Fuller from "Pour les Ressources Alternatives en eau" Sécheresse, vol. 11, No. 4, December 2000.
10. Bolonkin, A. "Extraction of Freshwater and Energy from Atmosphere," arxiv:0704.2571.
11. Bolton, D. "The Computation of Equivalent Potential Temperature," Monthly Weather Review. Vol.108, No. 7, July 1980, pp. 1046–1053 .
12. Davtalab, R. and Salamat A. "Water Harvesting from Fog and air Humidity in the Warm and Coastal Regions in the South of Iran," ICID 21st, International Congress on Irrigation and Drainage, 15-23 October 2011, Tehran, Iran.
13. Directorate of Meteorology, Civil Aviation, Kingdom of Bahrain, verbal communication.
14. Eindhoven University of Technology. "Cotton with special coating collects water from fogs in desert," Science Daily, January 2013.

15. Esfandiarnejad, A., Ahangar, R., Kamalian, U. and Sangchouli T., "Feasibility Studies for Water Harvesting from Fog and Atmospheric Moisture in Hormozgan Coastal Zone (south of Iran)" Proceedings of the 5th International Conference on Fog, Fog Collection and Dew, Münster, Germany, July 2010 .

16. Gaware, N., Nadar, M. and Vedpathak M. "Harvesting of Water from Air Using Helical Coil," International Journal of Engineering Sciences & Research Technology, Vol. 4, No. 2, Feb. 2015, pp. 178-182.

17. Grand View Research "Atmospheric Water Generator Market Analysis, Market Size, Application Analysis, Regional Outlook, Competitive Strategies And Forecasts, 2015 To 2022" A Report by Grand View Research, Inc. San Francisco, California, U.S.A,

18. Hamed, A., Aly, A and Zeidan E. "Application of Solar Energy for Recovery of Water from Atmospheric Air in Climatic Zones of Saudi Arabia," SCRIP Open Access Natural Resources Journal, Vol. 2, No. 1, March 2011, pp. 8-17,

19. Hamed, A., Kabeel, A., Zeidan, E. and Aly, A. "A Technical Review on the Extraction of Water from Atmospheric Air in Arid Zones," JP Journal of Heat and Mass Transfer, Vol. 4, No. 3, 2010, pp 213-228.

20. Kumar ,M. and Yadav, A. "Experimental Investigation of Solar Powered Water Production from Atmospheric Air by Using Composite Desiccant Material "CaCl2/saw Wood"," Elsevier Desalination Journal, Vol. 367, No. 1, July 2015, pp. 216-222.

21. Lindblom, J. and Nordell, B. "Water Production by Underground Condensation of Humid Air," Elsevier Desalination Journal, Vol. 189, March 2006, pp. 248-260.

22. Nandy, A., Saha, S., Ganguly, S. and Chattopadhyay S." A Project on Atmospheric Water Generator with the Concept of Peltier Effect," International Journal of Advanced Computer Research, Vol. 4, No 2, June 2014, pp 481-486.

23. Ngaina J, Muthama J., Opere A., Ininda J., Ng'etich C., Ongoma V. and Mutai B. "Potential of Harvesting Atmospheric Water over Urban Cities in Kenya," International Journal of Physical Sciences, Vol. 2, No. 5, May 2014, pp. 69-75.

24. Pauli, G. "The Blue Economy: 10 Years, 100 Innovations, 100 Million Jobs," Paradigm Publications, New Mexico, U.S.A., 2010, ISBN 9780912111902

25. Reznikov, M., Salazar, M. and Lopez, M. "Electrically Enhanced Harvesting of Water Vapor from the Air" Proceedings of The 2015 Electrostatics Society of America (ESA) Annual Meeting on Electrostatics, Paloma, CA, U.S.A. June 2015, pp. 1-11.

26. World Health Organization: Guidelines for drinking water quality, Incorporating First Addendum, Recommendations, 2006.

27. Yang, H., Zhu, H., Hendrix, M., Lousberg, N., de With, G., Esteves, C. and Xin J. "Temperature-Triggered Collection and Release of Water from Fogs by a Sponge-Like Cotton Fabric," Advanced Materials, 2013; DOI: 10.1002/adma.201204278

 DOI: 10.1051/e3sconf/20172303002

Hydrogen production from bio-fuels using precious metal catalysts

Joachim Pasel[*1], *Sebastian Wohlrab*[2], *Mikhail Rotov*[1], *Katrin Löhken*[1], *Ralf Peters*[1], *Detlef Stolten*[1,3]

[1]Forschungszentrum Jülich GmbH, Institute of Energy and Climate Research, IEK-3: Electrochemical Process Engineering, 52425 Jülich, Germany

[2]Leibniz Institute for Catalysis at the University of Rostock (LIKAT), Albert-Einstein-Str. 29a, 18059 Rostock, Germany

[3]Chair for Fuel Cells, RWTH Aachen University, Germany

Key words: catalyst activity; catalyst characterization; autothermal reforming; commercial diesel fuel; fuel cells

* Corresponding author: tel: +49 2461 615140; fax: +49 2461 616695; e-mail: j.pasel@fz-juelich.de

Abstract

Fuel cell systems with integrated autothermal reforming unit require active and robust catalysts for H_2 production. Thus, an experimental screening of catalysts for autothermal reforming of commercial bio-diesel fuel was performed. Catalysts consisted of a monolithic cordierite substrate, an oxide support (γ-Al_2O_3) and Pt, Ru, Ni, PtRh and PtRu as active phase. Experiments were run by widely varying the O_2/C and H_2O/C molar ratios at different gas hourly space velocities. Fresh and aged catalysts were characterized by temperature programmed methods and thermogravimetry to find correlations with catalytic activity and stability.

Introduction

According to the current policies scenario, by 2030 the global CO_2 emissions will be more than 50% higher than today. In addition, the world population is supposed to grow to more than 9 billion of people in 2050. As a consequence, the energy demand strongly increases, especially driven by China and India. The Institute of Energy and Climate Research (IEK-3) at Forschungszentrum Jülich has been developing a two-step approach to contribute to facing these environmental problems. On the one hand, efficient fuel cell systems based on the autothermal reforming of diesel fuels have been developed, which can be used as auxiliary power units for efficient on-board power supply in aircraft or long-haul trucks, for example. On the other hand, a renewable Bio-To-Liquid (BTL) fuel is used as feedstock for these systems.

Biodiesel fuel synthesized by hydro-treating of triglycerides

Transesterification is one of the most important routes for the production of bio-fuels [1-3]. During this process, triglycerides from vegetable oil or animal and plant fats are converted with alcohol molecules (methanol or ethanol) yielding mono-alkyl esters and glycerol. The mono-alky ester molecules then make up the bio-diesel. Another option for the production of bio-diesel is hydro-treating. This process can be run in existing petroleum refineries [4]. Figure 1 shows a simplified scheme for the conversion of triglycerides contained in Jatropha oil, algae oil or plant and animal fats to straight-chain and branched alkanes [4]. In the first two steps of this reaction scheme, the triglyceride molecule is hydrogenated and split into a number of intermediates which can be monoglycerides, diglycerides and carboxylic acids. These intermediates are then converted into alkanes on three different routes: decarboxylation, decarbonylation and a combination of hydrogenation with dehydration. Reaction equations (1)-(3) show exemplarily the different routes for alkane production starting from carboxylic acids. Decarboxylation involves converting a carboxylic acid into a methyl group and CO_2 as shown in equation (1). CO is produced by decarbonylation where the carboxylic carbon is reduced by hydrogen to yield a methyl group, CO and water appear as side products, as shown by equation (2). During the combination of hydrogenation with dehydration the carboxylic acid reacts with hydrogen to produce an alkane and water as shown by the third equation [4].

$$R\text{-}CH_2\text{-}COOH \longleftrightarrow R\text{-}CH_3 + CO_2 \quad (1)$$

$$R\text{-}CH_2\text{-}COOH + H_2 \longleftrightarrow R\text{-}CH_3 + CO + H_2O \quad (2)$$

$$R\text{-}CH_2\text{-}COOH + 3\,H_2 \longleftrightarrow R\text{-}CH_2\text{-}CH_3 + 2\,H_2O \quad (3)$$

For the investigations in this work, so-called NExBTL diesel fuel produced by the Finnish company Neste was used as fuel for the autothermal reforming reactor. NExBTL diesel fuel has a sulfur mass proportion of less than 1 ppm and a fraction of aromatics far lower than 1 vol%.

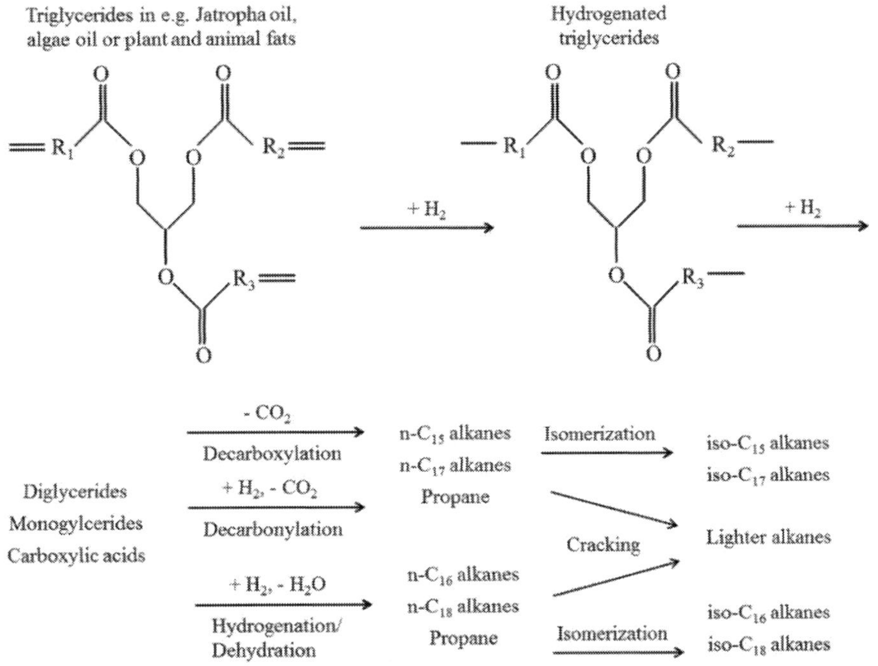

Figure 1 Simplified scheme for the conversion of triglycerides contained in in Jatropha oil, algae oil or plant and animal fats to straight-chain and branched alkanes [4]

DOI: 10.1051/e3sconf/20172303002

Catalyst preparation

The catalysts were synthesized in mass fractions related to the cordierite substrates, whereby for the oxide support and the active metal phase approx. 10 wt.% and 1 wt.%, respectively, were obtained. Each value referred to the total mass of the catalyst. The synthesis of the support γ-Al_2O_3 was identical to that described in previous work [5] except for the temperature, at which the impregnated cordierite substrate was dried after impregnation. This temperature was decreased from 110 °C to 80 °C in order to reach a more homogeneous coating of the support on the substrate. For the synthesis of the bimetallic catalyst PtRu/γ-Al_2O_3 with 0.5 wt.% Pt and 0.5 wt.% Ru, the cordierite substrate, which was already coated with γ-Al_2O_3, was at first impregnated with a precursor solution of $Ru(O_2C_5H_7)_3$ (tris(acetylacetonato)ruthenium (III), dissolved in tetrahydrofurane). This impregnation step was performed twice from the top and the bottom side of the cylindrical substrate, respectively, to ensure a homogeneous coating. The substrate was then dried at 60 °C for 2 hours and calcined at 700 °C for 1 hour. In a second step, this substrate was impregnated with a precursor solution of $[Pt(NH_3)_4](NO_3)_2$ (tetraammineplatinum (II) nitrate). Then, the substrate was dried at 80 °C for 2 hours and calcined at 700 °C for 1 hour. For the synthesis of the bimetallic catalyst PtRh/γ-Al_2O_3 with 0.5 wt.% Pt and 0.5 wt.% Rh, the γ-Al_2O_3 coated cordierite substrate was at first impregnated with a precursor solution of $Rh(NO_3)_3$, then dried at 80 °C for 2 hours and calcined at 700 °C for 1 hour. Again, the impregnation step was performed twice from the top and the bottom side of the cylindrical substrate, respectively, to ensure a homogeneous coating. In a second step, the substrate was impregnated with a precursor solution of $[Pt(NH_3)_4](NO_3)_2$ (tetraammineplatinum (II) nitrate), dried at 80 °C for 2 hours and calcined at 700 °C for 1 hour. The preparation of the monometallic catalyst Pt/γ-Al_2O_3 with 1.0 wt.% was performed by impregnation of the γ-Al_2O_3 coated cordierite substrate with a precursor solution of $[Pt(NH_3)_4](NO_3)_2$ (tetraammineplatinum (II) nitrate). The substrate was then dried at 80 °C for 2 hours and calcined at 700 °C for 1 hour.

Analogously, the monometallic catalyst Ru/γ-Al_2O_3 with 1.0 wt.% was prepared by impregnating the γ-Al_2O_3 coated cordierite substrate with a precursor solution of $Ru(O_2C_5H_7)_3$ (tris(acetylacetonato)ruthenium (III), dissolved in tetrahydrofurane). The substrate was then dried at 80 °C for 2 hours and calcined at 700 °C for 1 hour. The catalyst Ni/γ-Al_2O_3 was synthesized using a $Ni(NO_3)_3$ * 6 H_2O precursor solution. The γ-Al_2O_3 coated cordierite substrate was impregnated with this solution, dried at 80 °C for 2 hours and calcined at 700 °C for 1 hour.

Results and discussion

In the above mentioned previous work by our group [5], it was demonstrated that among the investigated samples Rh/γ-Al_2O_3, Rh/La-Al_2O_3, Rh/CeO_2, Rh/Gd-CeO_2, Rh/ZrO_2 and Rh/Y-ZrO_2 the catalyst Rh/γ-Al_2O_3 had the highest initial catalytic activity and showed the lowest catalytic deactivation in the course of the applied evaluation pattern under rough reaction conditions. For that reason, it was decided to choose γ-Al_2O_3 as catalyst support for the experiments of this contribution. In this respect, Figure 2 shows H_2 concentrations of catalysts Pt/γ-Al_2O_3, Ru/γ-Al_2O_3, Ni/γ-Al_2O_3, PtRh/γ-Al_2O_3 and PtRu/γ-Al_2O_3 as a function of the H_2O/C molar ratio at an O_2/C molar ratio of 0.47 and different GHSVs. NExBTL diesel fuel was used. For means of comparison, the corresponding trend of the H_2

DOI: 10.1051/e3sconf/20172303002

concentration of catalysts Rh/γ-Al_2O_3 from previous work taken under identical reaction conditions was added to Figure 2. The first sequence was started at a H_2O/C molar ratio of 2.1 and a GHSV of 59,000 h^{-1}. That means, Figure 2 has to be interpreted from right to left.

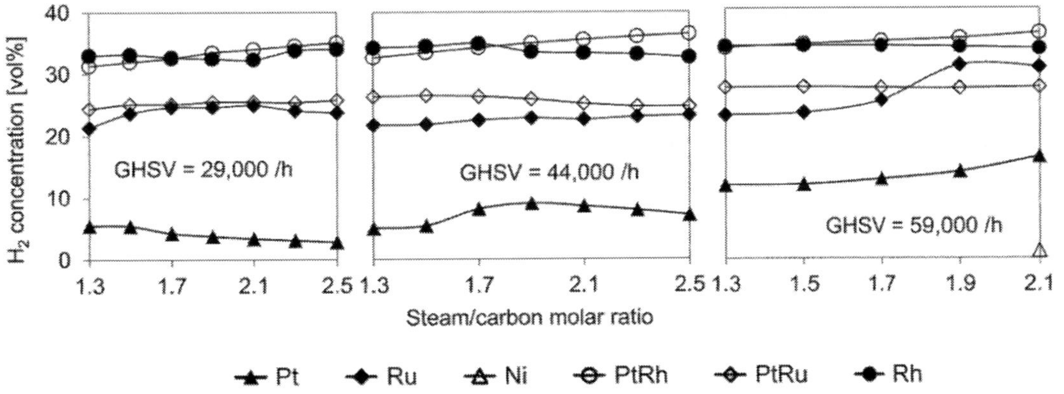

Figure 2 H_2 concentration on all six catalysts as a function of the H_2O/C molar ratio, O_2/C molar ratio = 0.47, at different mean GHSVs, NExBTL diesel fuel

Figure 2 shows a very low H_2 concentration in the range of only 3 vol% for the catalyst Ni/γ-Al_2O_3 at the beginning of the first sequence (H_2O/C molar ratio of 2.1). The experimental evaluation of this catalyst had to be stopped after this first experiment. Otherwise, the large residual mass stream of unconverted hydrocarbons in NExBTL diesel fuel would have severely contaminated and destroyed many components of the test rig downstream the catalytic test reactor including the gas analytical system. Also the sample Pt/γ-Al_2O_3 showed comparatively low H_2 concentrations between 16 and 11 vol% during the first sequence in Figure 2. The values of catalyst Ru/γ-Al_2O_3 started at a higher level of approx. 31 vol% but then decreased with decreasing H_2O/C molar ratio to approx. 23 vol%. The bimetallic catalyst PtRu/γ-Al_2O_3 revealed very stable H_2 concentrations of approx. 27 vol% independent of the H_2O/C molar ratio. Thus, the addition of Ru to the Pt-based catalyst had a very positive effect on the catalytic activity. The bimetallic sample PtRh/γ-Al_2O_3 and the reference catalyst Rh/γ-Al_2O_3 showed highest H_2 concentrations ranging from 36 to 34 vol% for the bimetallic catalyst and from 33.5 to 34 vol% for the reference catalyst from previous work.

Figure 2 in the middle and on the left show the results of the second and third experimental sequence with respect to H_2 concentrations of the five investigated catalysts and the reference sample. The catalysts were not replaced between the first and second or between the second and third sequence. This means that the series of sequences run with each catalyst can be considered an accelerated aging experiment. It can be seen that during sequences 2 and 3 the H_2 concentrations of the catalyst Pt/γ-Al_2O_3 further decreased to values between 9 vol% and only 3 vol%. An additional deactivation of this catalyst for autothermal reforming occurred. As a consequence of the deactivation step from the first sequence, the H_2 concentrations of the monometallic catalyst Ru/γ-Al_2O_3 were always lower than those of the bimetallic sample PtRu/γ-Al_2O_3 during the second and third sequence. At a GHSV of 44,000 h^{-1}, they were almost constant at 22 – 23 vol% and thus in the same order magnitude as after the deactivation step in the first sequence. During the third sequence at a GHSV of 29,000 h^{-1}, the values for the H_2 concentration of Ru/γ-Al_2O_3 slightly increased to 25 vol% at most favorable H_2O/C molar ratios between 2.1 and 1.7. Obviously, the lower residence time during the third sequence had a positive effect on the catalytic activity of Ru/γ-Al_2O_3. As mentioned above the H_2 concentrations of the bimetallic sample PtRu/γ-Al_2O_3 were always slightly higher than those of Ru/γ-Al_2O_3 during the second and third sequence. The values for PtRu/γ-Al_2O_3 were quite stable between 25 and 26 vol% thus showing no sign of deactivation. However, it can be concluded from the first three sequences that both catalysts, Ru/γ-Al_2O_3 and PtRu/γ-Al_2O_3, were also not suitable for efficient autothermal reforming of BTL diesel fuel. The observed H_2 concentrations were simply too low compared to those given in the literature. This would lead to poor H_2-productivity of the reforming step within a fuel cell system and thus also significantly reduce the overall efficiency of the whole fuel cell system. Also during the second and third sequence the H_2 concentrations of the bimetallic catalyst PtRh/γ-Al_2O_3 were highest among the investigated samples and varied between 34 and 36 vol% at most favorable H_2O/C molar ratios between 2.3 and 1.9. Under these reaction conditions these values were even higher than those of the reference catalyst Rh/γ-Al_2O_3, which was considered suitable for autothermal reforming of diesel fuels in previous work [5]. So, from the first three sequences with widely varying H_2O/C molar ratios it can be concluded that the bimetallic catalyst PtRh/γ-Al_2O_3 from this work is promising for the autothermal reforming of BTL diesel fuel. It showed a high initial catalytic activity and no detectable signs of deactivation.

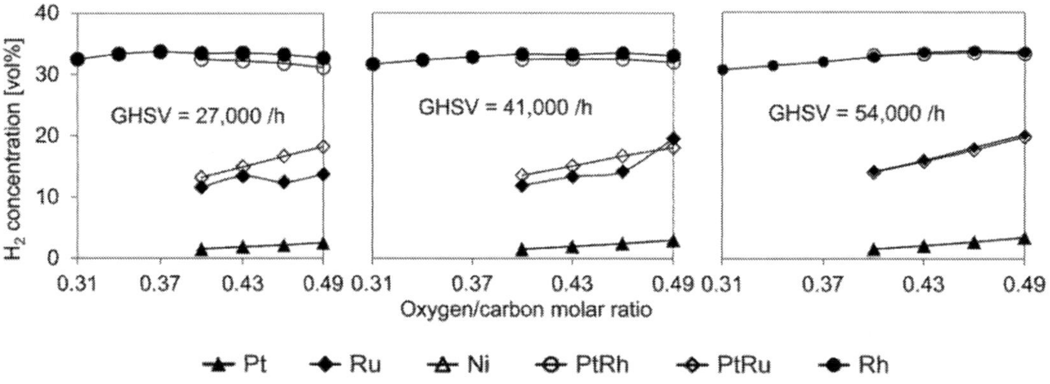

Figure 3 H_2 concentration on all six catalysts as a function of the O_2/C molar ratio, H_2O/C molar ratio = 1.9, at different mean GHSVs, NExBTL diesel fuel

In Figure 3 the influences of the O_2/C molar ratio at three different mean GHSVs on the catalytic activity and the stability of the catalysts from this work and the reference catalyst are shown. The pre-aged catalysts from the first three sequences were not replaced. So, the accelerated aging experiment continued. It becomes obvious from Figure 3 on the right that the bimetallic catalyst PtRh/γ-Al_2O_3 remained as active as the reference catalyst during the fourth sequence. H_2 concentrations were in the range between 34 and 33 vol% and almost identical to those of the reference catalyst Rh/γ-Al_2O_3. The addition of Rh by Pt did not lead to any decrease in the catalytic activity although the trend for the monometallic catalyst Pt/γ-Al_2O_3 in Figure 3 would suggest this conclusion. As during the first three sequences the H_2 concentrations of Pt/γ-Al_2O_3 were very low showing only values between 3 and 2 vol%. In Figure 3 on the right the H_2 concentrations of the bimetallic sample PtRu/γ-Al_2O_3 and the monometallic catalyst Ru/γ-Al_2O_3 were very similar. They started at approx. 20 vol% and decreased to 14 vol% at an O_2/C molar ratio of 0.40. Obviously, a significant deactivation of these samples occurred under the rough conditions of the fourth sequence. As can be seen in Figure 3 in the middle and on the left, the deactivation of the catalyst Ru/γ-Al_2O_3 proceeded. It revealed H_2 concentration gradients between 20 and 12 vol% at a GHSV of 41,000 h^{-1} and between 14 and 12 vol% at a GHSV of 27,000 h^{-1}, respectively. The H_2 concentrations of the sample Pt/γ-Al_2O_3 remained at their very low level in the range of 3 to 2 vol%. In both figures (middle and left), a slight deactivation of the bimetallic catalyst PtRh/γ-Al_2O_3 could be observed. H_2 concentrations slightly decreased from the range of 34 to 33 vol% to the level of 32 vol% and 31 to 32 vol%, while the concentrations of the reference catalyst remained constant. Nevertheless, catalyst PtRh/γ-Al_2O_3 confirmed the conclusion drawn from the results of the first three sequences that it is a promising candidate for the autothermal reforming of BTL diesel fuel with high initial catalytic activity and only marginal signs of deactivation under rough reaction conditions.

In the following, the reducibility of the catalysts from this work will be analyzed with the aid of measurements from temperature-programmed reduction. Figure 4 on the left shows the profiles from the temperature programmed reduction of the fresh catalysts Ni/γ-Al_2O_3, Pt/γ-Al_2O_3, Ru/γ-Al_2O_3, PtRh/γ-Al_2O_3 and PtRu/γ-Al_2O_3, before they were subjected to the evaluation pattern. Again for comparison reasons, the corresponding TPR profile of the reference catalyst Rh/γ-Al_2O_3 is added. The profile **a** of fresh catalyst Ni/γ-Al_2O_3 shows a very broad signal in the temperature range between 300 and 900 °C, which can attributed to the formation of solid-state solutions of nickel-aluminum-oxide [6]. This signal has a comparatively high intensity, which points at a high reducibility of the fresh sample. The finding stands in sharp contrast to the very low catalytic activity of Ni/γ-Al_2O_3 for the autothermal reforming of NExBTL diesel. The profile **b** of the catalyst Pt/γ-Al_2O_3 does not show any reduction signal at all. This means that this sample does not possess any reducibility, which can be detected by temperature programmed reduction, and thus that it is only marginally able to catalyze the autothermal reforming of NExBTL diesel fuel by adsorbing and releasing electrons. This result is well consistent with the comparatively low initial catalytic activity of Pt/γ-Al_2O_3. In the literature [7], the reduction of platinum oxide is reported at a temperature of approx. 250 °C. In the case of the catalyst Ru/γ–Al_2O_3, the TPR profile **c** shows a weak shoulder at 105 °C and a more pronounced peak at 145 °C. In the literature [8], the weak shoulder is assigned to the conversion of RuO_3 to RuO_2, while the main peak is explained by the reduction of RuO_2 to elementary Ru. In the profile **e** of the fresh catalyst PtRu/γ-Al_2O_3 very similar signals for the consumption of H_2 were detected at temperatures of 103 and 139 °C, respectively. As in the case of catalyst Ru/γ-Al_2O_3, they can be explained by the conversion of RuO_3 to RuO_2 and the reduction of RuO_2 to elementary Ru [9]. No interaction between Ru and Pt particles possibly forming a new joint phase was detected. This is in accordance with results from the literature [10]. The overall intensity of both peaks of PtRu/γ-Al_2O_3 in profile **e** is slightly lower than that of the monometallic sample Ru/γ-Al_2O_3 in profile **c**. That means that the addition of Pt to the Ru-based catalyst lowered the reducibility in the fresh state. This graduation is in good accordance with the initial H_2 concentrations of these both catalysts showing initial values of 31 vol% (Ru/γ-Al_2O_3) and 27 vol% (PtRu/γ-Al_2O_3), respectively. The largest consumption of H_2 during the TPR experiments with the fresh catalysts was observed in the case of sample PtRh/γ-Al_2O_3. In profile **d** two very pronounced signals at temperatures of 83 and 150 °C were identified, which can be attributed to the reduction of a three-dimensional Rh_iO_x phase or a dispersed two-dimensional Rh_iO_x phase on γ-Al_2O_3 [11]. The overall intensity of these signals is even significantly higher than that of the signals of the monometallic catalyst Rh/γ-Al_2O_3 (cf. profile **f**). Also in this case no interaction between the two metals Pt and Rh forming a joint phase could be detected in profile **d**. The observed outstanding reducibility of the fresh sample PtRh/γ-Al_2O_3 is in good accordance with its very high initial catalytic activity with H_2 concentrations in the range of 36 to 35 vol%. Figure 4 on the right shows the profiles from the temperature programmed reduction of the aged catalysts Ni/γ-Al_2O_3, Pt/γ-Al_2O_3, Ru/γ-Al_2O_3, PtRh/γ-Al_2O_3 and PtRu/γ-Al_2O_3, after they were subjected to the evaluation pattern as described above. A comparison of the figures on the left and on the right makes obvious that in any case the profiles of the fresh and aged catalysts are very similar with respect to the position of the respective peaks. Profile **a** of the aged catalyst Ni/γ-Al_2O_3 again shows the broad signal between 300 and 900 °C. Again, no H_2 was consumed by the catalyst Pt/γ-Al_2O_3. In the profiles **c** and **e** of the catalysts Ru/γ-Al_2O_3 and PtRu/γ-Al_2O_3 the weak shoulder at approx. 130 °C and the more pronounced peaks at 160 and 145 °C, respectively, can be rediscovered. Again as in the case of the fresh samples, the overall intensity of the two peaks of the aged sample PtRh/γ-Al_2O_3 (98 and 144 °C) is by far highest. The only difference is that for the aged catalyst the peak at 98 °C is less pronounced than it was in the case of the fresh catalyst PtRh/γ-Al_2O_3. For the catalyst Ru/γ-Al_2O_3 a significant decrease in the overall intensity of the weak shoulder and the main peak could be detected after the sample was run in the evaluation pattern indicating a loss in reducibility. This is in good agreement with the observed deactivation of this catalyst in the course of the evaluation pattern. However with all other catalyst, the overall intensities of fresh and aged samples were comparable. The rough reaction conditions applied during the evaluation pattern did not significantly lower the reducibility of the active phases of these catalysts. This is in contradiction to the deactivation phenomena, which were observed with the catalysts Pt/γ-Al_2O_3, PtRu/γ-Al_2O_3 and PtRh/γ–Al_2O_3.

 DOI: 10.1051/e3sconf/20172303002

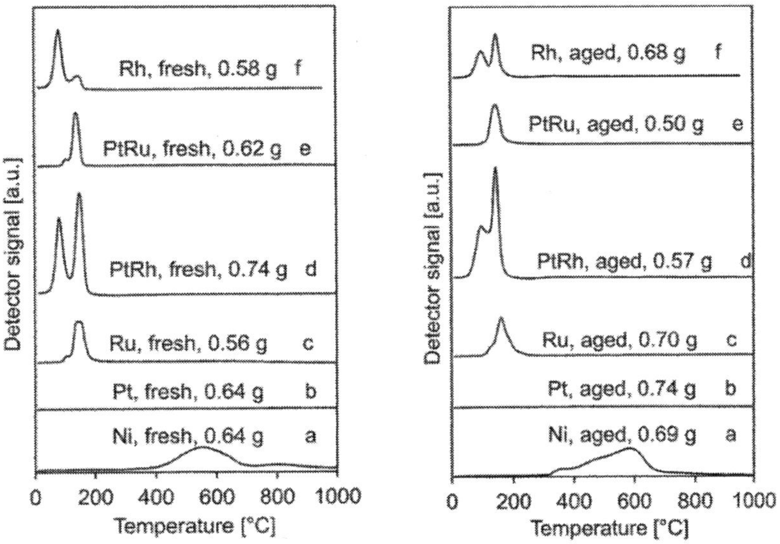

Figure 4 TPR profiles of fresh (left) and aged (right) catalysts Ni/γ-Al_2O_3, Pt/γ-Al_2O_3, Ru/γ-Al_2O_3, PtRh/γ-Al_2O_3, PtRu/γ-Al_2O_3 and Rh/γ-Al_2O_3, H_2 consumption is displayed as a function of temperature

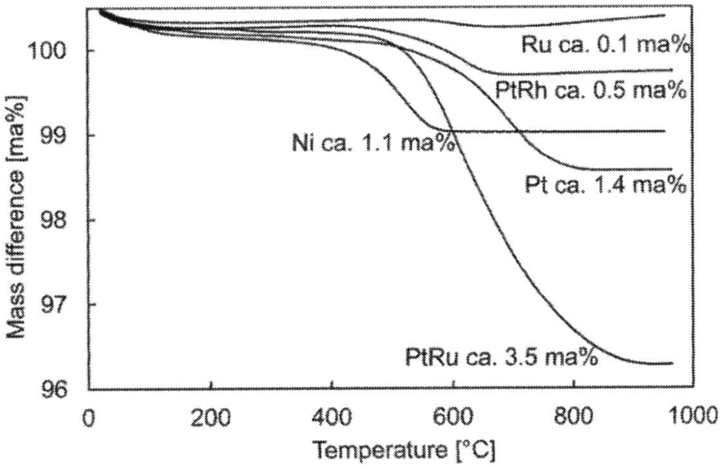

Figure 5 Diagram for the thermogravimetric investigation of aged catalysts: Pt/γ-Al_2O_3, Ru/γ-Al_2O_3, Ni/γ-Al_2O_3, PtRh/γ-Al_2O_3 and PtRu/γ-Al_2O_3

In the next subsection, the results of thermogravimetric methods try to shed additional light on causes for the reduced H_2 concentrations of the samples in this contribution. The catalysts were heated to temperatures of almost 950 °C in a stream of air while the change in mass was simultaneously determined. The goal was to quantitatively investigate whether carbon-containing deposits were formed on the surfaces of the five catalysts in this study. It becomes obvious in Figure 5 that in the cases of the catalysts Pt/γ-Al_2O_3, Ni/γ-Al_2O_3 and PtRu/γ-Al_2O_3 significant mass losses were detected during the thermogravimetric analyses in the temperature range between 500 and 800 °C. They amounted to 1.4 ma% for Pt/γ-Al_2O_3, 1.1 ma% for Ni/γ-Al_2O_3 and even 3.5 ma% for PtRu/γ-Al_2O_3. The analysis of the off-gases by means of mass spectrometry revealed high concentrations of carbon dioxide in the temperature range between 500 °C and 800 °C. Cheng et al. [12] found in their work a combustion temperature of coke on Rh/γ-Al_2O_3 of 600 °C. Therefore, it can be derived that the observed mass losses are due to the burn-off of carbon-containing components from the surface of the catalysts which were formed during experiments on autothermal reforming. Forzatti and Lietti [13] describe how the carbon-containing deposits diminish the catalytic activity, sometimes strongly, since they accumulate on the active catalyst centers so that the latter are no longer accessible for the reactants in autothermal reforming. For the catalysts Pt/γ-Al_2O_3 and PtRu/γ-Al_2O_3 from this work, the detected carbonaceous deposits can explain the observed loss in catalytic activity of these samples during the evaluation pattern. The thermogravimetric investigation of the catalysts Ru/γ-Al_2O_3 and PtRh/γ-Al_2O_3 showed only minor carbon-containing deposits on the catalysts' surfaces. Therefore, correlations between these findings and the catalytic activities of the catalysts cannot be derived.

Conclusions

Of the samples tested, PtRh/γ-Al_2O_3 was found to be the most active and robust catalyst for the autothermal reforming of NExBTL diesel fuel. H_2 concentrations were in the range of 35 - 36 vol% and decreased only slightly when the evaluation pattern was applied. In contrast, Pt/γ-Al_2O_3 showed the strongest deactivation. Ni/γ-Al_2O_3 was not active at all. As described above in detail, decreased reducibilities of the active phases and coke depositions on the catalyst surfaces can be adduced to explain the deactivation phenomena observed with the catalysts from this work.

References

[1] Adewale P, Dumont M-J, Ngadi M. Recent trends of biodiesel production from animal fat wastes and associated production techniques. Renewable and Sustainable Energy Reviews 2015;45:574-88. DOI: 10.1016/j.rser.2015.02.039

[2] Dai Y-M, Wu J-S, Chen C-C, Chen K-T. Evaluating the optimum operating parameters on transesterification reaction for biodiesel production over a LiAlO2 catalyst. Chemical Engineering Journal 2015;280:370-6. DOI: 10.1016/j.cej.2015.06.045

[3] Maneerung T, Kawi S, Wang C-H. Biomass gasification bottom ash as a source of CaO catalyst for biodiesel production via transesterification of palm oil. Energy Conversion and Management 2015;92:234-43. DOI: 10.1016/j.enconman.2014.12.057

[4] Huber GW, O'Connor P, Corma A. Processing biomass in conventional oil refineries: Production of high quality diesel by hydrotreating vegetable oils in heavy vacuum oil mixtures. Appl. Catal., A 2007;329:120-9. DOI: 10.1016/j.apcata.2007.07.002

[5] Pasel J, Wohlrab S, Kreft S, Rotov M, Löhken K, Peters R, et al. Routes for deactivation of different autothermal reforming catalysts. J. Power Sources 2016;325:51-63. DOI: 10.1016/j.jpowsour.2016.06.005

[6] Li C, Chen Y-W. Temperature-programmed-reduction studies of nickel oxide/alumina catalysts: effects of the preparation method. Thermochimica Acta 1995;256:457-65. DOI: 10.1016/0040-6031(94)02177-P

[7] Ebitani K, Hattori H. Combined Temperature-Programmed Reduction (TPR)- Temperature-Programmed-Desorption (TPD) Study od Supported Platinum Catalysts. The Chemical Society of Japan 1991;64:2422-7.

DOI: 10.1051/e3sconf/20172303002

[8] Bianchi CL. TPR and XPS investigations of Co/Al_2O_3 catalysts promoted with Ru, Ir and Pt. Catalysis Letters 2001;76:3-4.

[9] Nurunnabi M, Murata K, Okabe K, Inaba M, Takahara I. Performance and characterization of Ru/Al2O3 and Ru/SiO2 catalysts modified with Mn for Fischer–Tropsch synthesis. Appl. Catal., A 2008;340:203-11. DOI: 10.1016/j.apcata.2008.02.013

[10] Bi J-L, Hong Y-Y, Lee C-C, Yeh C-T, Wang C-B. Novel zirconia-supported catalysts for low-temperature oxidative steam reforming of ethanol. Catal. Today 2007;129:322-9. DOI: 10.1016/j.cattod.2006.11.027

[11] Yao HC, Japar S, Shelef M. Surface interactions in the system $RhAl_2O_3$. J. Catal. 1977;50:407-18. DOI: 10.1016/0021-9517(77)90053-7

[12] Cheng CK, Foo SY, Adesina AA. Carbon deposition on bimetallic Co–Ni/Al2O3 catalyst during steam reforming of glycerol. Catal. Today 2011;164:268-74. DOI: 10.1016/j.cattod.2010.10.040

[13] Forzatti P, Lietti L. Catalyst deactivation. Catal. Today 1999;52:165-81. DOI: 10.1016/S0920-5861(99)00074-7

Pine Needles as Potential Energy Feedstock: Availability in the Central Himalayan State of Uttarakhand, India

L.D. Kala*, P.M.V. Subbarao

Mechanical Engineering Department

Indian Institute of Technology Delhi, Hauz Khas, New Delhi – 110016 (India)

**Corresponding author. Email - ldkala@yahoo.com;*

Phone - +91-9313899865; Fax - +91-11-26581121

WREC17 Conference track: Biomass and Waste-to-energy

Abstract

The amount of pine needles (*pinus roxburgii*) potentially available for use as energy feedstock in the Central Himalayan state of Uttarakhand in India has been estimated. It involves estimating the gross annual amount of pine needle yield followed by a comprehensive identification and quantification of the factors that affect the net annual pine needle yield available as energy feedstock. These factors include considerations such as accessibility, alternative uses, forest fires, other losses, etc., that are influenced by aspects ranging from physical constraints to traditional societal traits. Tree canopy cover method has been used for estimating the gross annual pine needle yield. The information on canopy density is obtained from remote sensing data, that forms the basis for forest classification. The annual gross pine needle yield has been estimated at 1.9 million tonnes while the annual net pine needle yield at 1.33 million tonnes. The annual primary energy potential of pine needles available as energy feedstock has also been estimated. For annual net energy potential estimation, thermal and electrical routes are considered. Electrical energy generation from pine needles using thermochemical conversion has been examined and the corresponding potential for electricity generation been estimated. An installed capacity of 789 MW can be supported with pine needles feedstock for supplying electricity in rural areas for five hours a day. For round the clock generation, an installed capacity of 165 MW can be supported by the pine needle energy feedstock.

Keywords

Pine needle gasification, Gross annual yield , Factors affecting availability as energy feed-stock, Primary and net energy delivery potential.

DOI: 10.1051/e3sconf/20172304001

Introduction

Uttarakhand in Central Himalayas, a northern hill state of India, is rich in flora and fauna and the source of the major river systems – Ganga and Sharda. This study brings in focus energy situation in the state where demand exceeds supply. The year 2014-15 recorded the peak electricity demand at 1930 MW and the peak energy shortage between 8 and 46 percent [1]. The state has power generation installed capacity of 2361 MW (in 2015) including state allocation of central generating stations [2]. Almost 17 percent of the population of the state do not have access to electricity, though 99 percent of the villages have been electrified [3]. The demand for electricity has steadily grown since formation of the state in year 2000. This is understandable as the new state had reasonable thrust on industrialisation.

The options available for growth of power generation in Uttarakhand exist in conventional as well as in non conventional sectors. Hydro based generation offers major potential in conventional power generation [4]. But the large dams come with their enormous socio-environmental consequences. In non conventional, solar photovoltaic (PV) has good potential as the state receives enough sunshine [4]. Microhydel based power generation again offers enormous opportunites for harnessing. This technology has advantages over large hydro in terms of environmental and socio-economics. There are pockets where wind energy generation offers good scope and thorough site specific life cycle analysis for such locations is needed before deciding in their favour.

Biomass offers good potential as energy feedstock in the state [5]. Biomass can be used for obtaining thermal energy and electrical energy both. Other uses of biomass can be for domestic applications such as cooking. Uttarakhand has relatively cold climate and wider biomass uses can be exercised, such as for house warming.

Forest based biomass is available in plenty in the state owing to its large forest area. Pine needles are forest biomass and have no major usage as energy feedstock so far. Their useful application is highly desirable. Pine needles are known to be prone to fire, causing loss of forests and forest biomass. The fire catching property can be utilised for gasification of pine needles, generating fuel gas that can be used for

 DOI: 10.1051/e3sconf/20172304001

energy generation, electrical power being the most preferred in such circumstances of remotely scattered energy feedstock.

Pine forests are spread across the length and breadth of the state and pine needles as energy feedstock are therefore suited for decentralised power generation. This energy can also be transmitted through grid if the local demand is not adequate. Physical availability of grid in near vicinity is essential for this. Use of pine needle for energy generation is most likely to have positive impact on energy access in the villages of the state.

Gasification of pine needle makes use of thermochemical conversion technology in a gasifier equipment. The producer gas so generated is cleaned for operating an engine coupled to a generator. A downdraft gasifier is best suited for electricity generation [6,7]. Pine needle is compared with some commonly used gasification feedstocks in Table-1. The calorific value of pine needle is comparable to the most common energy feedstock for a gasifier i.e. wood. It is in fact a shade higher. It however, has lower bulk density. This may render it unsuitable for gasifier unless densified as all low density feedstocks such as agricultural wastes and loose biomass should be [8,9]. Figure-1 gives a flow chart for pine needle utilisation for electricity generation using a gasifier.

Table-1: Comparision of pine needle as gasification feedstock

Name feedstock	**Calorific value (MJ/kg)**	**Bulk density (kg/m^3)**	**Source**
Fuelwood	7.2-14.4	188-277	[10]
Pine needle	19.44	94.33	[11]
Bagasse	14.6	935.6	[12]
Rice husk	13.24-16.2	86-114	[13]

DOI: 10.1051/e3sconf/20172304001

Electricity generation through pine needles can also be done for captive power by industrial units in the vicinity of pine forests. These units would mosty lie in industrial segment of small and medium enterprises as large industries are not located in hill areas of the state. This would necessitate dealing with logistics for pine needle collection, transportation and storage.

Densification of pine needle biomass yields briquettes (or pellets) that can be fed in a gasifier for conversion into produer gas fuel. The density may now lie above 400 kg/m^3 depending upon the type of densification process and equipment used [14]. The positive results of pine needle gasification after densification is evident from from some efforts in India and Nepal [15,16].

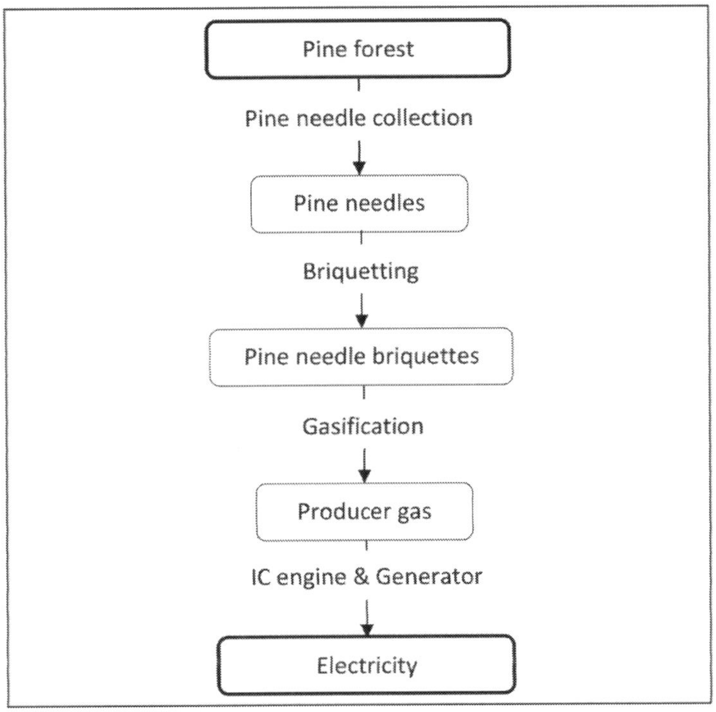

Figure-1: Flow Chart depicting process of power generation from pine needles

Such large scale usage of pine needles need enabling environment in terms of policy thrust and availability of credible associated information such as matrix of pine needle availability with particulars of pine forests including yield, location, area, accessibility, proximity to road, etc. This study is an attempt towards evolving such information to promote usage of pine needles as energy feedstock in the state of Uttarakhand. The factors affecting net annual pine needle yield as energy feedstock have been thoroughly discussed, analysed and quantified. Estimates of potential net pine needles available as energy feedstock has been made. The extent of thermal and electrical energy generation using pine needles have also been estimated.

Methodology for estimation of pine needle availability for use as Energy Feedstock

The estimation of pine needle availability in Uttarakhand as energy feedstock involves assessment of (i) gross annual pine needle yield (ii) pine needles lost as energy feedstock due to various losses, alternative uses, limitations etc., and (iii) net annual pine needle yield as energy feedstock, obtained as the difference of the gross pine needle yield and the amount of pine needles lost.

Tree density based approach is used to estimate pine needle availability potential.The estimation of gross pine needle yield required data on forest cover, forest class or type, tree canopy density, tree canopy cover, tree density, pine needle yield per tree, etc. Two types of forests – very dense forests (VDF) and moderately dense forests (MDF) – are included for the estimation. The forest types along with their classification is given in Section 2.1. Forests with pine trees alone (pure pine forests) are considered for the estimations here as the aim was to obtain first approximation of the pine needle yields. Pine trees however, are also present in mixed pine forests such as pine-sal, pine-oak, etc., which are ignored for this study. It can therefore be argued that these estimates are conservative in nature.

Forest classification

Forest statistics are generated using satellite based remote sensing technology. This technique classifies forests on the basis of tree canopy density, which is percent land area covered by the overall tree canopy as viewed from top (by the satellite). Forest Survey of India (FSI) has used digital data of LISS-III satellite with spatial resolution of 23.5 metre for India State of Forest Report (ISFR) 2013 [17].

Table-2 presents forest classification and related tree canopy density. The forests are classified as very dense forest (VDF), moderately dense forest (MDF), open forest (OF) and scrub forests. For very dense forest tree canopy density is 70 percent and above while for open forest it lies between 10 and 40 percent.

The state of Uttarakhand has total forest cover of 3,799,900 hectares (ha) and this includes reserve forests, protected forests, civil soyam, private forests and special forests [18]. Tree canopy based forest distribution is available for 2,449,200 hectares of reserved forest land. For altitude between 500 - 3000 metres, where pine mainly grows, VDF consists of 413,700 hectares (21 percent), MDF consists of 1,106,900 hectares (56 percent) and OF consists of 455,100 hectares (23 percent) [17].

The area of pine forests in the state is 394,383 hectares (Data for 2011). The class distribution of pine forest is assumed to be same as for the total forest, as such distribution for pine forests is not available.

Table-2: Tree canopy density based categorization of Forests [17]

S. No.	Forest class or type	Tree canopy density
1	Very dense forest (VDF)	Canopy density of 70 % and above.
2	Moderately dense forest (MDF)	Canopy density of 40 % and more but below 70 %.
3	Open forest (OF)	Canopy density of 10 % and more but below 40%.
4	Scrub	Tree canopy density less than 10 %.

Estimating gross annual pine needle yield in the state of Uttarakhand

The pine needle yield of a pine forest may depend upon factors like type of forest, location, etc. All these factors are assumed to be encompassed in the tree canopy density of the forest. For estimation of gross annual pine needle availability, OF may not justify economic pine needle collection owing to small pine needle quantity per unit area. The area under VDF and MDF forests arrived on the basis of the above discussion are given below.

 DOI: 10.1051/e3sconf/20172304001

Area of pine forest, A_{pf} = 394383 ha

Area of very dense pine forest, A_{vdf} = 82821 ha

Area of moderately dense pine forest, A_{mdf} = 220855 ha

Tree canopy density of a forest can be a used to calculate tree density of that forest. Tree canopy density is canopy cover (the sum of tree canopy cover) per unit forest area as seen from top [19]. Tree canopy cover is the area covered by canopy of an individual tree which may nearly be same for each tree in a grown up forest. A VDF forest therefore has more tree canopy density for greater number of trees and MDF has lower tree canopy density and lower number of trees as evident from Table-2. Annual pine needle yield of the forest per unit area can be estimated if along with tree density, per tree pine needle yield is available. Canopy cover of a unit area forest is estimated by Wu [20]. Therefore these parameters can be given as:

(1) Canopy cover = $\frac{\pi}{4}\sum_{i=1}^{n} di^2$ ....(1)

(2) Tree canopy density = $[\frac{\pi}{4}\sum_{i=1}^{n} di^2]/[Forest\ Area]$ ....(2)

where di is the average effective diameter of the tree crown as seen from top.

The effective diameter is smaller than the real diameter of the tree canopy and it may vary as per the tree packing in the forest or the tree density. Closer tree packing may have smaller effective diameter. Gill etal. [19] have reported minimum crown radius as 2.2 m for conifers. The number of trees obtained for a hectare VDF pine forest is 680 and for MDF forest, 440 per hectare for effective canopy diameter as 4 metre.

Pine needle yield per tree reported in literature varies widely as given in Table-3. The variation may be due to factors such as location, age of trees, method of measurement, etc. These measurement most probably may have been made for the pine forests falling in VDF and MDF forests. The average value therefore has been taken for further estimation.

Table-3: Reported estimates of pine needle yield per tree per year in Uttarakhand

 DOI: 10.1051/e3sconf/20172304001

S. No.	Source reference	Needle yield (kg/ha/yr)	Pine tree per ha	Needle yield per tree (kg/yr)	Remarks
1	[21]	5994	575	9.2	-
2	[22]	5371	278	19	-
3	[22]	5700	1630	3.5	Tree density high.
4	[22]	6800	657	10.4	-
5	[22]	7400	820	9	-
6	[22]	7600	540	14.1	-
7	[23]	8900	700	12.7	-
Average value of pine needle per tree = 12.4 kg					

Annual Pine Needle Yield (APNY) for the state now can be obtained as following.

$$APNY = (A_{vdf}) \times (N_{tree,vdf}) \times (PNY_{tree}) + (A_{mdf}) \times (N_{tree,mdf}) \times (PNY_{tree}) \quad \text{..........(3)}$$

where N refers to number of pine trees per hectare, PNY indicates pine needle yield per year and others refer to their usual representation.

Estimation of pine needles available for use as energy feedstock

The methodology developed above will give the gross annual pine needle yield. This amount may not be entirely available as energy feedstock because of pine needles lost due to reasons ranging from physical access to cultural behaviour. These losses may be due to poor accessibility, alternative traditional uses of pine needles, forest fires and various losses such as due to rains, winds, storage, transportation, etc. These are analysed in the following paragraphs.

Pine needles for use as energy feedstock would entail collection from the forests and taken to the site of the gasifier power generation unit. Pine forests exist from the height of 500 metres to 3500 metres and physical accessibility being an issue at several locations may impact collection of pine needles. Accessibility of pine forests may be affected due to steep slopes, gorges, difficult terrains, etc., making it humanly difficult to collect pine needles. Forests at higher reaches, say above 3000 metre height, may pose greater difficulty and 8 percent of forest lies above this height [17].

Accessibility also means distance from inhabitation where it is economically not feasible to collect pine needles. This includes isolated patches of forests. Presence of wild dangerous animals may also hamper accessibility. Forests have thorny shrubs and other undercover vegetation that makes traversing for pine needle collection difficult. An estimated 20 percent of pine forest may be difficult due to accessibility related reasons causing loss in pine needle collection for energy feedstock.

Pine needles do find some use in villages close to pine forests. This mostly pertains to livestock bedding material [24]. Other uses are small and can be ignored. Cattle population in hills of Uttarakhand is estimated to be 2.43 million [25]. About 15 percent of families in the hills of the state are expected to be using pine needles for animal bedding. An animal may be needing about 1.5 kg of pine needle as bedding material per night. The total annual usage of pine needles for animal bedding may be about 24337 tons.

No alternate use of pine needles other than animal bedding is reported. Attempts are being made to make use of pine needles using new technologies for paper making, chemical extraction, as fuel in cookstoves, etc., but such efforts will take some time before their wide scale acceptability. These therefore have no impact on current estimates.

Forest fires also render loss of pine needles. Forest fire data for last 15 years reveal that the forest area affected by forest fire has been between a low of 231 hectares to a high of 4983 hectares [18]. On an average 4000 hectares can be taken as the forest

DOI: 10.1051/e3sconf/20172304001

area affected by fire and loss of its pine needle yield. This is about 1 percent of the total pine forest area in Uttarakhand.

Loss of pine needles may occur due to winds and rains. Excessive wind may blow pine needles to inaccessible surface and rains strong water current may wash pine needles lying in the stream path. The loss due to both these can be minimised by periodic pine needle collection. Pine needle availability may also be affected by reasons owing to inefficient collection.

Pine needles have low bulk density require large space for storage. The storage is inevitable for ensuring constant availability of pine needles for their utilisation. The storage is best located at the site of utilisation of pine needles and necessitates transportation from pine forests. Losses in storage and transportation are also likely to occur. Together, all these losses may be taken as about 10 percent.

The discussion above yields values of loss factor affecting pine needle availability as energy feedstock in Uttarakhand. These factors enable estimation of the shortfall in net pine needle collection as energy feedstock. Table-4 gives a summary of these in terms of loss factors.

Table-4: Loss factors for estimating annual pine needle yield

S. No.	Loss factor type	Value	Reasons
1	Accessibility	0.2	Difficult terrain, distance, undercover deterrence
2	Alternative uses	0.01	Mainly animal bedding
3	Forest fire	0.01	Burnt by forest fire
4	Various other losses	0.1	Washed away by rains, blown by winds, transportation and storage loss
	Total	**0.32 say 0.3**	

Loss due to accessibility, estimated at 20 percent, carries highest significance in estimation of pine needle lost as energy feedstock. About 10 percent loss of pine needles occur on account of various other losses together such as loss due to rains, winds, during transportation and storage. Alternative uses and forest fire account for loss of pine needles at 1 percent each. The net loss factor can be taken as 0.3. This leaves 70 percent of annual gross pine needle yield available for use as energy feedstock and can be called net annual pine needle yield for the state.

Primary and net energy from pine needles

Primary Energy available annually is the energy available in pine needles for conversion into other energy forms. Estimation of Net Energy has been carried out in terms of electric energy and thermal energy. For electric energy, producer gas generated in a gasifier can operate a modified IC engine coupled with an electric generator [7,26,27]. Producer gas is a fuel gas containing CO, H_2, CH_4, CO_2 and N_2 gases. The conversion is generally possible at an efficiency of 0.2 [28]. The installed capacity for electricity generation will depend upon daily generation hours decided by the application of electricity. The energy in pine needles can be used for process heating and its estimation has also been carried out. In industrial sector, boiler is employed for process heating applications. Boiler efficiency of 66% have been reported for biomass boiler by Chau etal. [29].

Results and Discussion

The estimation of gross annual pine needle yield has been carried out. Losses in pine needle as energy feedstock based on loss factors are also estimated. Using these figures, net annual pine needle yield is estimated. Table-5 presents a summary of the input parameters for these estimations.

Table-5: Input parameters for gross annual pine needle yield estimation

S.No.	Parameter	Units	Value

 DOI: 10.1051/e3sconf/20172304001

1	Area pine forest, A_{pf}	ha	394383
2	Area very dense forest, A_{vdf}	ha	82821
3	Area moderately dense forest, A_{mdf}	ha	220855
4	Number of tree VDF, $N_{tree,vdf}$	No.	680
5	Number of tree MDF, $N_{tree,mdf}$	No.	440
6	Pine needle yield per tree, PNY_{tree}	kg/tree	12.4

The results of gross annual pine needle yield (APNY) and the net annual pine needle yield (NAPNY) for the state of Uttarakhand are given in Table-6. The estimated value of gross annual pine needle yield is 1.90 million tonnes. The net annual pine needle yield is 1.33 million tonnes.

Table-6 also gives results obtained for potential annual primary energy (APE) and annual net energy (ANE) when pine needles are used as energy feedstock. The annual primary energy obtained from pine needles is 25.9 peta joules (PJ). The annual net energy for thermal applications is estimated at 17.1 PJ. For electricity generation, annual net energy (ANE) potential using pine needles is 1438.9 million kWh. This estimation is based on electricity generation through gasification route using gasifier-engine-generator combination.

Table-6: Pine needle gross and net yields and energy potential

S.No.	Parameter	Unit	Value
	Pine needles		
1	Gross annual pine needle yield	million tons	1.90
2	Net annual pine needle yield	million tons	1.33
	Energy potential		

DOI: 10.1051/e3sconf/20172304001

3	Annual primary energy	peta joules	25.90
4	Annual net energy (Thermal)	peta joules	17.10
5	Annual net energy (Electrical)	million kWh	1438.90

The annual net energy (ANE) potenial of pine needles may be seen in the background of electricity consumption for the state. For the commercial sector alone, state's consumption of electricity was 1411 million kWh in the year 2013-14 [3]. For realising the net annual energy potential of pine needles as given in Table-6, installed capacities of the generation units have been estimated. Table-7 gives cumulative generation capacity for two different user specific scenarios. In the event of supplying power for 5 hours a day, say for rural households in the vicinity of pine needle based generation units, 789 MW intalled capacity is needed. Whereas for round the clock electricity generation, an installed capacity of 165 MW is required.

Table-7: Cumulative installed capacity of pine needle power generation units

S.No.	**Generation hours**	**Cumulative installed capacity (MW)**
1	Five hours per day electricity for rural households	789
2	Round the clock generation	165

Concluding Remarks

The central Himalayan state of Uttarakhand in India generates large amount of pine needles on forest floors. Estimation of pine needles produced has been made. The gross annual pine needle yield in the state obtained using tree canopy density method has been estimated at 1.9 million tonnes. Net annual pine needle yield as energy feedstock is affected by factors such as accessibility, alternative uses, rains and winds, various other losses, etc. The discussion and subsequent estimation here puts these losses at nearly 30 percent. The net annual pine needle yield is estimated at 1.33 million tonnes. The annual pimary energy potential of pine needles in the state have been estaimated at 25.9 peta joules. The annual net energy is estimated at 17.1 PJ for thermal applications. Alternatively, an estimated amount of 1438.9 million kWh of electrical energy can be obtained. This is a significant amount as commercial sector in Uttarakhand consumed 1411 million kWh in the year 2013-14.

References:

[1] Daily System Energy Report. Uttarakhand Power Corporation Limited (UPCL), Government of Uttarakhand, Dehradun; 2015. Website: https://www.upcl.org/wss/ViewDailySupplyPos.htm [Accessed 20.01.2015].

[2] Uttarakhand 24x7 Power for All (UPFA). A Joint Initiative of Government of India and Government of Uttarakhand, India; 2015.

[3] Statistical Diary Uttarakhand. Directorate of Economics and Statistics Uttarakhand (DESU), Dehradun, India; 2015.

[4] Chauhan A, and Saini RP. Renewable energy based off-grid rural electrification in Uttarakhand state of India: Technology options, modelling method, barriers and recommendations. Renewable and Sustainable Energy Reviews 2015; 51: 662–681.

[5] Chauhan S, and Silori CS. Assessment of biomass availability for power Generation in selected talukas of Uttranchal state. ENVIS Bulletin: Himalayan Ecology 2004; 12(2).

[6] McKendry P. Energy production from biomass (part 2): Conversion technologies. Bioresource Technology 2002; 83(1): 47–54. http://doi.org/10.1016/S0960-8524(01)00119-5

[7] Martinez JD, Mahkamov K, Andrade RV, and Silva Lora EE. Syngas production in downdraft biomass gasifiers and its application using internal combustion engines. Renewable Energy 2012; 38(1): 1–9.

DOI: 10.1051/e3sconf/20172304001

[8] Erlich C, and Fransson TH. Downdraft gasification of pellets made of wood, palm-oil residues respective bagasse: Experimental study. Applied Energy 2011; 88: 899–908.

[9] Kala LD, Susastriawan AAP, Subbarao PMV, and Subrahmanyam JP. Performance investigations on gasification of pongamia shells in a conventional downdraft gasifier, In: Proceedings of International Symposium on Aspects of Mechanical Engineering and Technology for Industry (AMETI); 2014.

[10] FAO. Wood Fuels Handbook. Food and Agriculture Organisation of United Nations; 2015.

[11] Dhaundiyal A, and Gupta VK. The Analysis of Pine Needles as a Substrate for Gasification. Hydro Nepal 2014; 15: 73 - 81.

[12] Sharma A. Assessing the Suitability of Various Feedstocks for Biomass Gasification. M.Sc. Thesis, Dept. of Biological & Ag. Engg., Luisiana State University; 2006.

[13] Mansaray KG, amd Ghaly AE. Physical and thermochemical properties of rice husk. Energy Sources 2007; 19(9): 989-1004, DOI: 10.1080/00908319708908904.

[14] Clarke S, and Preto F. Biomass densification for energy production. Ministry of Agriculture Food and Rural Affairs, Govt. of Canada; 2011.

[15] Dahal M, Baral B, Karki S, Dahal S, and Banjara R. Pine needle briquette as a promising fuel for gasifier. In: Rentech Symposium Compendium 2014; Vol. 4: pp 31 – 34.

[16] Jain VK, and Srinivas SN (Eds). Empowering Rural India the RE Way. Ministry of New and Renewable Energy, Government of India, New Delhi, India; 2012

[17] India State of Forest Report (ISFR 2013). Forest Survey of India, Dehradun; 2013.

[18] Uttarakhand Forest Statistics (UFS). Uttarakhand Forest Department, Dehradun, India; 2013.

[19] Gill Samantha J, Biging GS, and Murphy EC. Modeling conifer tree crown radius and estimating canopy cover. Forest Ecology and Management 2000; 126(3): 405–416.

[20] Wu W. Derivation of tree canopy cover by multiscale remote sensing approach. ISPRS Workshop on Geospatial Data Infrastructure: from data acquisition and updating to smarter services; 2011.

[21] Rana BS, Singh SP, and Singh RP. Biomass and net primary productivity in central Himalayan forests along an altitudinal gradient. Forest Ecology and Management 1989; 27: 199-218.

[22] Kaul ON, Negi JDS, Sharma DC and Srivastava PBL. Organic matter and plant nutrient distribution in a chir (Pinus roxburgii) plantation. Indian For. 1981; 107(2): 86-95.

[23] Singh SP, Adhikari BS, and Zobel DB. Biomass, Productivity, Leaf Longevity, and Forest Structure in the Central Himalaya. Ecological Monographs 1994; 64(4): 401-421.

[24] Kala CP. Indigenous uses and structure of chir pine forest in Uttaranchal Himalaya, India. International Journal of Sustainable Development & World Ecology 2004; 11(2): 205-210. <DOI: 10.1080/13504500409469824>

[25] Uttarakhand at a Glance. Directorate of Economics and Statistics Uttarakhand (DESU), Dehradun, India; 2015b.

[26] Reed TB, and Das A. Handbook of Biomass Downdraft Gasifier Engine Systems, Solar Energy Research Institute (SERI), Colorado; 1988.

[27] Basu Prabir. Biomass Gasification and Pyrolysis: Practical Design and Theory, Academic Press Elsevier, USA; 2010.

[28] McKendry P. Energy production from biomass (part 1): overview of biomass. Bioresource Technology 2002; 83(1): 37–46.

[29] Chau J, Sowlati T, Sokhansanj S, Preto F, Melin S, and Bi X. Techno-economic analysis of wood biomass boilers for the greenhouse industry. Applied Energy 2009; 86: 364–371.

EVALUATION OF SOLAR THERMAL DRIVEN COOLING SYSTEM IN OFFICE BUILDINGS IN SAUDI ARABIA

Majid T. Linjawi, Qazi Talal, Fahad A. Al-Sulaiman

Abstract

In this study solar driven absorption chiller is used to reduce the peak cooling load in office buildings in Saudi Arabia for different selected cities. The study is conducted for six cities of Abha, Dhahran, Hail, Jeddah, Nejran and Riyadh under three operating durations of 4, 6, and 8 hours using flat plate or evacuated tube collectors. The energy analysis concluded that flat plate collectors are better than evacuated tube collectors. However, the results from economic analysis suggest that while proposing a gas fired absorption chiller will reduce running costs, further reduction by using solar collectors is not feasible because of its high initial cost. At the best case scenario the Net Present Value of a 10 Ton Absorption chiller operated by natural gas boiler and two large flat plate collectors (12m^2 each) running for 8 hours/day, 5days/week has a value of 117,000 $ and Internal Rate of Return (IRR) of 12%. Solar driven absorption chiller could be more feasible if the gas prices increases or the solar collector prices decreases significantly. Finally, government economic incentives and taxes are recommended to provide a boost for the feasibility of such projects.

Keywords: absorption chiller, solar cooling, energy analysis, economic analysis

Introduction

The growth in economy and advancement in technology depends on the amount of energy that is available and can be utilized in that country. Fossil fuels are the primary source of energy that is utilized in almost all places throughout the world. Nevertheless fossil fuels are a non-renewable source of energy and its resources are getting depleted. In addition, burning fossil fuel causes an adverse effect on the environment such as pollution, global warming and depletion of ozone layer due to the greenhouse gas effect. Hence the need arises to explore alternative sources of energy that reduces the burning of fossil fuels. In the present day renewable energy sources are the best alternative for supplying the required energy and decrease the combustion of fossil fuels. Examples of renewable energy include solar energy, wind energy, geothermal energy, tidal energy etc. Solar energy has the most potential among all these different types of energy because it's high availability in nature. At present the energy consumed by the world is about 1.84×10^{13}W [1] and the amount of solar radiation that is transmitted to the earth is around 1.74×10^{17}W [2]. Hence solar energy has vast potential as an energy source.

DOI: 10.1051/e3sconf/20172305001

We cannot however completely eliminate utilization of conventional energy sources as it is currently impossible, but the focus should be on reducing the dependency on conventional energy sources. Focusing on the Gulf region, primarily in Saudi Arabia, high amounts of energy are consumed by the air conditioning and refrigeration applications. According to a report from 2010 [3] the amount of electrical energy consumed by the air conditioning sector in KSA was more than 36% of the total generated electricity, moreover this demand for energy is increasing. Hence we should try to satisfy these energy requirements in the air conditioning sector by the utilization of the renewable energy i.e. primarily by solar energy. The advantage of solar energy is that the peak demand for air conditioning matches with the peak of solar energy radiation [4]. Hepbasli and Alsuhaibani [5] carried out research reviewing the various applications and studies on solar energy in Saudi Arabia. Research in various applications has been carried out in Saudi Arabia. These include solar water desalination, solar powered irrigation, solar photovoltaic systems, solar stills, exergetic solar radiation and also some studies in solar cooling/refrigeration etc. However, less studies has been made for utilization of solar absorption systems for providing air conditioning in Saudi Arabia.

A project to reduce the cost and improve the efficiency of air cooling system was carried out by SOLARES in which Saudi Arabia participated. [6, 7]. Solar cooling can have a huge impact in Saudi Arabia and hence the number of projects for development and demonstration in this field should be increased. A variety of different solar refrigeration technologies have been developed and are now available at comparatively cheaper prices [8]. Two kinds of system can be used to produce cooling: solar thermal and solar electrical cooling systems. In the case of solar electrical system, photovoltaic system is used to obtain electrical energy from solar energy, which is then used to produce the refrigeration like the conventional vapor compression system. In the case of solar thermal system, solar energy is used to run the absorption refrigeration system. According to a study by Kim and Ferreira [8] solar thermal systems are cheaper than solar electrical systems. Another study by Otanicar et al. [9] shows that up to 95% of incident solar radiation can be utilized by solar thermal systems compared to up to 35% of the incident solar radiation that is converted to electrical energy by photovoltaic system. Therefore, solar thermal systems have potential in producing the required cooling and more research is being done on this field; however, they have some challenges. A recent research paper by Siddiqui and Said [10] reviewed solar absorption systems and summarized the works done in this field.

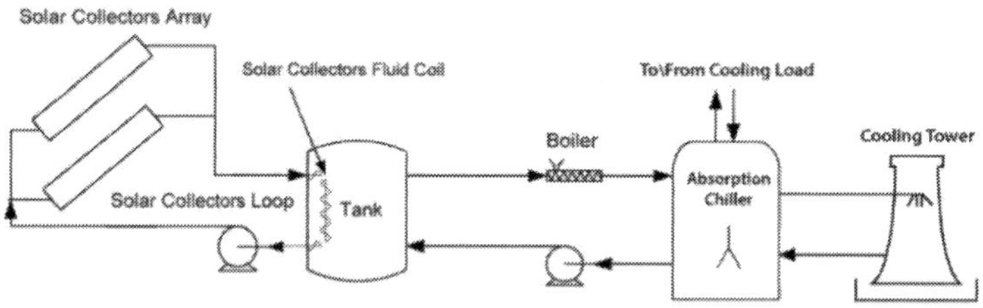

FIG. 1. SCHEMATIC DRAWING OF SOLAR POWERED ABSORPTION CHILLER SYSTEM

Some of the previous studies on solar absorption systems were carried out by Trombe and Foex [11] and Chinnappa [12]. Research was carried out by Kundu et al. [13] to study the performance of solar absorption systems using different types of solar collectors and the conclusion was made that the concave parabolic collector provided the best performance. Later, many components were added to absorption system to increase its COP (coefficient of performance). Aqua-ammonia and LiBr–water are the most common working pairs in the absorption refrigeration systems [10]. Various experimental and simulation studies showed that single effect and double effect solar powered absorption chiller have COP of around .7 and 1.3 respectively [14-17] with double effect chillers being cheaper by 50% for the same COP [18].

Florides et al. [19] carried out the simulation to study the performance of LiBr– water system in Cyprus. The system was equipped with 15 m^2 compound parabolic collector and also consisted a 600 L hot water storage tank. For this system he also performed the cost optimization. Their optimization results indicated a life cycle savings of C£ 1376 without subsidized fuel cost. Fumo et al. [20] studied a hypothetical solar thermal driven cooling system for data center in Albuquerque, New Mexico. Their results suggest using flat plat collectors are superior to evacuated tubes. In their conclusion the researchers suggest using the system in typical office building operating (8 am to 5 pm) and they expect that the system will have better energy performance.

Eicker and Pietruschka [21] performed the cost analysis of solar powered LiBr–water absorption system for office buildings. Their cost analysis indicated that for long operation hours in Southern European locations, cooling costs are around 200 euros per MWh. They prepared a full simulation model of the absorption cooling system along with storage tank and collector model. Their results revealed that varying the load time

DOI: 10.1051/e3sconf/20172305001

for same maximum cooling load results in a variation of solar collector area by more than a factor of 2 for getting the same solar fraction.

The literature review revealed that there are limited studies on solar-driven absorption chiller for office buildings. However, no study has been conducted for such system in office building for Saudi Arabia weather conditions. Therefore, this study will fulfil this gap.

SYSTEM DESCRIPTION

For this study, an office building will be considered. The building's air conditioning system will run for 5 days a week. The cooling load of the building is satisfied with vapor compression air conditioning system having a COP of 4. This vapor compression system also works as a heat pump when heating is required in the building with a COP of 3.

TABLE 1. SPECIFICATIONS OF THE MODEL

Parameter	Magnitude	Units	Parameter	Magnitude	Units
Tank solar collector coil effectiveness	0.8		Chiller heat medium inlet temperature	88 (190)	°C (°F)
Boiler efficiency	0.8		Chiller heat medium outlet temperature	83 (181)	°C (°F)
Chiller nominal cooling capacity	35 (10)	kW (Ton)	Heating water mass flow rate	2.4	kg/s
Absorption chiller cooling COP	0.7		Absorption Chiller ump	210	W
Chiller nominal heating capacity	48.5 (14)	kW (Ton)	Cooling tower onsumption	1500	W
Absorption chiller heating COP	0.97				

Solar collectors azimuth angle	0	degree			
Solar collectors tilt	Latitude of the selected city	degree			
Flat-plate solar collectors			Evacuated Tube Collectors		
FR ($\tau\alpha$)	0.726	W/(°Cm²)	FR ($\tau\alpha$)	.55	W/(°Cm²)
FR (UL)	3.35	W/(°Cm²)	FR (UL)	1.71	W/(°Cm²)
Collector gross area	12.5	m²	Collector gross area	4.6	m²
Collector mass flow rate	0.0161	kg/s	Collector mass flow rate	.0161	kg/s
Storage Capacity / Solar Collector Area	75	L/m²	Storage Capacity / Solar Collector Area	75	L/m²
Number of Solar Collectors	(2,4,8,12 ,16,20)	variabl e	Number of Solar Collectors	(10,20,4 0,60,80,1 00)	variable

However, during the peak cooling load excess electricity is consumed which transfers into high demand on the electricity grid. Also, for meeting the peak summer cooling demand the cooling system of a much higher capacity is required, compared to the capacity of the cooling system throughout the year. Therefore, introducing a solar powered absorption chiller to support at peak cooling loads to minimize the cooling required from the vapor compression system is considered to be a good option. The solar powered absorption chiller will be investigated when being run by two different types of solar collectors: flat plate collectors and evacuated tube collectors. The system consists of solar collectors, storage tank with a heat exchanger, an auxiliary natural gas boiler to provide the required inlet temperature to the absorption chiller, and a cooling

 DOI: 10.1051/e3sconf/20172305001

tower for cooling the condenser coils of the absorption chiller. Chilled water from the absorption chiller is pumped to the air conditioning distribution system. Figure 1 shows the schematic diagram for the system under study.

Cooling and Heating Loads

The absorption chiller is assumed to be working at steady state condition and provide constant full capacity cooling or heating power for various cases at the specified duration of operation. When the average temperature of the month is above 20°C, the absorption chiller system will operate during peak cooling demand of the office building, the absorption chiller is assumed to provide a constant 10 Ton of cooling, and the rest will be met by the conventional cooling system which also satisfies the cooling demand during the remaining hours of the day. Study has been made for 3 scenarios of peak load durations namely 4, 6, and 8 hours.

Similarly when the average temperature of the month is below 20°C, the absorption chiller system will operate during peak heating demand of the office building, the absorption chiller is assumed to provide a constant 14 Ton of heating, and the rest will be met by the conventional heating system which also satisfies the heating demand during the remaining hours of the day. Study has been made for 3 scenarios of peak load durations namely 4, 6, and 8 hours.

Location Selected

Six cities from different provinces in KSA were chosen. The cities and their metrological data are presented as follows.

TABLE 2. SELECTED CITIES AND THEIR LOCATIONS

City	Abha	Dhahran	Hail	Jeddah	Nejran	Riyadh

DOI: 10.1051/e3sconf/20172305001

Location	18°13′N 42°30′E	26°16′N 50°09′E	27°31′N 41°41′E	21°32′N 39°10′E	17°29′N 44°7′E	24°38′N 46°43′E

All data in this section were obtained from weather data obtained from NASA through RetScreen Software [22]. Figure 2 shows the general curve of ambient air temperature for all cities. Hail city has the lowest ambient temperatures during the winter; therefore, it is recommended that peak load heating is mostly required for this area. Its average temperatures are around 15 degrees Celsius in winter. Jeddah on the other hand has the lowest temperature deviations between summer and winter with average temperatures are around 25°C in winter and around 31 degrees in summer. Hence it does not require heating throughout the year. However, the summer peak daily average temperatures are in Riyadh and Dhahran with temperatures 35 and 36°C respectively. This peak suggest that peak load cooling is mostly needed in these two cities. On the other hand, during summer Abha has the lowest ambient temperature among all cities with an average temperature of around 22°C. This suggests that Abha requires the least peak load cooling.

Figure 3 shows the daily Horizontal solar radiation in kWh/m^2/d. Similar to the air temperature curve, the solar radiation curve has a maximum in summer and minimum in winter for all cities, Nejran is closer to the equator and, hence, has the highest peaks in summer as well as winter with average values around 7.5 kWh/m^2/d during summer and 6.4 kWh/m^2/d during winter. During the summer Abha and Riyadh has the lowest peaks of solar radiation with average values around 5.6 kWh/m^2/d. Hence Abha has the least potential to use solar energy compared to the other cities during summer. During the winter the three locations with lowest radiation are Dhahran, Hail and Riyadh with average values around 4.3 kWh/m^2/d. Thus, these areas are less suitable for solar energy application in winter. However during the summer, Dhahran and Jeddah have Radiation values on the higher side with average values around 6.85 kWh/m^2/d. Hence during summer these areas are more feasible for solar energy.

ENERGY ANALYSIS

Cooling and heating air conditioning loads from the office building were transformed into hot water load or demand by dividing the load by its corresponding absorption chiller COP. Then, hot water demand is entered into RetScreen, with different combinations of cities, number and type of solar collectors, number of operational hours. The input variables into the energy analysis were given in Table 1. The following

DOI: 10.1051/e3sconf/20172305001

values of solar fraction were obtained as illustrated in Table 3. The results indicate the number of flat plate solar collectors and evacuated tube solar collectors required to achieve solar fraction 1 for various operating hours for different cities.

For the case of flat plate collectors for 4 hours of operation for all cities approximately 12 collectors are required. In case of 6 hours of operation at-least 16 collectors are required and for 8 hours of operation more than 20 collectors are required. When evacuated tube collectors are used for 4, 6, and 8 hour operations approximately 40, 60, and 80 collectors are required for achieving solar fraction of 1, that is, for operating the absorption chiller independently by solar thermal energy only.

The results clearly shows that flat plate collectors are superior to evacuated tubes in terms of delivering the required energy for this specific application. This coincides with the findings of Fumo et al. [20], and the reason is that since the temperature rise in the collectors are small, around 5°C only, flat plate collector's efficiency is higher for that range. Another observation in the data is that Nejran, with the lowest latitude, has the highest solar fraction, then Jeddah, Riyadh, and Dhahran which have similar intermediate latitude have similar values of solar fraction. Finally, Hail city, the city with the highest latitude has the lowest values of solar fraction. However, Abha is not included in that pattern due to its low temperatures and different cooling/heating demand pattern.

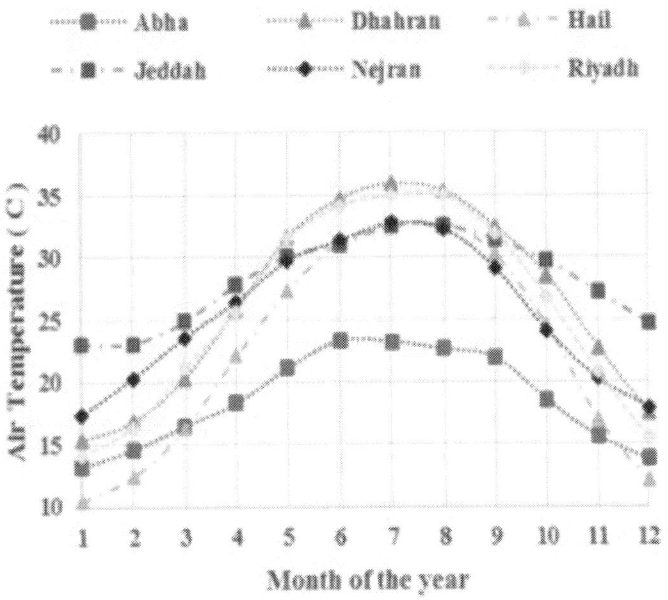

DOI: 10.1051/e3sconf/20172305001

FIG. 2. TEMPERATURE PROFILES

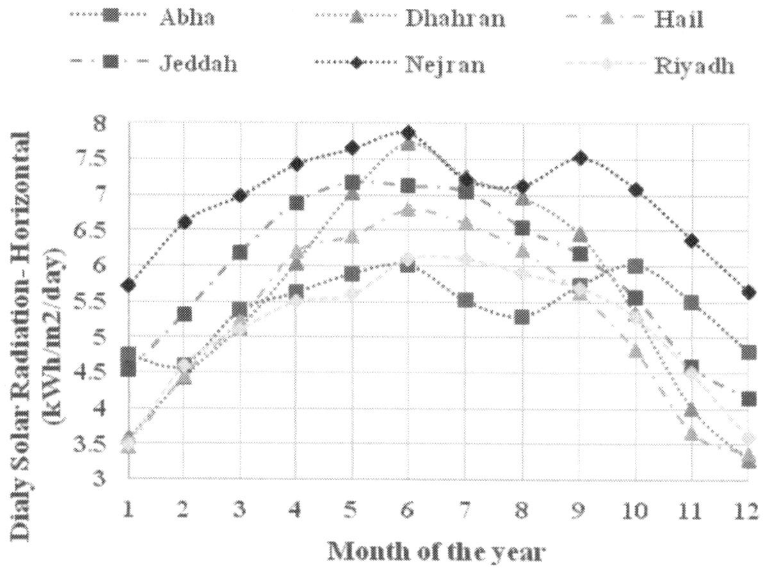

FIG. 3. SOLAR RADIATION PROFILES

TABLE 3. SOLAR FRACTION

		Abha			Dhahran			Hail			Jeddah			Nejran			Riyadh		
		Operating Hours			Operating Hours			Operating Hours			Operating Hours			Operating Hours			Operating Hours		
		4	6	8	4	6	8	4	6	8	4	6	8	4	6	8	4	6	8
# of Flat Plate Collectors	2	0.24	0.17	0.13	0.27	0.19	0.15	0.24	0.17	0.13	0.30	0.21	0.16	0.37	0.26	0.26	0.28	0.20	0.12
	4	0.43	0.31	0.24	0.47	0.34	0.27	0.43	0.31	0.24	0.51	0.37	0.30	0.60	0.46	0.46	0.48	0.36	0.23
	8	0.75	0.54	0.43	0.76	0.57	0.47	0.74	0.53	0.43	0.79	0.61	0.51	0.87	0.71	0.71	0.78	0.59	0.42
	12	1	0.75	0.59	1	0.76	0.62	1	0.74	0.58	1	0.79	0.66	1	0.87	0.87	1	0.78	0.57
	16	1	0.98	0.75	1	0.98	0.76	1	0.97	0.74	1	0.99	0.79	1	1	1	1	0.99	0.72
	20	1	1	0.93	1	1	0.92	1	1	0.91	1	1	0.94	1	1	1	1	1	0.89
		Abha			Dhahran			Hail			Jeddah			Nejran			Riyadh		
		Operating Hours			Operating Hours			Operating Hours			Operating Hours			Operating Hours			Operating Hours		
		4	6	8	4	6	8	4	6	8	4	6	8	4	6	8	4	6	8
# of Evacuated Tube Collectors	10	0.43	0.31	0.24	0.46	0.34	0.26	0.43	0.31	0.24	0.50	0.36	0.28	0.59	0.44	0.34	0.42	0.30	0.24
	20	0.68	0.53	0.43	0.71	0.57	0.46	0.67	0.53	0.43	0.75	0.60	0.50	0.84	0.70	0.59	0.66	0.52	0.42
	40	0.92	0.78	0.68	0.93	0.80	0.71	0.91	0.77	0.67	0.95	0.84	0.75	1	0.91	0.84	0.90	0.76	0.66
	60	1	0.92	0.81	1	0.93	0.84	1	0.91	0.81	1	0.95	0.87	1	1	0.94	1	0.90	0.80
	80	1	1	0.92	1	1	0.93	1	1	0.91	1	1	0.95	1	1	1	1	1	0.90
	100	1	1	1	1	1	1	1	1	1	1	1	1	1	1	1	1	1	1

ECONOMIC ANALYSIS

Riyadh was selected as the city to make the detailed economic analysis because it is the capital and has the highest number of office buildings. Additionally, resources and their availability is high. Finally, since it has similar cooling/heating demand as Dhahran and it also has similar solar fraction, it is expected that the economic analysis for both cities will be very close to each other. Table 4 lists the main inputs used for the 25 years economic analysis of the project. Natural Gas prices and electricity prices are today's prices for commercial sector. The Net Present Value (NPV) for the project in Riyadh are given in figures 4 and 5 below for flat plate collectors and evacuated tube collectors respectively for three different running time. The results from the figures clearly indicate that it is feasible to run the absorption chiller with natural gas instead of having vapor compression system since it has positive NPV at 0 Collectors. In addition, the introduction of solar collectors is not economically feasible because it decreases the NPV of the project as number of collectors is increased. However, another observation made from comparing the figures indicate that utilization of flat plate collectors are more feasible than evacuated tube collectors. Also it is shown that the longer the time the absorption chiller runs per day, the more economical it becomes.

The results indicate that using 20 flat plate solar collectors for 8 hour operation of absorption chiller has a positive NPV, and hence is economically feasible. But as the number of hours of operation are decreased the NPV becomes negative. Utilizing up to 12 flat plate collectors for all 3 durations of operation has a positive NPV, hence with lower number of collectors the operation is feasible. For evacuated tube collectors, utilizing more than 20 collectors results in a negative NPV, hence evacuated tube solar collectors are not economically feasible. The low prices of natural gas and electricity and the high initial cost of solar collectors are the key elements for this result of NPV. To make this system feasible many approaches may be taken. An increase in fuel prices, introduction of greenhouse emissions tax, a subsidy in the collectors prices are only few of many other possible governmental solutions. However, from consumer point of view, running the system for longer periods of time, 7-days a week will make the project more feasible. Another possible approach is the use of two stage absorption chillers which will lead to higher NPV.

DOI: 10.1051/e3sconf/20172305001

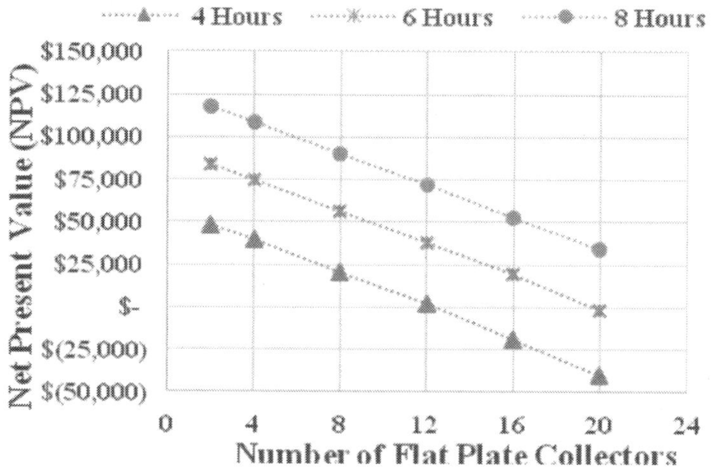

FIG. 4. NPV OF FLAT PLATE COLLECTORS

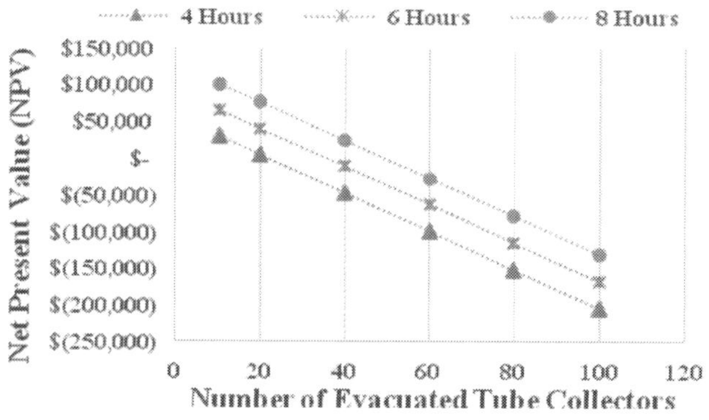

FIG. 5. NPV OF EVACUATED TUBE COLLECTORS

TABLE 4. ECONOMIC ANALYSIS ASSUMPTIONS

Initial Costs	Unit price	Units

 DOI: 10.1051/e3sconf/20172305001

Flat plate solar collector	$	300	m^2
Evacuated tube collector	$	500	m^2
Absorption chiller	$	16,000	10 Ton
Thermal storage tank	$	1,000	m^3
Boiler cost	$	102	kW
Electric chiller	$	1,000	ton
Natural gas price	$	1.25	USD/MMBTU
Peak electricity price	$	202.67	$/MWh
Normal Electricity Price	$	50.45	$/MWh
Electricity escalation		10%	
Natural Gas escalation		5%	
Escalation Rate		0%	

CONCLUSION

This study presents results by performing simulations with different arrangements of solar thermal driven cooling systems as a cooling plant in order to reduce the peak cooling and heating loads in office buildings in Saudi Arabia. The different arrangements considered were the number of collectors, type of collectors, and different duration of operation during the day when the peak cooling load is required varying from 4 to 8 hours. Flat plate collectors and evacuated tube collectors were compared. Our results with energy analysis suggests that flat plate collectors are better than evacuated tube collectors in terms of delivering the required energy for the office building with electric chillers since that temperature rise in the collectors are small, around 5°C only, flat plate collector's efficiency is higher for that range. With increasing the number of operation hours the number of collectors required also increases to have the same solar fraction. Also the results from economic analysis suggest that while proposing a gas fired absorption chiller will reduce running costs, further reduction by using solar collectors is not feasible because of its high initial cost. At the best case scenario NVP of a 10 Ton Absorption chiller operated by natural gas boiler and 2 large flat plate collectors (12m^2 each) running for 8 hours/day, 5days/week has a value of 117,000 $ and Internal Rate of Return IRR of 12%. Using solar collector for the absorption chiller can be more feasible solution if it used for seven days a week rather than five days a week. Further significant reduction in solar collectors prices and/or increase in the fossil fuel prices, will make solar driven cooling system a feasible alternative. Finally, government economic incentives and taxes can serve as a boost for the feasibility of such projects.

REFERENCES

[1] Solanki, C. S., .Solar Photovoltaic Technol. Syst. 2013;1:23–4.

[2] Balaras, C. A., Grossman, G., Henning, H-M., Infante Ferreira, C. A., Podesser, E., Wang, L., and Wiemken, E., 2007, "Solar Air Conditioning in Europe—an Overview," Renewable and Sustainable Energy Reviews, 11(2), pp. 299-314.

[3] Ministry of Water & Electricity, KSA annual report; 2010.

[4] Said, S. A. M., Habib, M. A., and Iqbal, M. O., 2003, "Database for Building Energy Prediction in Saudi Arabia," Energy Conversion and Management, 44(1), pp. 191-201.

[5] Hepbasli, A., and Alsuhaibani, Z., 2011, "A Key Review on Present Status and Future Directions of Solar Energy Studies and Applications in Saudi Arabia," Renewable and Sustainable Energy Reviews, 15(9), pp. 5021-5050.

[6] SAW, Saudi Aramco World. A solar village. September/October print edition; 1981. p. 16–29.

[7] Vincent-Barwood A. A solar village. Saudi Arabia and solar energy—a special section; 2011.http://www.saudiaramcoworld.com/
issue/198105/saudi.arabia.and.solar.energy.a.special.section.html

[8] Kim, D. S., and Infante Ferreira, C. A., 2008, "Solar Refrigeration Options—a State-of-the-Art Review," International Journal of Refrigeration, 31(1), pp. 3–15.

[9] Otanicar, T., Taylor, R. A., and Phelan, P. E., 2012, "Prospects for Solar Cooling—an Economic and Environmental Assessment," Sol. Energy, 86(5), pp. 1287–1299.

[10] Siddiqui, M. U., and Said, S. A. M., 2015, "A Review of Solar Powered Absorption Systems," Renewable and Sustainable Energy Reviews, 42, pp. 93-115.

[11] Trombe, F., and Foex, M., 1957, "The Production of Cold by Means of Solar Radiation," Sol. Energy, 1(1), pp. 51–52.

[12] Chinnappa, J. C. V., 1962, "Performance of an Intermittent Refrigerator Operated by a Flat-Plate Collector," Sol. Energy, 6(4), pp. 143–150.

[13] Kundu, B., Mondal, P. K., Datta, S. P., and Wongwises, S., 2010, "Operating Design Conditions of a Solar-Powered Vapor Absorption Cooling System with an Absorber Plate having Different Profiles: an Analytical Study," Int. Commun. Heat Mass Transfer, 37(9), pp. 1238–45.

[14] Raghuvanshi, S., and Maheshwari G., 2011, "Analysis of Ammonia–Water (NH3–H2O) Vapor Absorption Refrigeration System Based on First Law of Thermodynamics," Int. J. Sci. Eng. Res., 2(8), pp. 1–7.

[15] Gomri, R., 2009, "Second Law Comparison of Single Effect and Double Effect Vapour Absorption Refrigeration Systems," Energy Conversion and Management, 50(5), pp. 1279-1287.

[16] Balghouthi, M., Chahbani, M. H., and Guizani, A., 2005, "Solar Powered Air Conditioning as a Solution to Reduce Environmental Pollution in Tunisia," Desalination, 185(1–3), pp. 105-110.

[17] Balghouthi, M., Chahbani, M. H., and Guizani, A., 2008, "Feasibility of Solar Absorption Air Conditioning in Tunisia," Building and Environment, 43(9), pp. 1459-1470.

[18] Sumathy, K., Huang, Z. C., and Li, Z. F., 2002, "Solar Absorption Cooling with Low Grade Heat Source — a Strategy of Development in South China," Solar Energy, 72(2), pp. 155-165.

[19] Florides, G. A., Kalogirou, S. A., Tassou, S. A., and Wrobel, L. C., 2002, "Modelling and Simulation of an Absorption Solar Cooling System for Cyprus," Solar Energy, 72(1), pp. 43-51.

[20] Fumo, N., Bortone, V., and Zambrano, J. C., 2011, "Solar Thermal Driven Cooling System for a Data Center in Albuquerque New Mexico," ASME. J. Sol. Energy Eng., 133(4), pp. 041008-041008-7.

[21] Eicker, U., and Pietruschka, D., 2009, "Design and Performance of Solar Powered Absorption Cooling Systems in Office Buildings," Energy and Buildings, 41(1), pp. 81-91.

[22] N. R. C. E. S. C. E. T. C. – V. Rets. I. Government of Canada, "RETScreen International Climate data."

DOI: 10.1051/e3sconf/20172305002

An investigation of the Performance of a Conical Solar Water Heater in the Kingdom of Bahrain

Nessreen Gaaliche

Department of Mechanical Engineering,

College of Engineering, University of Bahrain,

PO Box 32038, Kingdom of Bahrain

ngaaliche@uob.edu.bh

Teoman Ayhan

Department of Mechanical Engineering,

College of Engineering, University of Bahrain,

PO Box 32038, Kingdom of Bahrain

tayhan@uob.edu.bh

Raouf Fathallah

Department of Mechanical Engineering,

National Engineering School of Sousse,

University of Sousse

PO Box 264, Sousse Erriadh 4023

raouf.fathallah@eniso.rnu.tn

Abstract

Domestic water heater corresponds to 25% of the house energy consumption and can play an important role to reduce energy house expenses. Solar energy offers a preferred renewable energy resource because of its economic and environmental advantages. It is considered the best alternative to reduce domestic water heater energy consumption cost. Converting solar energy into heat can be considered among the simplest used systems. Solar thermal conversion is more efficient than solar electrical direct conversion method. Solar water heater systems are particularly easy to use and to repair. The integrated conical solar collector water heater (ICSCWH) is so far the easiest among water heating systems. The ICSCWH converts directly and efficiently the solar flux into heat. In order to expand the utilization of ICSCWH systems, many design modifications have been examined and analyzed. This study provides an experimental investigation and mathematical simulation of an ICSCWH system equipped with a glass cover resulting in the increase of the maximum absorption. Integrating the cone-shaped heat collector with an aluminum spiral pipe flow system may enhance the efficiency of the proposed system. In order to maximize the solar radiation of the system, the solar water heater has been designed in a conical shape, which removes the need to change its orientation toward the sun to receive the maximum sun radiation during the day. In this system, the heating of water has been obtained using the spiral pipe flow without the use of the solar cells and mirrors in order to reduce the total cost. The storage water tank of this system is coupled with a conical solar collector. Based on the above design, the solar water heater has been fabricated and tested. In addition, an analytical modeling approach aiming

DOI: 10.1051/e3sconf/20172305002

to predict the flow rate within the conical integrated collector storage solar water heater (ICSSWH) and its efficiency, was developed. Modeling through a numerical simulation approach based on energy equations was performed. Considering the entire water amount and the total area of the cone, the amount of water (facing the sun per unit absorbing area in the two symmetrical parts of the system) is found to increase, which is expected to reach a maximum water temperature at a high performance. Our experimental findings show that the daily performance is around 32% and the highest water temperature of about 45°C is obtained in the system at 4 pm, according to seasons and weather conditions. An efficient and simple mathematical simulation approach for the new conical solar water heater is described then validates using experimental data.

Keywords: *Solar water heater, Spiral pipe flow, conical shape collector, Collector-storage, Thermal efficiency.*

1. Introduction

In order to enhance the performance of a heating system, we need to optimize firstly the optical performance of the system by maximizing the absorbing parameters of the surface [2], then the efficiency of energy storage tank by maximizing the density of thermal storage [4,5], this means smaller volume and low building cost, as well as a good heat transfer in the system [15,9]. Kalogirou [6] describes in details the existing solar collectors explaining by type its thermal transfer and its different applications, leading to many innovations in the area of solar collectors like the photovoltaic-thermal collectors. In addition, it is important to highlight the great importance of the role of the storage reservoir design. Tyagi et al. [14] showed that the convection heat between the receiver surface and the ambient air decreases the performance of the system. For example, Kumar and Rosen [7] and El-Sebaii [3] proposed the integration of an insulation cover for night application with the purpose of conserving the temperature of hot water inside the storage vessel. Also Zhao, Lu and Tian [16] propose the use of insulation made from transparent material. In addition, Sefa T. et al. [8] pointed out the application of different types of special phase change materials in storage ICSWH. Several researchers studied the cylindrical design of storage reservoir because they are more found suitable for high water pressure [13]. Whereas; they proposed the use of trapezoidal storage tank as a solution for the stratification phenomenon in the water reservoir [1].
An experimental study was carried out by Smyth et al. [10] to investigate a novel inverted absorbing ICSSWH installed in the parabolic concentrator composed of a second cylindrical reflector. Both optical and thermal transfer were analyzed in details. In a modified design [11], the water storage vessel consisted of two different sections, an upper part which presenting the one third of total volume of the water is

 DOI: 10.1051/e3sconf/20172305002

totally covered by a vessel insulation, while the bottom consisting of the rest of the total volume is subjected to the incident solar irradiation. This configuration was found to conserve very well the heat due to the use of the transparent insulation. Sridhar and Reddy [11] invented a new cuboid shape of water storage vessel totally insulated, in which the receiver and the storage vessel present the same section in order to reduce the thermal loss during the non-collection periods. Suhas Vet al. [12] designed and tested an inverse conical-type solar heater. In the conical water tank, the sand was placed in contact with the surface of the sheet metal and the conical frame which improve the efficiency of the system. The tubes of water were placed inside the water vessel. The highest heat was obtained between 2 to 3 pm. In the literature, many solutions can be found such as the use of one or more transparent covers, a selective absorbing surface, and excellent insulation of the non-illuminated vessel parts in order to solve the problems of the utilization of this type of solar water heater.

In our case, we proposed also the solution of integrating the cone-shaped heat collector with an aluminum helical pipe. To achieve the objectives, firstly a mathematical model and analytical simulation of the integrated conical solar collector water heater (ICSCWH) were developed.

The simulation method was used in order to examine the daily performance of an ICSCWH for the month of May, and then compared with the experimental data. The simulation of performance was found to be perfectly consistent with experiment results.

2. System design

A new solar water heating system is proposed for operation in the Kingdom of Bahrain climate conditions. The novel ICSCWH system is principally composed of a conical collector and a storage vessel. The solar collector has a collecting surface facing to solar radiation and water supply to be heated and thermally coupled with the collecting plate. A part of the incident solar radiation in the collecting surface is absorbed and increases its temperature; thermal energy circulates from the collecting surface into the water. In fact, the present system is a solar water heater containing a conical structure and an aluminum helical pipe disposed vertically within the conical structure.

It can decrease the use of oil particularly in developing countries. The experimental set-up of ICSCWH system was installed at the Mechanical Laboratory of the University of Bahrain. The Kingdom of Bahrain has a latitude of 26° N and a longitude of 50.5° E. In our system, the conical collector is fabricated to receive the solar irradiation by the absorbing surface then converts the sunlight into heat. The sun energy peaks at local noon and then decreases significantly. Insulated water vessel at the bottom side could minimize thermal losses particularly during night.

DOI: 10.1051/e3sconf/20172305002

Figure 1 describes in detail each component of the ICSCWH system. Figure 2 shows a photograph of the ICSCWH experimental set-up.

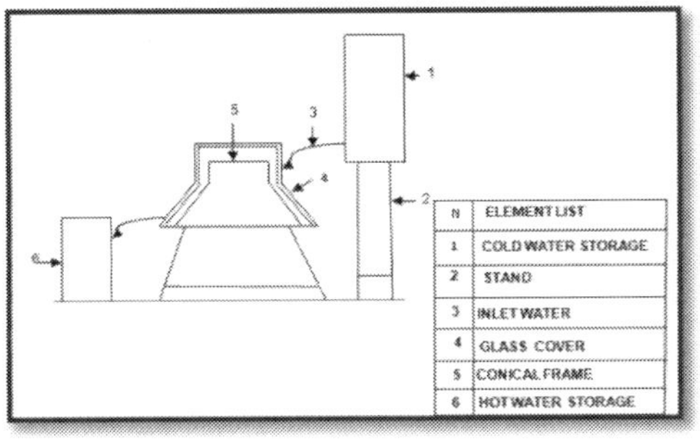

Fig.1 Line diagram of solar water heater.

Fig.2 Photograph of the Experimental Set-up of ICSCWH system at the University of Bahrain.

The water circulates in a closed loop system. By allowing the hot water to flow in a spiral-type aluminum pipe and dumped into the conical tank constantly, the water starts to receive some of the heat generated via solar radiation, thus increasing the water temperature.

3. Modeling and Simulation: Analytical Approach

3.1 Analytical Analysis of the System

Kumar and Rosen [7] showed that the modified integrated collector-storage solar water heater (ICSSWH) with extended storage unit will improve the heating efficiency. The water circulates from sections A to B and vice versa during collection periods. Natural convection in section A is developed as a result of buoyancy force. For simplification, each symmetrical-section of the system is studied separately. Given the symmetry of velocity field lines, the system is considered in the following approach, as shown in figure 3.

DOI: 10.1051/e3sconf/20172305002

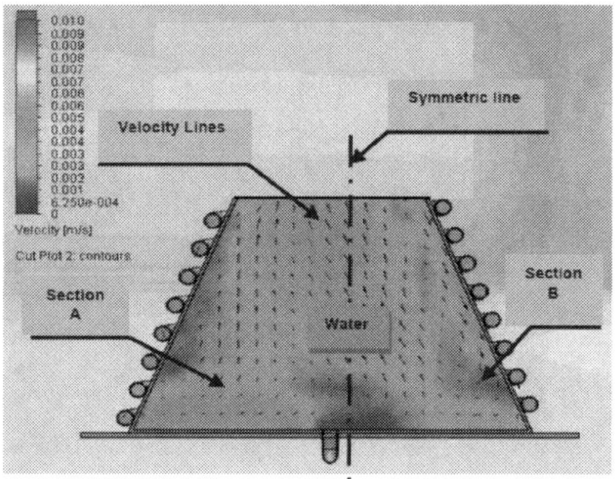

Fig.3 3-D Representation of velocity simulation of the system with natural flow.

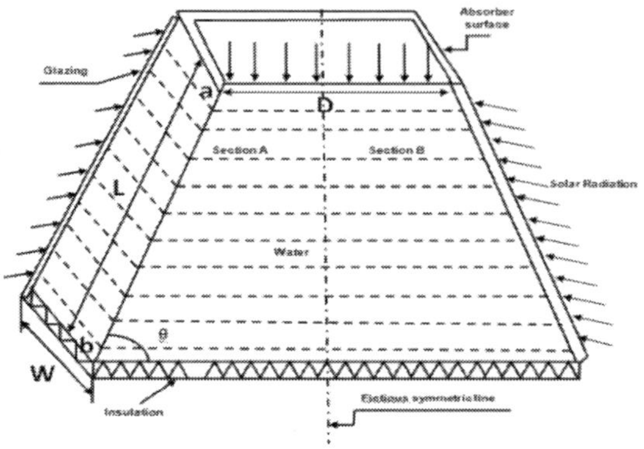

Fig.4 Description of approximate conical ICSCWH

(a):upper side of section A;(b): lower side of section A;(D): diameter of the upper side of the cone;(W): depth of section A ;(θ): inclination angle.

An equation is developed for natural convection flow rate in section A. In addition, to calculate the temperature of water and the efficiency, the energy equations are obtained for section A and solved using analytical simulation. The total water volumes of the two symmetrical-sections are designed to maximize thermal efficiency.

In order to study the influence of climatic and operation conditions on the efficiency of the simplified ICSCWH system described in Figure 4, energy equations are conducted for section A only without considering section B. But, for the entire volume of water and the whole surface, the heat is transferred in section A and section B via natural convection. To determine the thermal transfer in section A, an equation is established for natural convection flow rate resulting from buoyant forces. The mathematical calculation mentioned below for natural convection flow rate is the same of two parallel inclined surfaces separated by a fixed distance as reported by Kumar and Rosen [7].

DOI: 10.1051/e3sconf/20172305002

Section A is modeled as a rectangular section composed of two parallel surfaces inclined by an angle θ and separated by a constant distance W. For the calculation of natural convection flow rate, different hypothesis are taken into consideration:

1) The temperature of water is variable along the direction of length L;
2) The continuity and uniformity of the variation of water temperature from the bottom to the top;

Our modified design section A only with the top section without insulation;

3) The total water storage volume is distributed such that 50% is in section A and 50% in section B,
4) All the thermal and physical properties of water (except density) are not variable,

In our methodology, the calculation performed by Kumar and Rosen [7] is adopting by considering some appropriate modifications.

In Figure 4, the pressure difference in sections A caused by buoyancy force is noted Δp_{a-b} which represents the pressure difference in section A between two surfaces a and b.

The difference of pressure in section A, Δp_{a-b} is determined by:

$$\Delta p_{a-b} = g \sin\theta \int_0^L \{\rho 1 - \rho(y)\}\, dy \qquad (1)$$

where g is the gravitational acceleration, θ is the inclination angle of section A, ρ1 is the water density at the bottom, L is the total length of the section A and ρ(y) is the water density at a distance y from the bottom in the direction L.

The equation describing the natural convection mass flow rate $\dot{m}$ and the useful energy $\dot{Q}_u$ to section A which is equal to the difference between the input energy and the lost energy is developed as follow:

$$\dot{m} c_w \, (T_2 - T_1) = \dot{Q}_u = \dot{Q}_{in} - \dot{Q}_l \qquad (2)$$

$$\dot{m} = \left[\frac{W^3 g\, \beta \rho_0 (\dot{Q}_{in} - \dot{Q}_L)\, B_1}{12\, \upsilon c_w} \frac{\sin\theta}{2} \right]^{\frac{1}{2}} \qquad (3)$$

This equation for the natural convection flow rate in section A of a simplified ICSCWH heating system is obtained by heat transfer from section A to B from morning until noon time and vice versa from noon time until 4 p.m. Equation (3) indicates that the most significant factor influencing the flow rate on the depth W of section A. From noon time until 4 p.m, reverse convection thermal transfer is generated from section A to B, but at a slower rate, as a result of the drop of temperature gradient and gravity.

Energy Balance for Section A

DOI: 10.1051/e3sconf/20172305002

Section A is considered as a rectangular ICSSWH. Using the first law of thermodynamics, transient energy balances are developed for the absorbing plate and water.

In our case, an absorber surface is added in the top side of section A. As a result, energy equations for the absorber surface and water can be written respectively as follows:

$$m_p c_p \frac{dT_p}{dt} = (\alpha\tau)_e I_t A_p + (\alpha\tau)_e I_t \left(\frac{\pi D^2}{4}\right) - U_t(T_p - T_a) A_p - h_{pwA}(T_p - T_{wA}) A_p \quad (4)$$

$$m_{wA} c_w \frac{dT_{wA}}{dt} = h_{pwA}(T_p - T_{wA}) A_p \quad (5)$$

with, T_p the temperature of the absorbing surface, T_{wA} the water temperatures in section A, T_a the surrounding air temperature, h_{pwA} the thermal transfer coefficient from the absorbing plate to water, U_t the top thermal transfer coefficient from the absorbing plate to the surrounding air, I_t the total solar intensity, $(\alpha\tau)_e$ the effective transmittance–absorbance product for the absorbing plate, and $\dot{m}$ the natural convection flow rate. In addition, A_p the absorber area of the top side of section A, m_{wA} the mass of water in section A, m_p the mass of the absorber surface, and c_p and c_w are the specific heats of the absorber plate and the water, respectively.

The left parts of Eqs. (4) and (5) respectively represent the changes with time of energy inside of the absorber surface and the water.

Efficiency of Water Heater

The thermal efficiency of this conical solar water heater is defined as:

$$\eta = \frac{Q}{\int_0^t I_t dt \left(A + \left(\frac{\pi D^2}{4}\right)\right)} \quad (6)$$

The numerator of Eq. (6) describes the gain of useful thermal energy generating by this system, whereas the denominator describes the incident solar energy during the operating time by the total area of the system.

Results and Discussion

The simulations are performed for the climatic conditions of the Kingdom of Bahrain. The ICSCWH system is useful for seasonal applications in the Kingdom of Bahrain. The hourly variations of solar intensity for May in the Kingdom of Bahrain is considered for our numerical simulation (see Figure 5).To compute the water and absorbing surface temperatures, the two differential coupled equations (4) and (5) that describe the transient behavior of solar gain energy and heat loss of the system are resolved. The analytical model has been solved numerically. The numerical simulation of the novel conical solar integrated-collector-storage system has been realized. The analytical simulations with glass cover are carried out to the three different dimensional cases.

DOI: 10.1051/e3sconf/20172305002

Figure 6 shows that the flow rate increases as the solar radiation rises; which is faster in the period 9 a.m-11 a.m and slower at noon time. Based on figure 5 and figure 6, it is important to mention that the flow rate increases as the solar radiation increases. During solar irradiance 9 a.m-13 p.m, a continuous raise is examined in the natural convection flow rate; nevertheless the temperature in the absorbing plate diminished as the solar flux decreases from 13 p.m to 16 p.m.

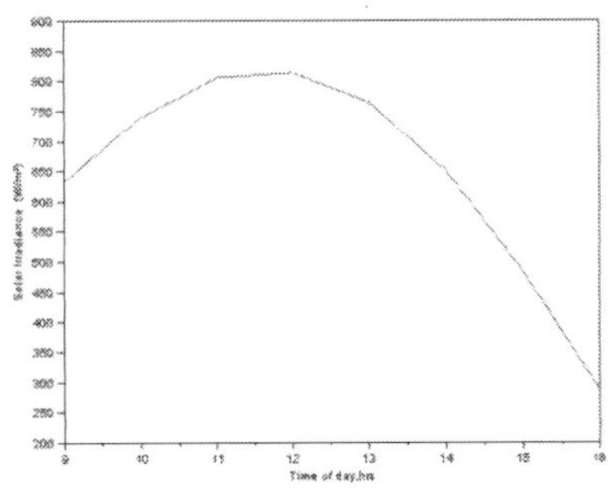

Fig.5 Hourly variation in solar intensity in Bahrain for 10 May.

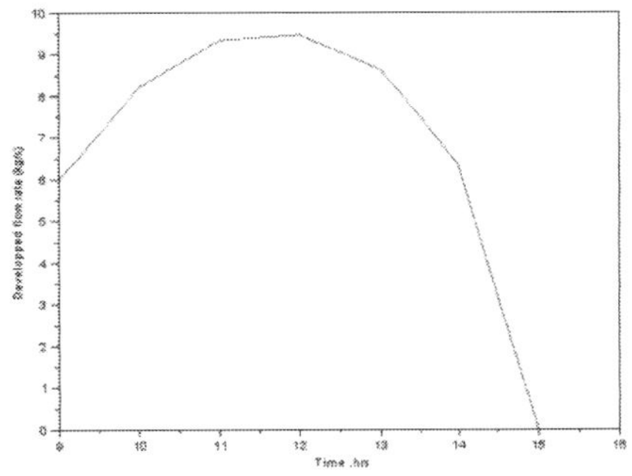

Fig.6 Simulation of natural convection flow rate.

From Figure 7, it can be concluded that as the volume ratio S/V increases, the estimated water temperatures increase. So, a case study 1 representing our investigated system has the highest output water temperature. Similarly, the volume ratio S/V increases, the inclination angle raises. The temperature of hot water increases of about 6°C as the inclination angle changesfrom 60° to 68° (study cases1 and 2). Thus, the tilt angle in our study case 1 for sections A and B generates the highest heat from the system for the climatic conditions in the Kingdom of Bahrain is 68°.

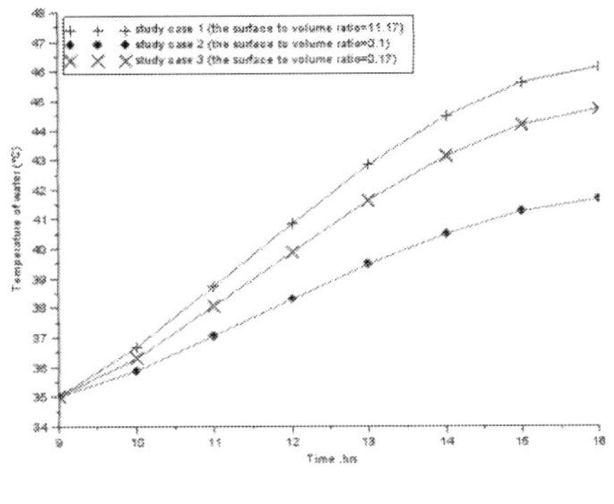

 DOI: 10.1051/e3sconf/20172305002

Fig.7 Comparison between simulated water temperatures for different surface to volume ratio (S/V).

Experimental Verification of the Mathematical Model

To verify the developed mathematical model, a comparison by means of an experimental approach is carried out. The difference between experimental and simulation results of water temperatures are resulted to the different approximations and error's of the instrument used in this research. Figure 9 shows that the temperature of collector at noon can reach 56°C. The difference between the water and surface temperature is resulted to the convection loss between the absorbing plate surface and water.

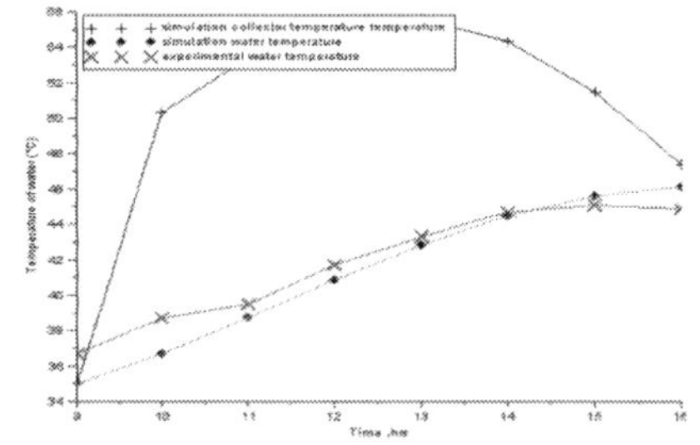

Fig.9 Validation of the numerical results with experiments.

In Figure 10, the variation of efficiency of the system between experimental and simulation results is plotted versus time on a selected day; i.e. 10, May 2015. It can be concluded that the highest efficiencies of the system are respectively around 32.75% and 30.57%, respectively.

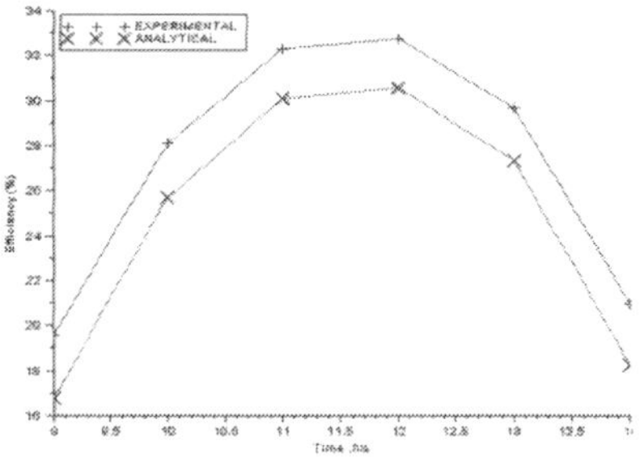

Fig.10 Comparison between experimental and simulation efficiencies for natural flow with glass.

Figure 11 presents the experimental and simulation useful energy which has the same behavior of the efficiency. The variation of experimental and simulation useful

DOI: 10.1051/e3sconf/20172305002

energy variation of the system is reported for a selected day 10 May 2015. It can be concluded that the optimum useful energy of the system is equal to 0.33 and 0.37 Kw, respectively.

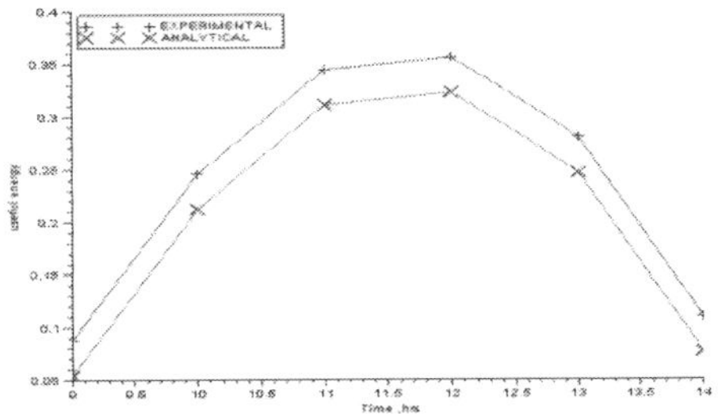

Fig.11 Comparison between experimental and simulation useful energy in Kw natural convection with glass.

From the above obtained results, it can be concluded that the developed analytical model is consistent. The analytical model is based on energy balances obtained from the first law of thermodynamic. The difference between the simulation and experimental results is mainly caused by instrumental errors (2°C). The natural convection flow rate increases with the solar intensity, reaching a maximum value at noon time. Also, the highest efficiency and thermal output are achieved at noon time when the solar irradiation reaches a maximum value. The highest efficiency for an experimental system is observed for S/V ratio 11.7.

4. Conclusion

The objective of this research work is developed to the investigation of thermal efficiency of an integrated conical solar collector water heater in the Kingdom of Bahrain. The numerical simulation of the system is performed for the determination of hourly variation. So, different parameters such as water and surface absorbing temperatures, natural convection flow rate, useful energy and efficiency.
Therefore, the following findings are very helpful for the design of solar water heater:

- ✓ Temperature of the ambient air, water and absorber surface rise with time of the day and achieve higher temperature after 7 hours of operation due to the rise in solar intensity.
- ✓ The temperature of absorber surface is greater than the water surface.

- ✓ Increasing the volume ratio S/V affects significantly the output water temperature.
- ✓ Raising the inclination of angle affects significantly the water temperature.
- ✓ The use of glass cover to enhance the optical coefficient of absorbing surface results in the enhancement of the thermal output of the system.
- ✓ It is found that the total daily performance is about 32%, according to weather and operating conditions.

References

(1) Cruz J.M.S., P.H. Geoffrey and J.P.S.R. Albino, "Thermal performance of a trapezoidal-shaped solar collector/energy store", *Applied Energy* , 2002; 73: 195-212.
(2) De Winter F. "Solar collectors, energy storage, and materials". Massachusetts: The MIT press; 1991.
(3) El-Sebaii A.A., "Thermal performance of a shallow solar pond integrated with a baffle plate", *Applied Energy;* 2005: 81- 33e53.
(4) Farid M.M., Khudhair A.M., Razack S.A.K. and Al-Hallaj S. "A review on phase change energy storage: materials and applications". *Energy Convers Manage*. 2004; 45:1597–615.
(5) Gil A, Medrano M, Martorell I, Lazaro A, Dolado P, Zalba B, et al. "State of the art on high
(6) temperature thermal energy storage for power generation. Part 1— concepts, materials and
(7) modellization". *Renew Sust Energy Rev* 2010; 14:31–55.
(8) Kalogirou SA. "Solar thermal collectors and applications".*Prog Energy Combust,* 2004; 30:231–95.
(9) Kumar R., M.A. Rosen, "Integrated collector storage solar water heater with corrugated absorber surface", *Applied Thermal Engineering* ,2010; 30 :1764-1768.
(10) Sefa T., S., A.M. Hakan, "Temperature distributions in trapezoidal built in storage solar water heaters with/without phase change materials", *Energy Conversion and Management*, 2006; 47, 2143-2154.
(11) Sharma A, Tyagi VV, Chen CR, Buddhi D. "Review on thermal energy storage with phase change materials and applications". *Renew Sust Energy Rev* 2009; 13:318–45.
(12) Smyth M., P.C. Eames, B. Norton, "Heat retaining integrated collector/storage solar water heaters", *Solar Energy,* 2003;75 :27–34.
(13) Sridhar A., and K.S. Reddy. "Transient analysis of modified cuboid solar integrated collector-storage Ahmet system", *Applied Thermal Engineering*. 2007:27-330e346.
(14) Suhas , SukeerthCalastawad, Sawan Kumar, Purujeeth and Shyam Sharma (2015)"performance of a conical shaped solar water heater with that of a conventional solar water heater", International Research Journal of Engineering and Technology. Vol.02 (4), pp 800-803.
(15) Tripanagnostopoulos Y. and P. Yianoulis, "Integrated Collector-Storage Systems With Suppressed Thermal Losses", *Solar Energ,* 1992 ; 48: 1: 31-43.
(16) Tyagi S.K., W. Wang, S.C. Kaushik, M.K. Singhal and S.R. Park, "Energy analysis and parametric study of concentrating type solar collectors", *International Journal of Thermal Sciences*. 2007;46 :1304-1310.

DOI: 10.1051/e3sconf/20172305002

(17) Zalba B, Marin JM, Cabeza LF, Mehling H. "Review on thermal energy storage with phase change: materials, heat transfer analysis and applications". *ApplThermEng,* 2003; 23:251–83.

(18) Zhao CY, Lu W, Tian Y. "Heat transfer enhancement for thermal energy storage using metal foams embedded within phase change materials (PCMs)". *Sol Energy,* 2010; 84:1402–12.

Optimum Design Parameters of Box Window DSF Office at Different Glazing Types under Sub Interval of Intermediate Sky Conditions (20-40 klux)

O.K Elayeb[a], M.A. Alghoul[b,c,], K. Sopian[a,]*, N.G. Khrit[a]*

[a] *Solar Energy Research Institute, Universiti Kebangsaan Malaysia, 43600 Bangi, Selangor, Malaysia*
[b] *Energy and Building Research Center, Kuwait Institute for Scientific Research, P.O. Box 24885, Safat 13109, Kuwait*
[c] *Center of Research Excellence in Renewable Energy (CoRe-RE), Research Institute, King Fahd University of Petroleum and Minerals (KFUPM), Dhahran 31261, Saudi Arabia*

**Corresponding authors: dr.alghoul@gmail.com (M.A. Alghoul), ksopian@ukm.edu.my (K. Sopian)*

ABSTRACT:
Despite Double skin façade (DSF) buildings are widely deployed worldwide, daylighting strategy is not commonly incorporated in these buildings compare to other strategies. Therefore, further theoretical and experimental studies would lead to adopting daylighting strategy in DSF office buildings. The aim of this study is to investigate the daylighting performance of office building at different design parameters of box window DSF using different glazing types under sub interval of intermediate sky conditions (20-40) klux using the (IES VE) simulation tool from Integrated Environmental Solutions - Virtual Environment. The implemented design parameters are window wall ratio (WWR) of internal façade (10-100) %, cavity depth (CD) of DSF (1-2.5) m and different glazing types. The glazing types were selected from the list available in the (IES VE) simulation tool. After series of evaluations, bronze tinted coating (STOPSOL) is implemented for the exterior façade while clear float, clear reflective coating (STOPSOL), grey and brown tinted coating (Anti-sun float) and blue coating tinted (SUNCOOL float) are implemented for the interior façade. In this paper, several evaluation parameters are used to quantify the optimum design parameters that would balance the daylighting requirements of a box window DSF office versus sky conditions range (20-40) klux. The optimum design parameters of DSF office building obtained under different glazing types are highlighted as follows. When using bronze tinted coating (STOPSOL) for the exterior façade, the glazing types of interior façade that showed superior daylighting performance of DSF office at (CD of 1.0m with WWR of 70%), (CD of 1.5m with WWR of 70%), (CD of 2.0m with WWR of 70%) and (CD of 2.0m with WWR of 70%) are grey tinted coating (Anti-sun float), clear reflective coating (STOPSOL), brown tinted coating (Anti-sun float), and clear float glazing respectively. Blue Coating tinted (SUNCOOL float) of interior façade glazing failed to meet outstanding daylighting performance at any cavity depth.

Keywords: sub interval of intermediate sky conditions (20-40) klux, office building, box window DSF system, different glazing types, daylighting performance

Abbreviations

DSF	*Double Skin Façade*
IES VE	*the Integrated Environmental Solutions - Virtual Environment simulation tool*
CD	*Cavity Depth*
WWR	*Window Wall Ratio*
DF	*Daylight Factor*
ADF	*Average Daylight Factor*
UDI	*Useful Daylight Illuminance*

1 Introduction

Full glazing façades are widely used for modern architectural projects, particularly commercial buildings, due to their aesthetics and daylight potential. In certain circumstances, single layer glass façades have common weaknesses such as poor thermal insulation and sound reduction index. To overcome these problems, application of DSFs is widely accepted to reduce energy consumption and thus to improve the sustainability of buildings [1]. Double-skin facades had been defined as multiple layer skins construction with an external skin, an intermediate space and an inner skin. The external and internal skins could be of either single glaze or double glazed glass panes and it generally could be grouped under Box Window facade, Shaft-box facade, Corridor facade and Multi-story façade [2].

Daylighting can influence the task involvement, productivity, mood and health and comfort therefore is a critical factor in office design in the commercial buildings [3]. At a time when the daylight technology is applied to reduce energy cost in lighting sector, it must be considered some of factors have an impact on daylighting performance such as, location of building, time, climate, size and material of window [4]. Windows are a serious factor in building energy, because thermal energy that is transferred through their transparent surface is crucial, and help to transfer natural daylight deep into building. During summer, windows allow heat to pass in the building and during winter, heat escapes and variable light transmission of daylight allows the presence of a solar gain and glare [5, 6] It is estimated that windows are responsible for 60% of the total building energy losses. Hence, improvement of the performance for windows with low thermal transmittance can virtually contribute in reducing energy costs of buildings, and can provide large energy savings [6].

In the modern windows, the use of two or three panes together with different selective coatings is very common. This leads to increase the use of artificial light and gives darker appearance to the window but using an antireflection coating can be increases the transmittance of daylight leading to a brighter window. The optical properties of a glazing are commonly summarized in transmittance of daylight (Tvis), transmittance of all solar radiation (Tsol) and the g-value [7]. Although the changing from a high to a low transmittance glazing reduces energy savings for the office and the UDI value increases with low transmittance glazing, but it decreases with increasing WWR and reduces visual comfort in a indoor environment [8]. Daylighting under clear and intermediate sky condition often is accompanied by high temperature and glare. To limit depleting energy and improve performance of the window system, there are a wide variety of different types of glass in the market that can be used to reach this goal.

2 Materials and methods

2.1 Model description and materials used

The box window DSF office is used for daylighting simulations with CD, adjustable internal façade opening, and different glazing types. The study is performed in a simulated office with dimensions of 4 m in width, 6 m in depth and 3 m in height, located at a height of 16 m over the floor of a building with south orientation. These models are double glazed facades with a cavity depth of 1, 1.5, 2.0 and 2.5, respectively. Ten window sizes are modeled with fenestration window-to-wall area ratios. The light reflectance of the ceiling, wall and floor of the model is 0.7, 0.5 and 0.2 respectively. The exterior façade openings for a box window DSF is full glass. A bronze tinted coating (STOPSOL) is implemented for the exterior façade. The range of internal façade openings for a box window DSF (WWR 10-80%) was covered in the analysis. The five glazing types to be used for the internal windows are determined with respect to its transmission, absorption and reflection for light. A grey (G-I) and brown tinted coating (Anti-sun float) (G-II), blue coating tinted (SUNCOOL float 40/50) (G-III), clear float (G-IV) and clear reflective coating (STOPSOL) (G-V), are implemented for the interior façade. The glazing transmittance for an external window is 29%, while the glazing transmittance for an internal window with using (G-I), (G-II), (G-III), (G-IV) and (G-V) is 42, 58, 32, 82 and 53%, respectively.

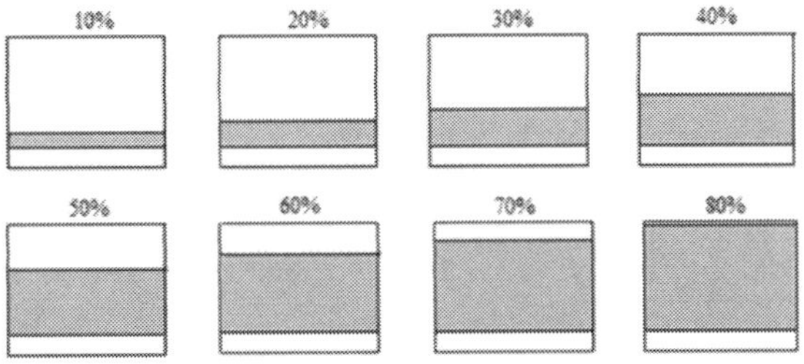

Fig 1: Implemented WWR of internal façade of DSF office

DOI: 10.1051/e3sconf/20172305003

2.2 Method

Integrated environmental solution <Virtual Environment> IES (VE) simulation tool is relatively unique amongst analysis tools which is intended to use during the earliest conceptual stages of design and provides daylighting calculations from first principles without any user programming. Therefore, in this study, this software was chosen and performed because it is widely used in building services and energy management industries. The indoor illuminance values are estimated at office table level (0.70 m) for all studied models under sky condition ranges (20, 30 and 40 klux).To ensure the accuracy of simulation results, the IES-VE simulation software has been validated by comparing the simulation results to the experimental ones. A 1:10 scale model has been fitted from a wood "sandwich cardboard" with an internal size of 75 cm length, 40 cm in width, 30 height and thick 12 mm. The external and internal walls were painted with white colour and light brown for the ceiling. The model was placed on a 1m high platform of a roof top of SERI, UKM, Selangor, Malaysia, (N 2.92° ; E 101.772°) with S orientation. The office has been tested under the real sky conditions.

3 Results and discussion

Simulating the effect of different internal window glazing types was carried out to determine the optimum design parameters of DSF system of office buildings for daylighting performance under outdoor illuminance ranges between 20 to 40 klx. The average daylight factor (ADF) and average indoor illuminance inside the office for each DSF model were commonly used in early stages of design to estimate the daylighting characteristics and the required window area. The ADF must meet the requirements of Malaysian Standard MS ISO/IEC 17025:2005 which limited it between 1 and 6%. Fig. 2 shows the methodology chart used to define and estimate the average indoor illuminance inside the office.

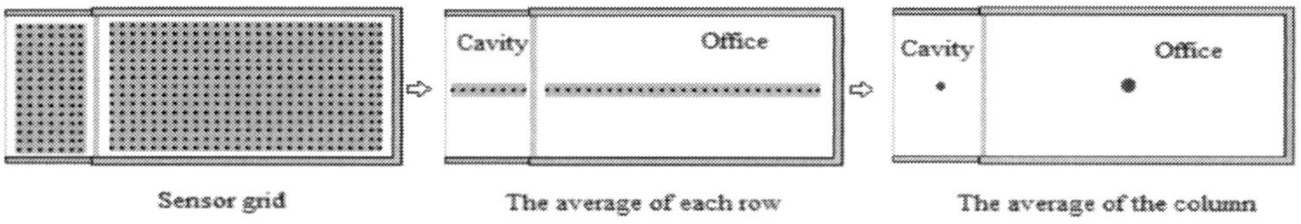

Fig 2: Method of determining and calculating of the mean office indoor illuminance values

In the cases the CD is 1.0m and at 30, 40, 50, 60, 70 and 80% WWR, the ADF is located mainly between 1-6% under outdoor illuminance (20,000-40,000 lx) for (G-I) and (G-II) and also under outdoor illuminance (20,000-30,000 lx) for (G-I), (G-II) and (G-V). In addition, it is important to mention that the models with high transmittance of glazing window such as (G-IV), fail to meet the standard ADF, this requires that the CD must be more than 1.0 m, as shown in Fig. 3. As seen in Fig. 4, in the cases CD is 1.5m and at (20- 80%) WWR, the models only succeed to meet the standard ADF under outdoor illuminance (20-40 klx) with using (G-V), while under outdoor illuminance (20-30 klx) and at (30- 80%) WWR, the models succeed to meet the standard ADF for all types of implemented glazing window, except (G-IV) showed a ADF more than 6% under outdoor illuminance 30 klx .

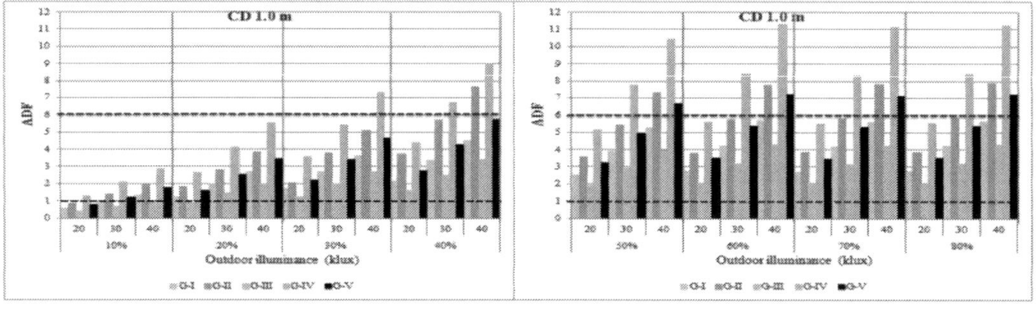

Fig 3: Average daylighting factor (ADF) versus outdoor illuminance (20-40klx) and different WWR at five types of implemented glazing window for CD 1.0m

DOI: 10.1051/e3sconf/20172305003

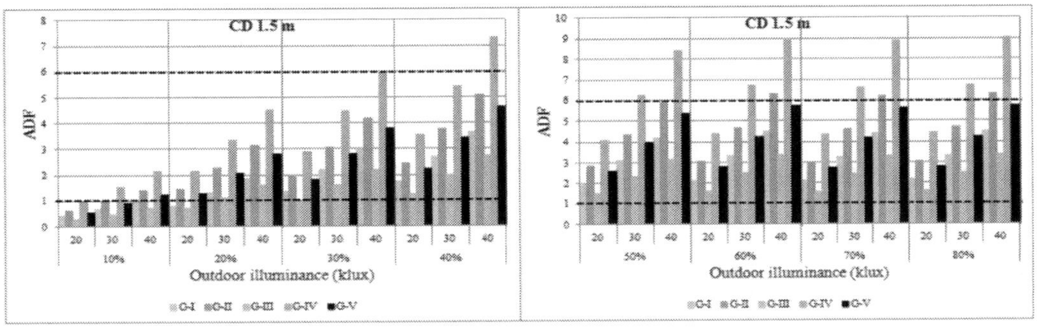

Fig 4: Average daylighting factor (ADF) versus outdoor illuminance (20-40klx) at different WWR and five types of implemented glazing window for CD 1.5m

Fig.5 shows that with increasing the cavity depth to be 2.0m and 2.5m, the indoor illuminance decreases inside the office, therefore, at (30-80%) WWR, the ADF is located between 1-6% under outdoor illuminance (20-40 klx) with (G-II), (G-IV) and (G-V). In the cases the CD is 2.5 m and at (50-80%) WWR, the ADF is located between 1-6% under outdoor illuminance (20-40 klx) with using all types of implemented glazing window (G-V) (G-I) (G-II) (G-IV) and (G-V), while the glazing which has low transmittance in these the case, at (10-30%) WWR, the ADF is less than a required minimum ADF under outdoor illuminance (20-30 klx), as shows in Figures 6.

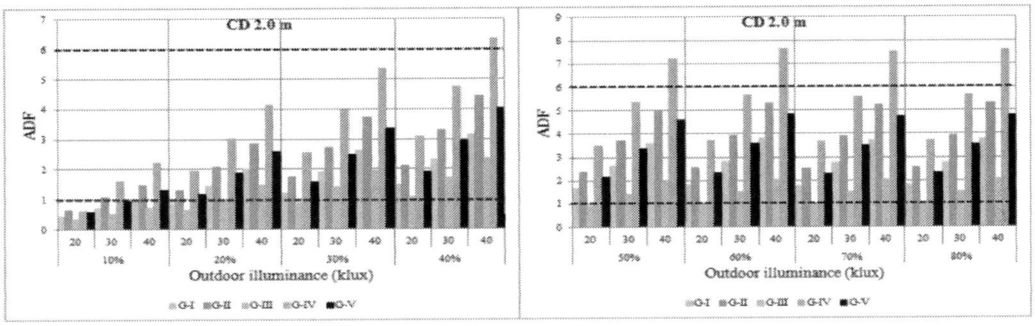

Fig 5: Average daylighting factor (ADF) versus outdoor illuminance (20-40klx) at different WWR and five types of implemented glazing window for CD 2.0m

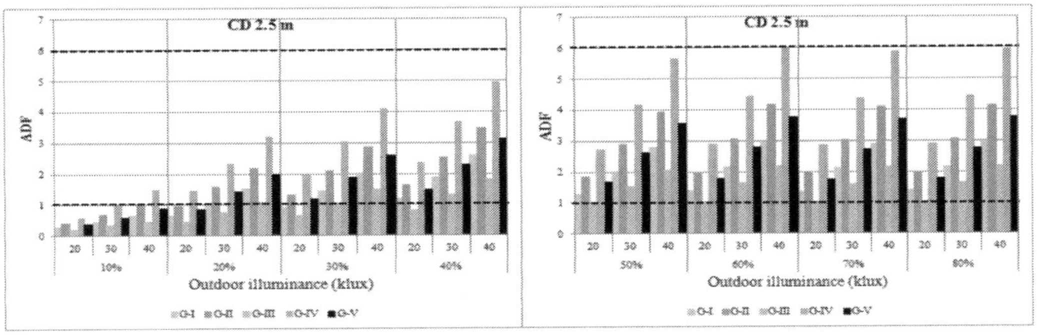

Fig 6: Average daylighting factor (ADF) versus outdoor illuminance (20-40klx) at different WWR for CD 2.5m and five types of implemented glazing window

DOI: 10.1051/e3sconf/20172305003

The average indoor illuminance also must meet the requirements of Malaysian Standard MS ISO/IEC17025:2005 which limited it between 300 and 400lx in commercial offices. Figure 7, 8, 9 and 10 illustrate the average indoor illuminance of double skin façade DSF models versus outdoor illuminance (20-40 klx) at a different window wall ratio WWR and cavity depth CD at five types of implemented glazing window. At CD 1.0m, the models have a recommended average indoor illuminance (300-400 lx) at a WWR range (60-80%) with using (G-II) and (G-V) under outdoor illuminance 30 klx, while with using all the implemented types of glazing window the average indoor illuminance is less than 300lx under outdoor illuminance (20-30 klx) at WWR 10%.

Also it can be noticed that the DSF model succeeds to reach to a recommended average indoor illuminance in the cases where the CD is 1.5m at a WWR range between (60-80%) under outdoor illuminance 30 klx with using (G-I) only and under at a WWR of 40% under outdoor illuminance 20 klx with using (G-IV) and under outdoor illuminance 30 klx with using (G-II) and (G-V). Increasing the cavity depth to be 2.0 and 2.5m, causes decreasing the average indoor illuminance inside the office, therefore, this requires using glazing has a high transmittance for high outdoor illuminance. In addition, the average indoor illuminance results when a WWR between (60-80%) is found with values that are almost similar with using same type of glazing as seen in the figures referred to previously.

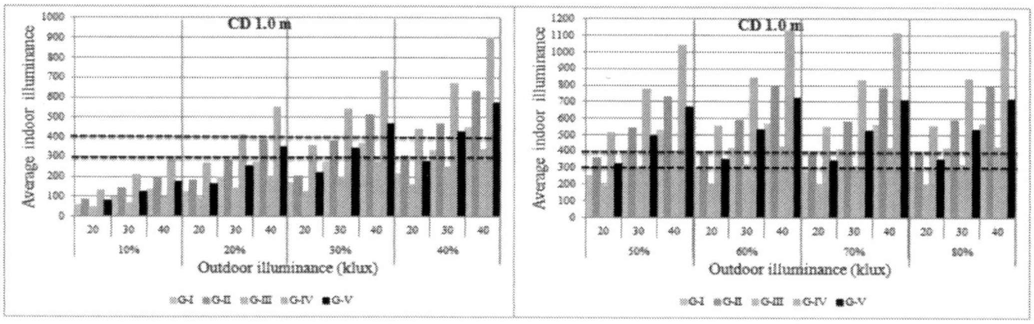

Fig 7: Average indoor illuminance versus outdoor illuminance (20-40klx) at different WWR and five types of implemented glazing window for CD 1.0m

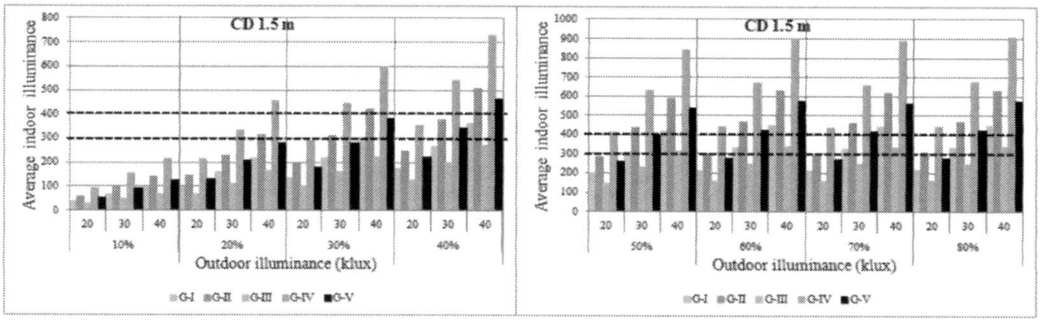

Fig 8: Average indoor illuminance versus outdoor illuminance (20-40klx) at different WWR and five types of implemented glazing window for CD 1.5m

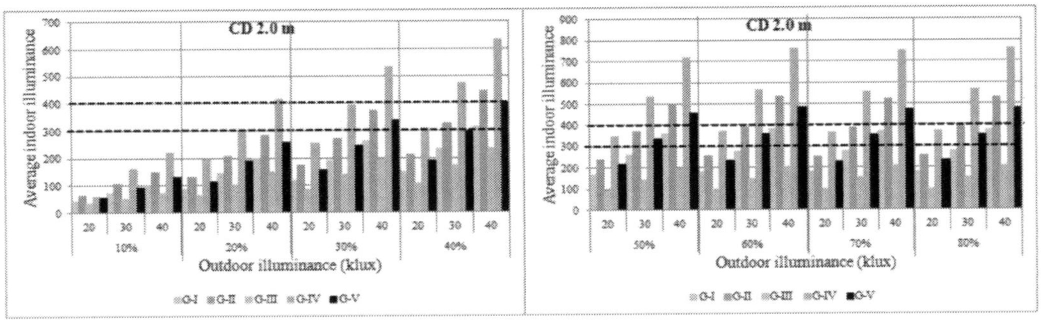

Fig 9: Average indoor illuminance versus outdoor illuminance (20-40klx) at different WWR and five types of implemented glazing window for CD 2.0m

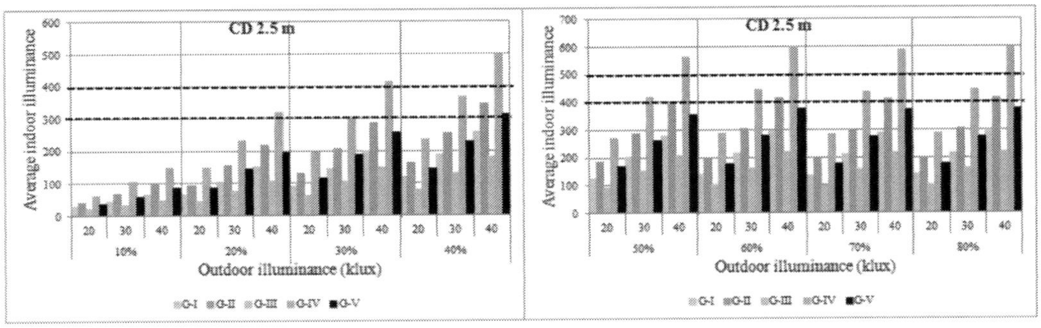

Fig 10: Average indoor illuminance versus outdoor illuminance (20-40klx) at different WWR and five types of implemented glazing window for CD 2.5m

As summarized in Table. 1, the DSF office with different glazing types are tested under outdoor illuminance range (20-40 klx) to determine the optimum glazing type for each CD. The optimum glazed type is the glazing that can supply the 2000 lux illuminance threshold near the window. It can be noted that, for a CD of 1.0m, the glazing type which is able to supply indoor illuminance close to 2 klx near the window, is a (G-I) and (G-III), but with using (G-I), only in the area that has a depth of 10 cm near of the window of office, the indoor illuminance is more than 2000 lx while the remaining area of the office has a good quality of the indoor illuminance. At a CD of 1.5m, the best choice of window glazing must be used which has a transmittance around 58 and 53% such as (G-II) and (G-V). Finally, on account of the need of the office to high indoor illuminance at high cavity depth, the models require glazing window with high transmittance light, this clearly shows at a CD of 2.0 and 2.5m, therefore in this case, the best glazing window glazing must be used which has a transmittance around 82% such as (G-IV).

Table 1: Daylighting characteristics of the DSF office daylighting at different glazing types for different CDs under outdoor illuminance range (20-40 klx)

CD (m)	Types of glazing	Outdoor illuminance (klux)	Illuminance level at the area near the window	Glare	Space depth with		
					100 < UDI< 2000 lux	Illuminance ≥ 200 lux (DF ≥ 2%)	Under-lit ≤199 lux DF < 2%
1.0	G-I	20	799	**0%**	**100%**	**48%**	**52%**
		30	1203	**0%**	**100%**	**69%**	**31%**
		40	1606	**0%**	**100%**	**100%**	**0%**
	G-II	20	1109	**0%**	**100%**	**66%**	**34%**
		30	1666	**0%**	**100%**	**100%**	**0%**

DOI: 10.1051/e3sconf/20172305003

		40	**2228**	**3%**	**97%**	**100%**	**0%**
	G-III	20	606	**0%**	**66%**	**66%**	**34%**
		30	914	**0%**	**100%**	**55%**	**45%**
		40	1222	**0%**	**100%**	**72%**	**28%**
	G-IV	20	1575	0%	100%	100%	0%
		30	2370	7%	93%	100%	0%
		40	3163	14%	86%	100%	0%
	G-V	20	1034	0%	100%	59%	41%
		30	1560	0%	100%	86%	14%
		40	2082	3%	97%	100%	0%
1.5	G-I	20	595	0%	76%	38%	62%
		30	898	0%	100%	59%	41%
		40	1200	0%	100%	83%	17%
	G-II	20	**828**	**0%**	**55%**	**55%**	**45%**
		30	**1249**	**0%**	**100%**	**100%**	**0%**
		40	**1668**	**0%**	**100%**	**100%**	**0%**
	G-III	20	451	0%	55%	28%	72%
		30	681	0%	100%	45%	55%
		40	911	0%	100%	62%	38%
	G-IV	20	1183	0%	100%	90%	10%
		30	1780	0%	100%	100%	0%
		40	2378	7%	93%	100%	0%
	G-V	20	**755**	**0%**	**100%**	**52%**	**48%**
		30	**1137**	**0%**	**100%**	**79%**	**21%**
		40	**1523**	**0%**	**100%**	**100%**	**0%**
2.0	G-I	20	478	0%	62%	31%	69%
		30	720	0%	100%	48%	52%
		40	965	0%	100%	69%	31%
	G-II	20	**663**	**0%**	**100%**	**55%**	**45%**
		30	**1003**	**0%**	**100%**	**76%**	**24%**
		40	**1342**	**0%**	**100%**	**100%**	**0%**
	G-III	20	361	0%	48%	55%	45%
		30	547	0%	76%	38%	62%
		40	731	0%	100%	100%	0%
	G-IV	20	**946**	**0%**	**100%**	**72%**	**28%**
		30	**1425**	**0%**	**100%**	**100%**	**0%**
		40	**1905**	**0%**	**100%**	**100%**	**0%**
	G-V	20	605	0%	100%	41%	59%
		30	914	0%	100%	66%	34%
		40	1225	0%	100%	100%	0%
2.5	G-I	20	341	0%	52%	24%	76%
		30	517	0%	100%	38%	62%
		40	693	0%	100%	76%	24%
	G-II	20	476	0%	76%	38%	62%
		30	722	0%	100%	59%	41%
		40	967	0%	100%	100%	0%
	G-III	20	256	0%	10%	10%	90%
		30	392	0%	62%	28%	72%
		40	525	0%	100%	52%	48%
	G-IV	20	**684**	**0%**	**48%**	**48%**	**52%**
		30	**1032**	**0%**	**100%**	**69%**	**31%**
		40	**1380**	**0%**	**100%**	**100%**	**0%**
	G-V	20	434	0%	100%	31%	69%
		30	656	0%	100%	38%	62%
		40	882	0%	100%	100%	0%

 DOI: 10.1051/e3sconf/20172305003

4 Conclusions

The amount of daylight that penetrates the window depends on the used glass type. The optimum WWR that should provide indoor illuminance inside the office without glare, maximum space within UDI, maximum space with an indoor illuminance of at least 200 lx is found 60-70%. Increasing the cavity depth, causes decreasing the average indoor illuminance inside the office, therefore, this requires using glazing has a high transmittance for high outdoor illuminance. The optimum glazing type for the interior façade opening can be seen in the follow cases: At CD of 1.0m, the best choice of window glazing is that has a transmittance around (32, 42 and 58) %; At CD of 1.5m, the best choice of window glazing is that has a transmittance around (53 and 58) %. Also, the best choice of window glazing for a CD of 2.0 m and 2.5m is that has a transmittance around 82%.

References

1. GHAFFARIANHOSEINI A., GHAFFARIANHOSEINI A., BERARDI U., TOOKEY J., LI D. H. W., KARIMINIA S. Exploring the advantages and challenges of double-skin façades (DSFs), Renewable and Sustainable Energy Reviews 2016: 60: 1052-1065.
2. AHMED M. M., ABEL-RAHMAN A. K., ALI A. H. H., SUZUKI M. Double Skin Façade: The State of Art on Building Energy Efficiency, Journal of Clean Energy Technologies 2016: 4.
3. ZOMORODIAN Z. S., KORSAVI S. S., TAHSILDOOST M. The effect of window configuration on daylight performance in classrooms: A field and simulation study, 24-15 :26 :2016 دانشگاه علم و صنعت ایران.
4. SRIPRAMONG S., ANANWATTANAPORN S., NGAOPITAKKUL A., JETTANASEN C. A Study an Impact of Daylight with Period Time on Illuminance and Uniformity for Building. Proceedings of the International MultiConference of Engineers and Computer Scientists; 2015.
5. GORGOLIS G., KARAMANIS D. Solar energy materials for glazing technologies, Solar Energy Materials and Solar Cells 2016: 144: 559-578.
6. JELLE B. P., HYND A., GUSTAVSEN A., ARASTEH D., GOUDEY H., HART R. Fenestration of today and tomorrow: A state-of-the-art review and future research opportunities, Solar Energy Materials and Solar Cells 2012: 96: 1-28.
7. GHOSH A., NORTON B., DUFFY A. Daylighting performance and glare calculation of a suspended particle device switchable glazing, Solar Energy 2016: 132: 114-128.
8. REINHART C. F. Effects of interior design on the daylight availability in open plan offices. 2002 ACEEE Summer Study on Energy Efficiency in Buildings; 2002, p. 309-322.

Profitability Evaluation of a Hybrid Geothermal and CO2 Sequestration Project for a Coastal Hot Saline Aquifer.

Tatyana Plaksina, [1] and Mohammed Kanfar[2]

[1]*Department of Chemical and Petroleum Engineering Schulic School of Engineering University of Calgary END 204E 2500 University Drive NW, Calgary, AB T2N 1N4 T: 403.220.5652 tatyana.plaksina@ucalgary.ca*

[2] *Department of Chemical and Petroleum Engineering Schulich School of Engineering University of Calgary 2500 University Drive NW, Calgary, AB T2N 1N4 kanfar@gmail.com*

Abstract

With growing interest in commercial projects involving industrial volume CO_2 sequestration, a concern about proper containment and control over the gas plume becomes particularly prominent. In this study, we explore the potential of using a typical coastal geopressured hot saline aquifer for two commercial purposes. The first purpose is to harvest geothermal heat of the aquifer for electricity generation and/or direct use and the second one is to utilize the same rock volume for safe and controlled CO_2 sequestration without interruption of heat production. To achieve these goals, we devised and economically evaluated a scheme that recovers operational and capital costs within first 4 years and yields positive internal rate of return of about 15% at the end of the operations. Using our strategic design of well placement and operational scheduling, we were able to achieve in our numerical simulation study the following results. First, the hot water production rates allowed to run a 30 MW organic Rankine cycle plant for 20 years. Second, during the last 10 years of operation we managed to inject into the same reservoir (volume of 0.8×10^9 m^3) approximately 10 million ton of the supercritical gas. Third, decades of numerical monitoring the plume after the end of the operations showed that this large volume of CO_2 is securely sequestrated inside the reservoir without compromising the caprock integrity.

Keywords: CO_2 sequestration; geothermal energy; dynamic plume control; hybrid energy systems

Introduction

At the beginning of the 1980s, when crude oil price soared to new historical highs, the US DoE started to investigate alternative energy sources that could provide abundant energy for electricity generation in highly populated coastal regions. As a part of the DoE initiative called "Wells of Opportunity" [1], the engineers established that geothermal potential of the Gulf of Mexico (GOM) coastal region was tremendous and highlighted that already drilled oil and gas wells can be used for economic production of hot geofluid for power generating plants. Though GOM geofluid tends to have low enthalpy [2], several pilot projects in this region demonstrated sustainably high geofluid production rates and promising power generation capacities. For example, in the 1990s, Riney [3] completed a report about the Pleasant Bayou project in Texas during which the engineers achieved production rates around 20,000 bbl of hot water per day and successfully converted mechanical and thermal energy into the total of 3,445 MWh using a hybrid binary cycle system.

After this successful attempt to assess viability of low-temperature geothermal development in hot saline aquifers, a number of authors investigated economic potential and profitable reservoir engineering design for geopressured geothermal systems. In their reservoir simulation study, Plaksina and White [4] provide a comprehensive overview of the GOM key projects including the discussion about hot water production rates achieved in the area, the hazards of land subsidence and loss of caprock integrity that become important during industrial size CO_2 sequestration activities.

Although the pilot projects demonstrated that it was possible to achieve high hot water production rates and thus, generate commercial electricity output, economic potential of a geopressured geothermal project can be further improved if the same reservoir is also used for secure CO_2 sequestration. Garapati et al. [5] proposed to use CO_2 as a secondary fluid that could displace hot reservoir water geofluid to the producing wells and placed these injection wells to mitigate the hazard of the uncontrolled CO_2 plume. With their numerical design, the authors managed to sequester approximately 2 to 4 Mtons of the supercritical gas without causing a significant damage to the caprock by the high gas concentration plume.

To eliminate the hazard of the plume and ensure caprock integrity, we propose to use strategic placement of cold water injection wells that can reduce gas concentration and keep the plume away from the caprock. In his thesis, Anchliya verified with numerical simulation that simultaneous injection of CO_2 in the middle or lower portion of the reservoir and cold water on top of the reservoir can diffuse the high concentration plume and keep it away from the caprock, thus, preventing the possibility of its fracturing and leakage of the gas to the upper formations and ultimately to the surface [6].

In this study, we use a typical hot saline aquifer geomodel from the Gulf of Mexico and offer an integrated approach to modeling and evaluation of the entire life cycle of this geopressured geothermal aquifer that can be found in many thick coastal sedimentary systems around the globe and used for simultaneous greenhouse gas sequestration and heat extraction [4]. Specifically, first, we develop a conceptual model of a strategic well placement and then a simulation flow model implementing this well arrangement. Second, we generate hot fluid production and CO_2 injection history based on this model and well placement, and evaluate the quantity of supercritical CO_2 gas that can be safely and permanently sequestrated inside the same reservoir. Third, we numerically monitor the reservoir for several decades to ensure that the gas plume is sufficiently dispersed and poses no threat to the caprock. Final, using the obtained production and injection history, we assess economic viability of this heat harvesting project using NREL CREST tool developed for economic assessment of geothermal projects [7].

Methodology

In this section, we present our approach to modeling of a geopressured geothermal sedimentary reservoir, the setup of the model, some practical considerations behind well placement, and description of the production and injection goals.

Simulation Model and Production Arrangement

To construct a simulation model that reflects features of aa actual aquifer, we conducted a survey of reports and scientific literature about potential low-enthalpy geothermal reservoirs in the US Gulf Of Mexico region. In the last several decades, the researchers have identified a number of potentially suitable geopressured formations, properties of which were derived from well logs and field well tests [8-9]. Note, that even though the presented geothermal prospects are found at depths of more than 4,000 m, their permeability is relatively high due to geopressure caused by fast burial of sand. Moreover, in such geopressured formations, overburden is mostly maintained by the incompressible pore fluid rather than touching rock grains, as a result we might observe both high porosity and permeability. Another important observation for these reservoirs is that their depth is not always the most influential factor in the formation temperature. For example, Camerina A sand due to its proximity to a salt dome has a thermal anomaly. Because of its thermal conductivity, salt transfers high amounts of heat from deeper hotter layers to shallower colder sand deposits [9]. Thus, it is possible to encounter anomalously high formation temperatures in this generally low temperature subsurface region.

 DOI: 10.1051/e3sconf/20172305004

Table 1 provides a summary of all relevant petrophysical, thermodynamic, and geometric properties surveyed from the literature and used in our geopressured geothermal reservoir model.

Table 1. Properties of the geothermal reservoir adopted in the simulation model [4].

Property	*Values*	*Unit*
Quarter model dimensions (width x length x thickness)	1000 x 1000 x 200	m
Initial formation pressure	80 or 11600	MPa or psi
Initial average formation temperature	140	°C
Reservoir rock density	2600	kg/m^3
Geothermal gradient	29	°C/km
Horizontal wellbore length	1000	m
Production period duration	20	years
Rock matrix porosity	20	%
Rock matrix compressibility	2.0×10^{-8}	1/Pa
Wet rock heat conductivity	2.0	W/m °C
Rock matrix permeability	200	md
Geofluid salinity	0	ppt
Re-injected cold water temperature	35	°C

Fig. 1 shows the geometry and initial pressure distribution inside the model going from approximately 80 MPa at the top of the reservoir to 81.5 MPa at the bottom. To adequately model constant heat flux from the lower hot layers to the producing formation, we assigned very high heat capacity to the bounding layers (shale layers with infinite heat capacity). This approach ensures constant heat conduction from the hotter lower lying layers to the reservoir without fluid influx.

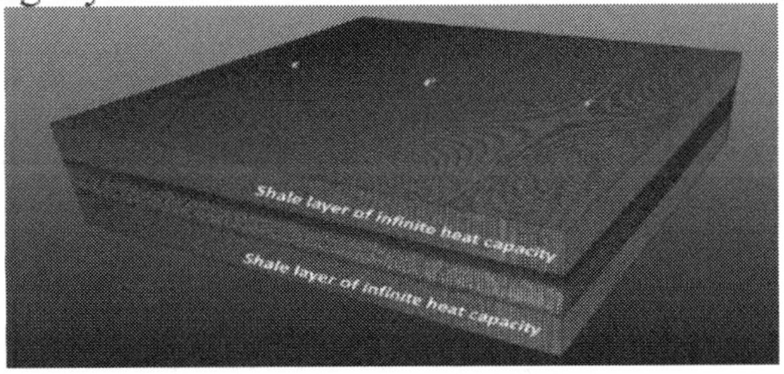

Figure 1. Geometry of a quarter of the geopressured geothermal reservoir (1000 m long, 1000 m wide and 200 m thick) (modelled in CMG STARS).

In his investigation of CO_2 plume control during industrial volume sequestration operations, Anchliya proposed to place a supercritical gas injection well underneath a cold water injector [6]. Such arrangement is reasonable from a fluid movement point of view and feasible from an engineering standpoint. **Fig. 2** provides visual demonstration of how cold water injector can help prevent formation of a high gas concentration plume, efficiently disperse the supercritical gas inside the reservoir, and avoid caprock integrity problems. Due gravity, cold re-injected water tends to move down toward the bottom of the reservoir and push supercritical gas plume away from the caprock and to the side of the formation. In the results section, we will demonstrate whether our arrangement performs according to this theoretical understanding and expectation of fluid flow and reservoir dynamics.

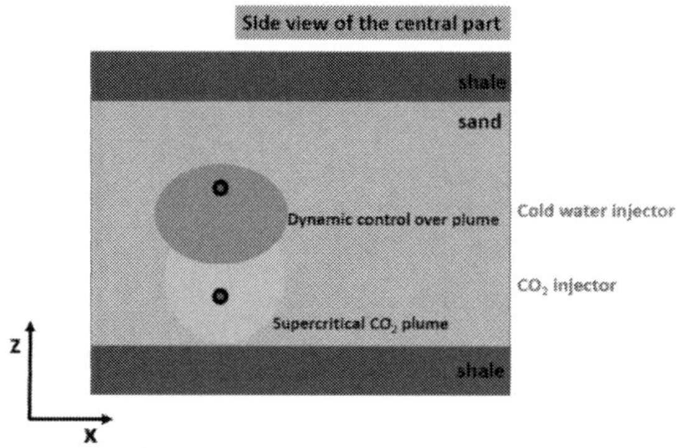

Figure 2. The central portion of the geothermal reservoir with CO_2 injection wells and cold water re-injection wells.

Coupling Subsurface Design with Surface Facilities and Power Output Considerations

Because geopressured geothermal systems produce geofluid of relatively low temperature (and therefore, low enthalpy), we cannot use flash conversion systems that require fluids of more than 200°C [2]. Instead, low-enthalpy geothermal development can take advantage of a binary cycle system such as a binary organic Rankine cycle (ORC) system. This energy conversion system works with geofluids of approximately 150°C (sometimes all the way down to 100 or up to 220°C) and has a number of economically attractive characteristics. Specifically, because of the low enthalpy drop inside a turbine, it can have much simpler design and, therefore, cost. Also the geofluid is not used directly like in typical flash systems. Instead, using a heat exchanger, the primary fluid (hot brine) heats the secondary one (isobutane or R-134a) and never enters the turbine or the condenser. This can be very important especially if the geofluid contains highly corrosive gas

 DOI: 10.1051/e3sconf/20172305004

impurities. The use of a secondary fluid also ensures superheating at the end of the expansion cycle, this helps avoid condensation and precipitation on the turbine and extend its lifetime [10].

The heat exchanger for the binary cycle is usually designed in the way that does not allow primary and secondary fluid to interact. Once the secondary fluid is superheated, it is passed to the turbine to generate electricity. After that the secondary fluid is cooled in the condenser and passed back to the heat exchanger. In the second part of the project (another ten years), two extra horizontal producer wells (wells **1** and **2**), CO_2 injection (location **3**) and cold water re-injection wells (location **4**) are drilled (or extended from the shallower wells). At this stage the formation pressure is reduced to a half of its initial value and is sufficiently depleted in its central portion. Now, the central part of the reservoir can be used for CO_2 sequestration and dynamic plume control while still maintaining commercial volumes of hot geofluid production from the sides of the reservoir. Now, produced and cooled water from all producer wells is pumped back through re-injection well to push the plume away from the caprock and toward the side of the reservoir.

To model behavior of this complex geopressured system on the subsurface level, we chose CMG STARS simulator because its EOS can handle extremely high pressures usually observed in deep saline aquifers [11]. EOS of CMG STARS can simulate fluid flow for pressures above 80 MPa. In addition to the active production phase of the project, we also modeled quiescent period after the heat extraction to monitor the plume behavior.

Results

After ten years of hot water production at 1200 m^3/day from the production wells (or quarter of the reservoir), the reservoir pressure is sufficiently lowered to start injection of CO_2 and cold water. **Fig. 3** shows three snapshot of the reservoir throughout the second phase of the project. The production wells are producing at 2400 m^3/day each, cold water re-injection wells pumps 2400 m^3/day back into the formation, and CO_2 injection wells injects 80,000 m^3/day of the supercritical gas. Note that the entire reservoir cools below 140°C at the end of the second phase of the project. We can also observe that high cold water injection rate creates a large front that engulfs the central portion of the reservoir with the stacked wells.

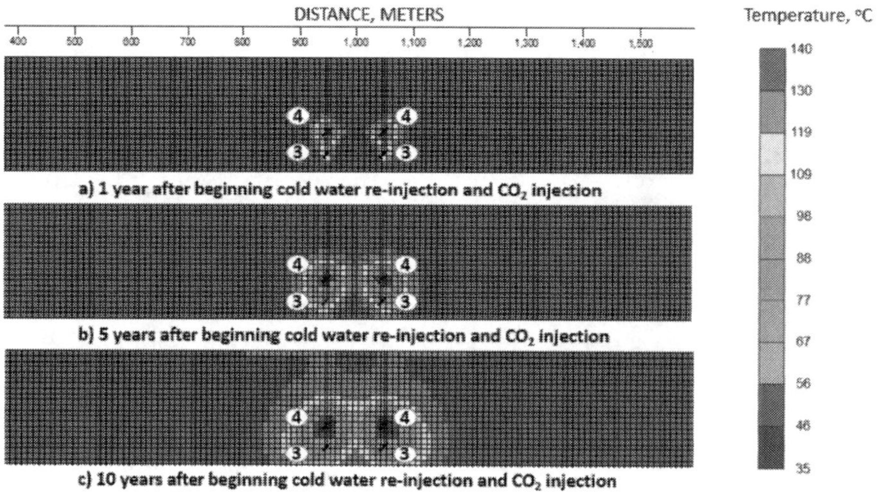

Figure 3. Temperature distribution inside the reservoir (sides with wells 1 and 2 are truncated to focus on the central part) after a) 1 year, b) 5 years, and c) 10 years of cold water re-injection and CO_2 injection. Locations 1 and 2 correspond to the production wells, location 3 is the CO_2 injection wells, and location 4 is cold water re-injection wells.

These simulation results confirm that the cold water front is large enough to push the gas plume down. **Fig. 4** presents snapshots of the gas saturation distribution during the second phase of the project corresponding to same time snapshot in **Fig. 3**. The juxtaposition of the two figures clearly illustrates the success of the dynamic control idea. Only toward the end of the project two small tails of low concentration gas plume reach the top of the reservoir. Moreover, the concentration of gas everywhere in the reservoir is less than 15% at the end of the operations. We can claim that dynamic control injection wells can be an effective tool to avoid well-defined gas plume and keep it away from the caprock.

DOI: 10.1051/e3sconf/20172305004

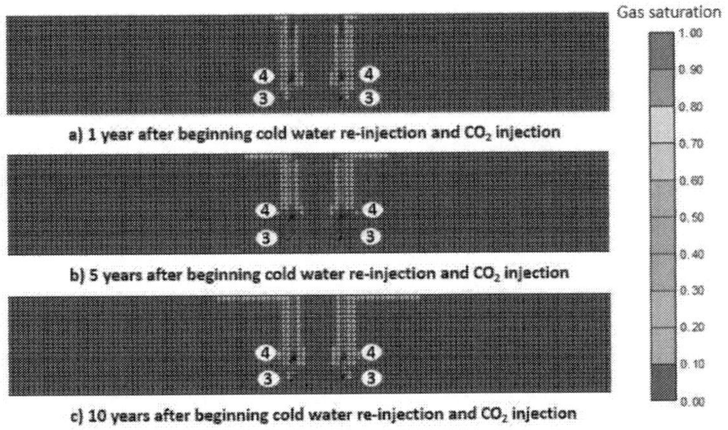

Figure 4. Distribution of CO_2 gas saturation inside the reservoir (sides beyond wells 1 and 2 are truncated to focus on the central part) after a) 1 year, b) 5 years, and c) 10 years of cold water re-injection and CO_2 injection.

It is widely known that projects involving CO_2 injection do not usually end with the end of injection operations. Such projects require extended monitoring of fluid movement and concentration inside the reservoir. **Fig. 5** offers the results of monitoring activities for 180 years after completion of the project. Note that even though the shape of the distorted gas plume does not change significantly over time, the concentration in the final snapshot (**Fig. 5**) is less than 1%.

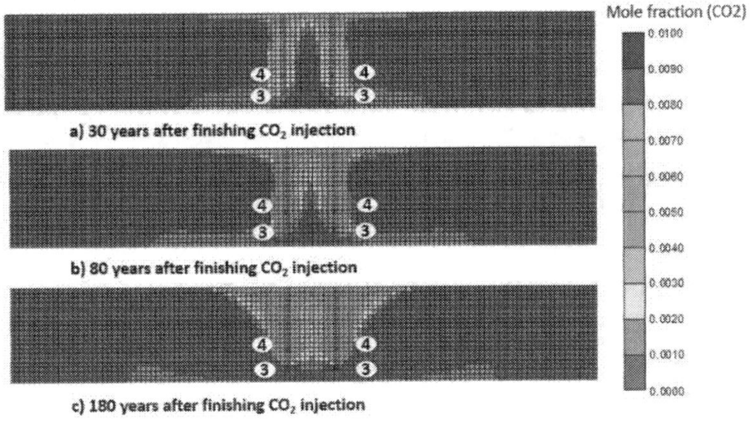

Figure 5. Monitoring concentration of CO_2 gas inside the reservoir after a) 30 years, b) 80 years, and c) 180 years after finishing cold water re-injection and CO_2 injection.

DOI: 10.1051/e3sconf/20172305004

The simulation results above reflect the subsurface aspect of our investigation. However, we posed a broader scope for this study and included the analysis of commercial potential of our geopressured geothermal project. To evaluate economic aspect of the project, we decided to use an elaborate model NREL CREST for cost analysis of renewable energy projects [7]. This model requires some essential input information that come both from our simulation results and literature survey of similar pilot projects. Note that in this model we accounted for tax credit for our CO_2 sequestration activities by adding US$40,000,000 as an additional federal grant [12]. Based on this US policy, the project can qualify for $20 per metric ton of CO_2 sequestered. Our simulation results show that we comfortably injected about 1 million metric tons of CO_2 during 10 years in the entire reservoir (the results above including rates are given for a quarter of the reservoir). Thus, US$40,000,000 can be added to the economic model.

Hot water production rates from the entire reservoir can maintain power generating capacities of about 30 MW. This value can also be achieved from scaling the pilot project at Pleasant Bayou which ran 10 MW capacity on rate of 1200 m^3/day of geofluid of comparable thermodynamic properties. After compiling a detailed cash flow model for 20 years of the project, we obtained internal rate of return after tax (IRR) is 15% in the last year which is good value for a type of project that has always been considered sub-commercial.

The most conservative estimates of the project performance are demonstrated in **Fig. 6** with cross-plot of expenses vs. revenue based on the NREL CREST economic analysis. Note that for the analyzed project based on the simulation model expenses never exceed revenue, which is a good indicator of commercial attractiveness of the project. These results are conservative, because we used very stringent tax regime and terms in NREL CREST model.

 DOI: 10.1051/e3sconf/20172305004

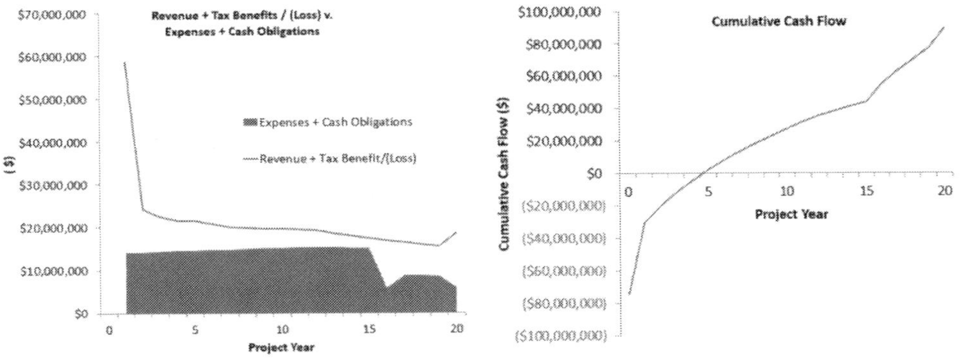

(a) **(b)**

Figure 6. Plot with (a) yearly revenue vs. expenses generated in NREL CREST geothermal model and (b) cumulative cash flow generated in NREL CREST geothermal model.

Discussion

The results of the geothermal reservoir simulation coupled with previous field pilots and the output from the economic model have demonstrated an encouraging prospect: it is possible to develop a geopressured geothermal reservoir economically and simultaneously achieve several other goals. One of such goals is secure and permanent sequestration of considerable amounts of CO_2 gas. Our simulations show that dynamic control injection wells can effectively disperse the plume and keep it away from the caprock to prevent leakage and/or fracturing. Although in our study we successfully pumped and monitored excess of 10 million tons of the gas, the formation pressure and our dynamic control strategy allows to increase this amount even further to exceed those of currently commercial CO_2 sequestration sites (4 to 10 million tons or 1 million tons per year). These results can be achieved even with the reservoir of choice for this study (bulk rock volume of 0.8 x 10^9 m^3). Higher gas injection volumes might also add extra leverage in the economic part of the project and help cover some of capital expenditures.

However, the biggest consideration in increasing volumes of CO_2 injection is the reservoir thickness and permeability. Thicker and tighter formations (but not necessarily less porous) might be good candidates for storing higher volumes of the supercritical gas. Thus, primary depletion of geopressured reservoirs is a necessary step before sequestration not only because of high injection cost, but also because of rock compaction which causes reduced permeability, higher capillary forces, and ultimately lower relative permeability. This will ensure proper and safe storage of high volumes of CO_2 without the plume rising dangerously close to the caprock.

DOI: 10.1051/e3sconf/20172305004

Conclusions and Future work

In our investigation, we pursued two main objectives: model the subsurface reservoir dynamics of a geopressured geothermal system under production and injection and evaluate commercial potential of such development project. Our simulation study solved two-fold problem of reconciling hot geofluid production with cold water re-injection and secure sequestration of substantial volumes of CO_2.

This study opens multiple possibilities for future investigations. Among further research directions are:

1. Automated optimization strategies for well placement controlled by geology and economic parameters (perhaps, an elaborate revenue function that includes elements of NREL CREST cash flow model);
2. Multi-objective optimization for sequestration projects with or without geothermal heat harvesting. This approach will help weight different and sometimes competing objectives (higher hot water volume and higher volume of injected CO_2 could be competing if the latter floods producing wells);
3. Incorporation of salinity effects, assessing its impact on density-driven convection inside the reservoir and potential for CO_2 sequestration (higher salinity leads to poorer sequestration potential).

References

[1] DOE. 1981. Wells of Opportunity Program. Final contract report. Testing geopressured geothermal reservoirs in existing wells. https://www.osti.gov/scitech/servlets/purl/5348034-YYECJY/

[2] MIT. 2006. The Future of Geothermal Energy: Impact of Enhanced Geothermal Systems (EGS) on the United States in the 21st Century. Technical Report, Renewable Energy and Power Department, Idaho National Laboratory. Accessed on February 18, 2016. https://www1.eere.energy.gov/geothermal/pdfs/future_geo_energy.pdf.

[3] Riney, T. 1991. Pleasant Bayou Geopressured-Geothermal Reservoir Analysis. *Journal of Energy Resources and Technology* **114** (4): 315-322.

[4] Plaksina, T. and White, C. 2016. Modeling Coupled Convection and Carbon Dioxide Injection for Improved Heat Harvesting in Geopressured Geothermal Reservoirs. *Geothermal Energy* **4**(1), 1-19 DOI: 10.1186/s40517-016-0044-x.

[5] Garapati, N., Randolph, J., and Saar, M. 2015. Brine Displacement by CO_2 Energy Extraction Rates, and Lifespan of a CO_2-limited CO_2-plume Geothermal (CPG) System with a Horizontal Production Well. *Geothermics* **55**: 182-194.

[6] Anchliya, A. 2009. Aquifer Management for CO_2 Sequestration. Master's Thesis. Texas A&M University, College Station, TX, USA.

[7] Gifford, J., and Grace, C. 2013. CREST Cost of Renewable Energy Spreadsheet Tool: A Model for Developing Cost-Based Incentives in the United States. User Manual, Version 4. NREL/SR-6A20-50374. Accessed on Feb 27, 2016 at https://financere.nrel.gov/finance/files/crest_user_manual_v-4.pdf.

[8] Bassiouni, Z. 1980. Evaluation of Potential Geopressured Geothermal Test Sites in Southern Louisiana. Progress Report. DOE Contract No. DE-AS05-76ET28465. April 1980.

[9] Gray, T. 2010. Geothermal Resource Assessment of the Guyedan Salt Dome and the Adjacent Southeast Guyedan Field, Vermilion Parish, Louisiana. MS Thesis, Louisiana State University.

[10] Quoilin, S., Van den Broek, M., Declaye, S., Dewallef, P., and Lemort, V. 2013. Techno-Economic Survey of Organic Rankine Cycle (ORC) Systems. *Renewable and Sustainable Energy Reviews* **22**: 168–186.

[11] STARS. 2014. Version 2014. User Manual, Computer Modeling Group, Calgary, Canada.

[12] C2ES. 2012. Recommended Modifications to the 45Q Tax Credit for Carbon Dioxide Sequestration. Accessed on Feb. 28, 2016 at http://www.c2es.org/docUploads/EOR-45Q.pdf.

 DOI: 10.1051/e3sconf/20172306001

STUDY OF THE USE OF A MICRO HYDRO IN KNOCKDOWN CONTAINER COMPLETED WITH A CYLINDRICAL FORM HOUSING OF FRANCIS HYDRAULIC TURBINE TO SUPPORT THE DEVELOPMENT PROGRAM OF ENERGY SELF-SUSTAINABILITY FOR REMOTE AND ISOLATED AREAS IN INDONESIA

Samsul Kamal, Prajitno, Janu Pardadi

Mechanical and Industrial Engineering, Engineering Faculty, Universitas Gadjah Mada

Jl. Grafika No. 2 Kampus UGM, Yogyakarta, INDONESIA, 55281

samsulugm@yahoo.com ; samsul@ugm.ac.id

ABSTRACT

With an intention to reduce the consumption of energy from fossil fuels and the CO_2 emission in relation with the climate change solution, Indonesian Government has targeted that the role of the New and Renewable Energy (NRE) resources reaches at least 23% , or about 27 GW , in 2025 and it is expected to increase to the extent of about 134 GW in 2050 within the scenario of mixed energy supply. Geographically Indonesia has many remote and isolated areas with lack of appropriate infrastructure supports. But in the most of the areas, huge potential of new and renewable energy are available such as hydro energy is about 75 GW, biomass energy is about 32 GW and bio fuel is about 32 GW. The total utilization of the energy from small hydro energy up to this year for example is only about 300 MW. The significant obstacle in optimizing the utilization of small hydro energy in the areas is mainly on the infrastructure conditions and the local manufacture capabilities. Difficulties in mobilization of experts, skill worker , parts and constructions material result in very time consuming and costly for site construction.

In this research a hydro turbine built in knock down container completed with a Francis turbine in cylindrical form housing is proposed and reported its performance on implementation. The hydro in a knock down container concept comes from the idea to manufacture hydro power solutions in a knock down container, readymade to be transported to installation sites. It can be easily manufactured in a quality controlled and cost effective environment, transported and installed in remote areas, to operate and maintain with minimal amount of equipment.

It shows that the implementation of the unit in a remote area has reduced the total site construction time by 1/3 compared to the predicted one with conventional unit. The performance of the Francis turbine with cylindrical form housing has shown comparable with the conventional one which has volute form housing. The superiority was even more pronounced by introducing a deflection insert in its input flow channel. The insertion was capable to create more distributed flow into the runner. Efficiency of the turbine up to 80% was performed in this

DOI: 10.1051/e3sconf/20172306001

research. Local manufacture for turbine usually use many craft work and welding to built a volute form housing for the Francis turbine. The cylindrical form housing here has also proven significantly in reducing the time and price for the turbine manufacturing process.

Keywords : Micro hydro, Francis turbine, container, isolated area.

INTRODUCTION

Indonesia is a large island country with around 17,000 islands spread over almost two million square kilometers, creating many remote areas. Most of the remote areas have minimum condition of utilities such as energy and also lack of adequate infrastructure such as roads . All of those conditions resulting the lack of economic development for most of the areas. In general the remote areas are very low dense population thus creating long distances between each hauses. This situation is also bring difficulties for government to construct a unified system of power generation to supply the need of energy on the scheme of nationwide scale [10]. The condition is now become even more uneasy and more expensive due to the decrease on the main national source of conventional sources of energy such as oil. It is fortunate, however, that the most of the remote areas have many type of renewable energy such as biomass, solar as well as hydro power potencies from small scale up to large scale. It is predicted that the potency of small hydro power in Indonesia is around 761 MW, but the utilization of the small hydro is only about 218 MW or just around 28% and the rate of the utilization process is also said some how to be very slow [9]. This situation does not support the government commitment to reduce GHG emission up to 20% by 2020 [2]. There are many causes can be noticed for this situation for example due to the lack of infrastructure such as access road to the location where the hydro potential is identified, the capability of local workshop as well as local craftsman in supplying some mechanical parts of the turbine and in some civil works activities. All of these are actually problems which are generally encountered in the development of micro hydro power [5].
The hydro in a knock down container concept , as shown in Fig. 1, comes from the idea to manufacture hydro power solutions in a knock down container, readymade to be transported to installation sites. It can be easily manufactured in a quality controlled and cost effective environment, transported and installed in remote areas, to operate and maintain with minimal amount of equipment.

 DOI: 10.1051/e3sconf/20172306001

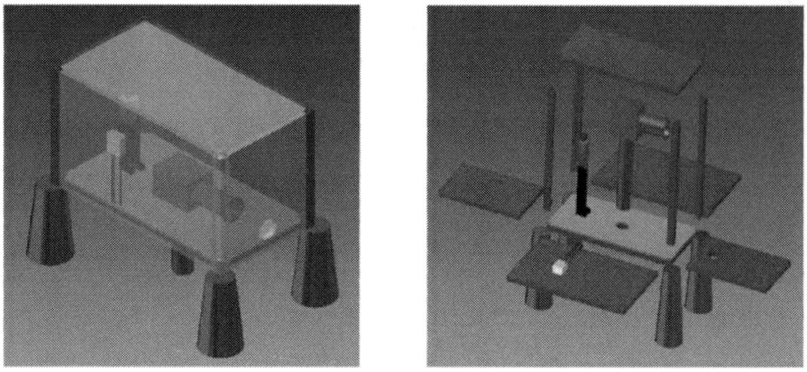

Figure 1. The concept of hydro in a knock down container.

For mechanical parts, there are two important parts in turbine that are runner and casing. For the runner, local foundry industries in developed part area of Indonesia such as in Java already have capability in producing runner turbine for Francis turbine. For the casing , however, the manufacturing process for the casing is still needed to be improved. They only manufacture a scrolled casing shape type which need longer time and work to be involved that cause this to be an expensive manufacturing process [1].

a. b.

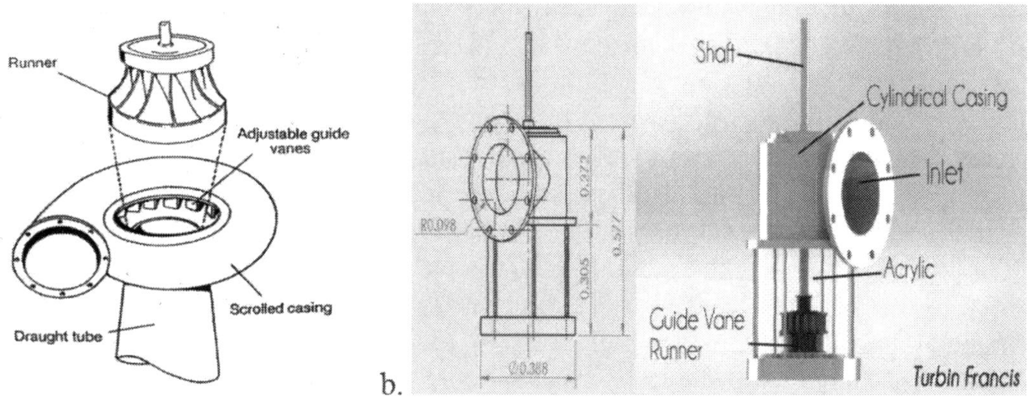

Figure 2. The scrolled (a) and cylindrical (b) casing for Francis turbine

A cylindrical casing shape is proposed here. This casing consists two cylindrical pipes, large and small, joined together by welding. The larger pipe is used as casing where the smaller one is functioned as the inlet port. The inlet pipe port is welded eccentrically to the casing pipe to produce swirl. The construction is shown

DOI: 10.1051/e3sconf/20172306001

in Fig. 2. For the reason of experimental uses, an acrylic glass wall is fitted in the pipe casing to examine flow patterns during the experimental.

MATERIAL AND METHODS

Figure 3. The test facility for the hydro in a container unit.

The micro hydro in knockdown container test facility has been developed in Mechanical and Industrial Engineering, Universitas Gadjah Mada. The test facility system is shown in Fig. 3. It uses an installation of high head water supply tank which is supplied by a pump. The turbine supply pipe (penstock) is constructed with adequate long and straight pipe to facilitate the pressure tank in the system to maintain flow steadiness. The turbine discharge directly into the sump which is completed with baffles to necessary prevent air from being drawn into the pump. Main instruments were installed in the facility , they are a stage marker for head measurement, an orifice plate and notches with weir for flow measurement, a mechanical tachometer for speed measurement and a torque meter and brake for torque measurement. Standard laboratory calibrations were used prior to the installation for the main instruments with acceptable accuracy. Test record sheets have been used where all experimental results are easily tabulated. The performance analysis follow the work of [6,7]. The recorded data consists : Valve setting (%), total brake load (kg), speed (rpm) , output power (W) , head (m) and flow rate (m^3/sec). The test was carried out for the two type of the turbine ; Francis with the scroll casing and Francis with the cylindrical casing. Valve setting in this test was varied from 100% (fully open) to 50% with decrement of 10%. From the

 DOI: 10.1051/e3sconf/20172306001

recorded data , they were then converted into turbine performance parameters for the following quantities to be determined ; output power *Po* (Watts), input power *Pi* (Watts), efficiency η, specific speed *Ns* (rpm . kW $^{1/2}$. m$^{-5/4}$), speed coefficient C_w (rads), flow coefficient C_Q and specific speed coefficient C_{ws} (rads).

Error analysis was performed for variables involved. The most significant errors were found in the flow rate but overall accuracy achieved in this test is about 3%.

A modification to improve the efficiency of the cylindrical casing Francis turbine was also examined by introducing an insertion profile, namely insertion profiles A and B, which was rotationally placed in the inlet port of the turbine to modify the inlet flow to the guide vane as shown in Fig. 7 which was inspired by the work of [8]. A computer simulation was also provided to examine the dynamic performance of the flow in the turbine due to the insertion which follow the work of [3] and [6].

Figure 4. The insertion profile A and B and the inlet port in where the profiles are fitted

The different between the insertion A and B is on the angle of the slope made by the surface of the insertion to the surface of the inlet port which are expected to modify the swirl velocity vector.

DOI: 10.1051/e3sconf/20172306001

PERFORMANCE AND EVALUATION

The results of the measurement and the calculations are plotted into some different correlations which represent the performance indicators, that are efficiency η versus speed coefficient C_w, flow coefficient C_Q versus speed coefficient C_w and efficiency η versus specific speed coefficient C_{ws} .

First, the effect of insertion was examined. Figure 5 shows the effect of the insertion A and B to the bare inlet port (no insertion) for the same valve setting 100% (fully opened) to the turbine performance. Variations of the valve setting were also fully examined (from 100% to 50% with percentages decrement) but the results are not shown here. All the results has the same trend that the A insertion shows the best effect. It seems that the insertion will affect the whirl velocity component of the fluid flows to the guide vane , as expected, and it finally strikes to the runner. This was also noticed by [8]. This phenomena can also be indicated by the simulation results as shown in Figures 6.a and 6.b which compare the flow in the scrolled casing (6.a) and in the cylindrical casing (6.b) of Francis turbine for three different conditions that are no insertion, A insertion and B insertion. The B insertion on cylindrical casing results a resemble stream line flow with the scrolled casing. The vortex flows will smoothly developed around the guide vane producing no disturbances to the flow through the guide vane. This means that performance of the cylindrical casing turbine with B type insertion is comparable with the one with scrolled casing but the first has gain on reducing the complexities of the manufacturing process.

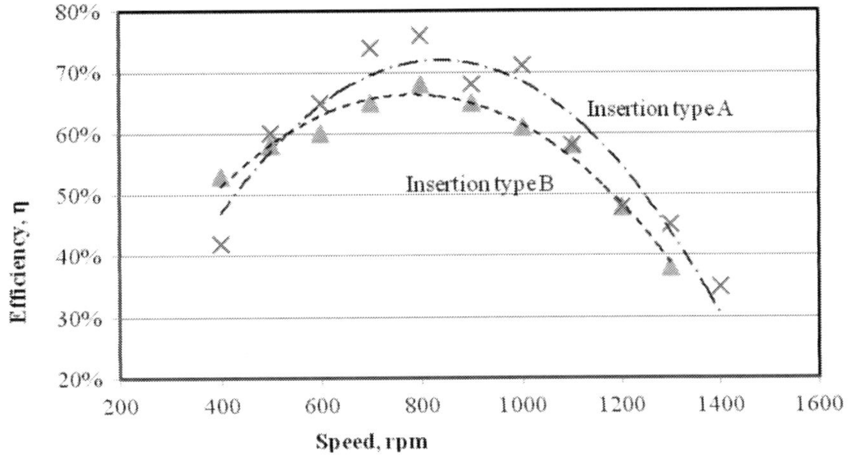

Figure 5. The effect of insertion type on the performance of the turbine

 DOI: 10.1051/e3sconf/20172306001

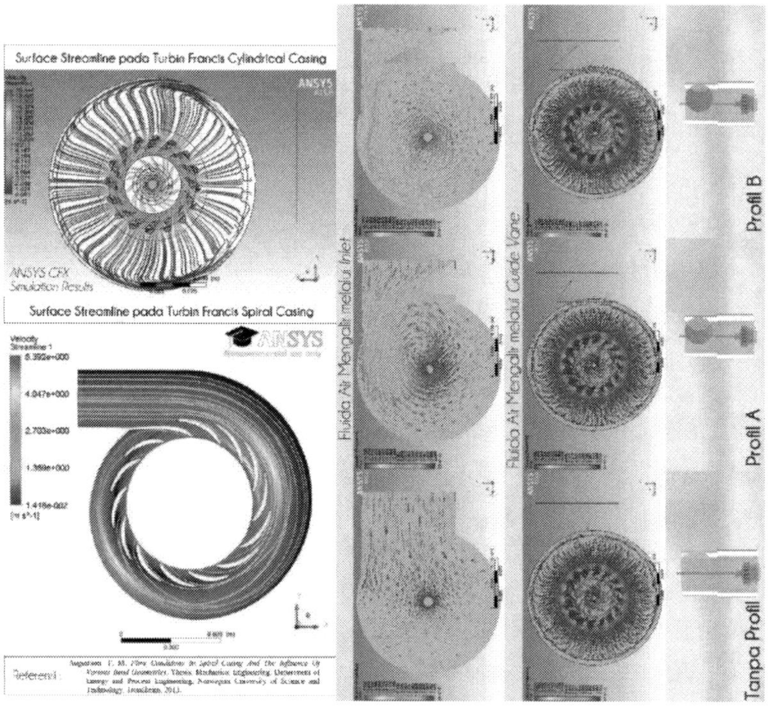

a b

Figure 6. a. Surface streamline configuration of water flow inside the cylindrical casing and the scrolled (spiral) casing.
b. The effect of profile insertion on the inlet port to modify the flow distribution in the guide vanes in the cylindrical casing of Francis turbine.

Figure 7.a and Figure 7.b are used to explore further the performance of the cylindrical casing turbine with B type insertion , relating the efficiency η and the specific speed, C_{ws} , for variation of valve settings 50%, 70% , 90% and 100%. The best efficiency for this turbine is occurred at the valve setting of 100% and it goes up about 75%.

DOI: 10.1051/e3sconf/20172306001

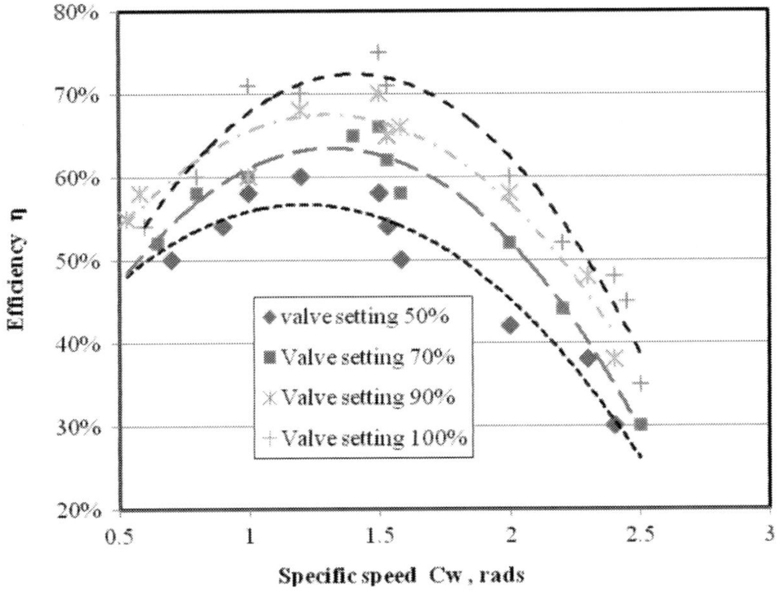

a

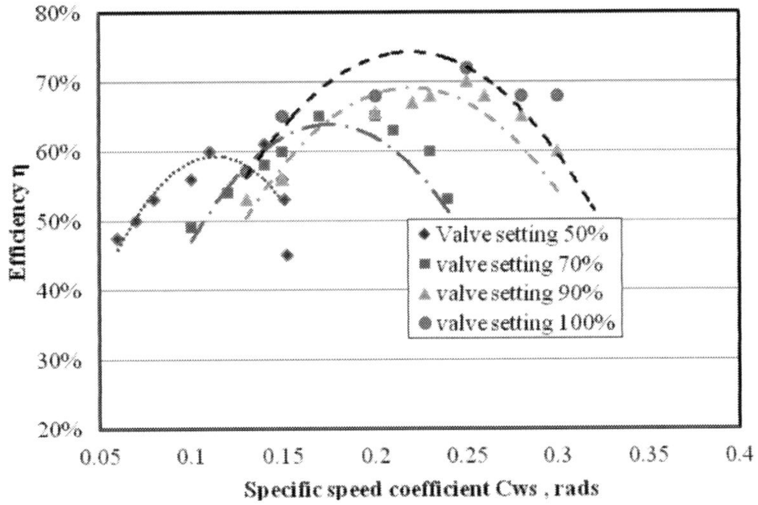

b

Figure 7. a. The performance of the turbine plotted against the specific speed with insertion type A.

b. The performance of the turbine plotted against the specific speed coefficient with insertion type A.

 DOI: 10.1051/e3sconf/20172306001

The performance of the cylindrical casing turbine here is dictated significantly by the design of insertion. The best type of the insertion , that is A type, performed very well for the variation of the valve setting.

Time for site construction was investigated by installing the knock down container unit in a place near by an existing micro hydro of a similar size unit which was already built by using conventional civil work method. The existing , greatly, has an adequate data of the construction. Finally the man hour used for the site construction of the two are then compared . Result shows that the total man hour can be reduced up to about 1/3. This result may also be used to predict the man hour needed for other location when the micro hydro in knockdown container is used although it must be also considered as well the level of the infrastructure difficulties in every specific location.

CONCLUSION

1. With a proper and specific design of the insertion, the performance of a cylindrical casing turbine could be more better than that of a scrolled casing turbine.
2. Cylindrical casing turbine could reduce the time and the difficulties in manufacturing process compared to the conventional type of casing that is scrolled type casing.
3. Micro hydro in a knockdown container completed with cylindrical casing turbine can significantly reduced the overall time for construction especially in an isolated area with lack of supporting adequate infrastructure.

DOI: 10.1051/e3sconf/20172306001

REFERENCES

[1] Adam Harvey, Andy Brown, Priyantha Hettiarachi and Allen Inversin (1993), Micro Hydro Design Manual, *Intermediate Technology Publications, London.*

[2] DEN , National Energy Council, (2015), National Energy Management, *Jakarta Indonesia* April 2015.

[3] Gohil, P. P., & Saini, R. P. (2014). *CFD:* Numerical Analysis and Performance Prediction in Fancis Turbine. *IEEE,* 94-97.

[4] Jain, S., Saini, R. P., & Kumar, A. (2010). CFD Approach for Prediction of Efficiency of Francis. *IGHM.*

[5] Kaunda, C. S., Kimambo, C. Z., & Nielsen, T. K. (2014). A Technical Discussion on Microhydro Power Technology and Its Turbines. *Renewable and Sustainable Energy Reviews*, 445-459.

[6] Khare, R., Prasad, V., & Kumar, S. (2010). CFD Approach for Flow Characteristics of Hydraulic Francis Turbine. *International Journal of Engineering Science and Technology* Vol. 2(8), 3824-3831.

[7] Kumar, M., Kumar, P., & Sinha, P. K. (2013). Performance Analysis For Francis Turbine – A Case Study. *International Journal of Emerging in Engineering and Development*, 377-383.

[8] Wei, Q., & Choi, Y. D. (2013). The Influence of Guide Vane Opening on The Internal Flow of Francis Turbine. *Journal of the Korean Society of Marine Engineering,* Vol. 37, No. 3, 274-281.

[9] Panigoro, A., (2013), Beyond 2014 Indonesia Jawara Energi Nabati Dunia Optimalisasi Biodiesel sebagai Solusi Krisis Energi Nasional, *Private communication.*

[10] Samsul Kamal, Prayitno, (2009), Pengembangan Gorlov Helical Turbine Untuk Pemanfaatan Arus Air Pada Pembangkit Listrik Terapung dengan Apungan Berbentuk Nosel. *Research Report* , sponsored by HKPSPN, DIKTI

 DOI: 10.1051/e3sconf/20172307001

Analyzing Long-run Relationship between Energy Consumption and Economic Growth in the Kingdom of Bahrain

Hanan Naser *

Assistant Professor, Faculty of Business Studies, Arab Open University, Bahrain

Abstract

Since the relation between energy consumption and economic growth is important to design effective energy policies that will promote economic growth, this study investigates the short run dynamics and causality among energy consumption, co2 emissions, oil prices and economic growth in Kingdom of Bahrain. To do so, annual data that covers the period from 1960 till 2015. Empirical work tests for unit root, co-integration relationship using Johansen (1988) approach and then estimate both long and short run dynamics using the vector error correction model (VECM). Results indicate that there is a long-run relationship between the suggested variables. Since economic growth has a predictive power to estimate the energy demand of Kingdom of Bahrain, it is recommended that the government of Bahrain and policy designers shed the light on energy efficiency strategies and carbon emissions reduction policy in the long run without impeding economic growth in order to move towards sustainability.

Key Words: emissions; energy consumption; economic growth; Bahrain

*Naser: Assistant Professor, Faculty of Business Studies, Arab Open University, Bahrain; *Email* : *hanan.naser@aou.org.bh*.

1 Introduction

Energy is not only essential for economic development, it is indeed fuels the economy from resource and material extraction to the technologies producing electricity. Energy carriers and end-use equipment are also needed to deliver the desired energy services. Accordingly, the amount of energy demand has witnessed a considerable boom in last ten years. The threat of global warming problem which is foremost a problem of excessive build-up of carbon dioxide (CO2) emissions in the atmosphere due to the burning of fossil fuels has raised a global concerns about energy consumption. A key central question is: how much energy do economies need to function smoothly and thus being able to augment social development and well-being.

Looking at BP annual report 2015, the GCC member countries are major oil and natural gas producers, holding about 40% and 24% of worldŠs proven oil and gas reserves, respectively (BP 2015). The GCC collectively produced higher CO2 emissions than EU-25 and Organization for Economic Co-Operation and Development (OECD) countries on both a per capita and per GDP basis Patlitzianas et al. (2006). As such, the general thought among international analysts and observers is that there is considerable scope for the GCC group to contribute to global emission reductions via energy efficiency improvement and/or development of renewable energies. All GCC member states had ratified the Kyoto Protocol by 2006. Being Non-Annex I parties, they are committed to reduce GHG emissions without any binding reduction targets. However, the GCC as a cohesive trade group has always acted concertedly in climate change negotiations and policies. The action of a single member country may derail or delay the decision-making process (Reiche, 2010). To date there has been lack of consistent and coherent policies and strategies across the GCC member states with respect to energy conservation, development of renewable energies and emission reductions. While no GCC country to date has a consistent national policy to deal with energy efficiency and renewable energy development, all GCC member states have created administrative capacities to address climate change issues (Reiche, 2010). Given increasing global concerns on climate change, it can be expected that the GCC member states will inevitably embark on promoting energy efficiency, emis-

DOI: 10.1051/e3sconf/20172307001

sion reductions and renewable energies in the near future. The vital question is this: Is it possible for GCC countries to realize a sustained economic growth in the long run without accumulation of pollution? In particular, if economic growth induces CO2 emission then growth policies may lead to greater degradation of the environment. However, depending on the nature of the long run relationship between CO2 emissions, economic growth and energy consumption, countries may resort to different policy options in contributing to the fight against global warming (Soytas and Sari, 2009).

Bahrain is a small GCC country which highly dependent on both oil and gas for boosting economic growth. However, acknowledging Bahrain's aspiration to limit the rise in CO2 emissions has mainly motivated this study. The problem of increasing energy efficiency in Bahrain by reducing energy consumption might slow down the economic growth where it depend on fossil fuels, in particular oil. Thus, analyzing the relationship between the economic growth as presented by gross domestic product (GDP) and energy consumption is very important in setting energy policies in Bahrain. Therefore, this paper attempts to shed some light on energy consumption, carbon emissions and economic growth relation in the case of Bahrain taking into account the fluctuation in oil price. To our knowledge the choice of Bahrain as a case study of the GCC group has not yet implemented using advanced econometric techniques that link energy consumption, CO2 emissions and economic growth taking into account the historical fluctuations in oil price.

In what follows, we first provide the material and methods used in this paper in Section 2. Section 3 describes the data and empirical methodology, where empirical results and discussion are represented in Section 4. Finally a conclusion and policy implications are provided in Section 5.

2 Background

2.1 Brief Literature Review

The literature concerning the relationship between energy consumption and economic growth has led to the emergence of a number of empirical investigations. A major strand has scrutinizes the linkage between energy consumption and output, suggesting that energy consumption and output may be jointly determined and the direction of causality between these two variables needs to be tested. Following the seminal work of Kraft and Kraft (1978), several others including Masih and Masih (1997), Yang (2000), Wolde-Rufael (2009), Apergis and Payne (2009) and Ozturk et al. (2010) have tested the energy consumption and economic growth nexus with a variety of techniques and for different panel of countries. Looking at the region of Gulf Cooporation Council (GCC), Al-Iriani (2006) investigates the causality relationship between gross domestic product (GDP) and energy consumption in the six countries of the Gulf Cooperation Council (GCC). Recently developed panel cointegration and causality techniques are used to uncover the direction of energyŨGDP causality in the GCC. Empirical results indicate a unidirectional causality running from GDP to energy consumption. Evidence shows no support for the hypothesis that energy consumption is the source of GDP growth in the GCC countries. Such results suggest that energy conservation policies may be adopted without much concern about their adverse effects on the growth of GCC economies. In the same context context, Hamdi et al. (2014) explores the relationship between electricity consumption, foreign direct investment, capital and economic growth in the case of the Kingdom of Bahrain. The Cobb-Douglas production is used over the period of 1980 - 2010. Using autoregressive distributed lag (ARDL), a cointegration relationship has been detected among the series. It is found that electricity consumption, foreign direct investment and capital add in economic growth. However, empirical works do not provide any precise answer, and there is still no consensus among economists whether there is a causal relationship or not and if it exists, there is no clear-cut answer about the direction of this causation (Ozturk, 2010). The contradictory results may occur due to the differences in data sets, characteristics of the investigated countries, variables that are included in the studies, and the diversification in using econometric methodologies (Ozturk, 2010).

Another stream of research has emerged, examine the dynamic relationship between carbon emissions, energy consumption and economic growth. Some of the recent studies using this approach are as follows: Akbostancı et al. (2009), Soytas and Sari (2009), Oz-

 DOI: 10.1051/e3sconf/20172307001

turk and Acaravci (2010), Apergis and Payne (2010) and Al-Iriani (2006). Arouri et al. (2012) investigate the relationship between carbon dioxide emissions, energy consumption, and real GDP for 12 Middle East and North African Countries (MENA) over the period 1981 - 2005. Results show that in the long-run energy consumption has a positive significant impact on CO2 emissions. More interestingly, we show that real GDP exhibits a quadratic relationship with CO2 emissions for the region as a whole. However, although the estimated long-run coefficients of income and its square satisfy the EKC hypothesis in most studied countries, the turning points are very low in some cases and very high in other cases, hence providing poor evidence in support of the EKC hypothesis. Thus, our findings suggest that not all MENA countries need to sacrifice economic growth to decrease their emission levels as they may achieve CO2 emissions reduction via energy conservation without negative long-run effects on economic growth.

Many researchers argue that if the estimated model does not account for other possible determinants such as that of energy prices, then results may be biased. For example, Glasure (2002) indicates that the real oil price is a major determinant of real national income and energy consumption. Hence, literature has included oil prices in many studies as an additional explanatory variable in energy growth models in order to improve the accuracy of the results. Yang (2000) proposes that the pattern of energy consumption may also influence the efficiency of the model as it differs over time. Looking at the aggregated energy consumption would accordingly ignore the change in the importance of different energy sources through time. Therefore, the performed studies end up with different results over different periods of time even for the same country.

2.2 Overview on energy sector in Bahrain

Although Kingdom of Bahrain is a small country within the GCC group, it appears to be an interesting case study due to the clear increase in it's CO2 emissions which have grown exponentially over the last 30 years. The levels of CO2 emission and fossil fuel consumption over the period 1980 - 2007 increased by 172% and 212%, respectively (calculated based on WDI, 2011). Currently, per capita energy consumption among Bahrainies is one

DOI: 10.1051/e3sconf/20172307001

of the highest in the world - it uses two folds more energy per capita than Japan. In 2007, an average Bahraini consumed 9456 kg oil equivalent of energy. By comparison, Japan and the US have per capita energy uses of 4033 and 7747 kg, respectively. Figure (1) shows per capita energy consumption in Bahrain as compared with neighboring countries (Oman and UAE) and selected developed countries. Such a high consumption rate of energy in Bahrain may have contributed to a high per capita emission level.

Looking at Figure (2), it is clear that energy sector still represents the biggest contributor to the kingdom's GDP, although its share of the economy has been falling as non-oil sectors continue to drive growth. The kingdom's financial sector represents the second-largest contributor to GDP, accounting for 16.5% in 2014, with Bahrain recognised as a pioneer in Islamic finance, having been the first country in the world to introduce and implement rules specific to Islamic banking in 2001. Manufacturing is the third-largest GDP contributor, at 14.4% of the total in 2014, with the kingdom home to one of the world's largest aluminuim smelters. Meanwhile, Bahrain continues to invest in considerable infrastructure upgrades, and these are expected to enhance the kingdom's logistics offerings, as well as help facilitate greater tourism numbers. For its part, tourism has been identified as an area with significant potential for growth, with the Supreme Council for Tourism created to help guide and develop the sector.

Although the global drop in crude oil prices is a concern, the early adoption of economic diversification efforts has created positive growth trajectories in non-oil sectors in Bahrain, such as construction and transport. Hydrocarbons continue to form the bulk of government revenues in Bahrain, with oil and gas accounting for 24.1% of Bahrain's GDP in 2014. Neighbouring Saudi Arabia remains a vital energy partner for the kingdom, and recent Saudi investments in its Eastern Province have benefitted both countries in terms of hydrocarbons processing. Meanwhile in June 2014 Bahrain opened its first solar energy facility in the town of Awali. The 5-MW facility relies on the concept of a smart grid that harvests energy through photo-votaic panels dispersed around the city.

The most important measure in the energy balance of Bahrain is the total consumption of 11.69 billion kWh per year, where it is with an average of 8,681 kWh per capita.

 DOI: 10.1051/e3sconf/20172307001

Bahrain could provide itself completely with self-produced energy. However, the local demand of energy has been increased recently due to many factors such as population growth and increase in electricity demand as summarized in Table (1). The total production of all energy producing facilities is 13 bn kWh. Thats 113% of the countries own usage. Despite this Bahrain trades its energy with foreign countries. Along with pure consumptions the production, imports and exports play an important role. Total per capita energy consumption is 10.8 toe (including about 19 400 kWh of electricity) (2014). Consumption has almost doubled since 2000 and reached 14.8 Mtoe in 2014. Looking at energy companies, the gas exploration and exploitation activities in the country are managed by Bahrain National Gas Company (BANAGAS), which is owned by the State (75%), Chevron Bahrain (12.5%) and Boubyan Petrochemical Company K.S.C. (BPC) (K.S.C, 12.5%). In general, energy Consumption per capita is 10.8 toe (including about 19 400 kWh of electricity) (2014). Therefore, government of Bahrain is considering developing renewable energies to diversify its power supply.

3 Data and Econometric Methodology

3.1 Data

The yearly data-set used in this paper cover the period from 1960 to 2015 for Bahrain. The set of variables are time-series variables that include real GDP per capita (GDP), energy consumption per capita (EC), carbon dioxide emission per capita (Co2) and real oil prices (OP). Oil consumption are obtained from BP Statistical Review of World Energy (2011) where OC is measured in thousand barrels daily. Real GDP per capita measured based on purchasing-power-parity (PPP) per capita in constant 2000 international dollars from the World Development Indicators (WDI, 2011). Real West Texas Intermediate (WTI) oil prices are defined as the US dollar prices of oil. Following Lee and Chiu (2011), oil prices are converted to the domestic currencies and then deflated by the domestic consumer price indices (CPIs), which is derived from International Financial Statistics (IFS, 2011) published by the International Monetary Fund (IMF). Descrptive statics are provided in

DOI: 10.1051/e3sconf/20172307001

Table (2).

3.2 Econometric Methodology

Before conducting cointegration analysis, stationary tests are essential for identifying the order of integration of the proposed variables. The most popular tests in literature are the Agmented Dickey and Fuller (1979) (ADF), Phillips and Perron (1988) (PP) and Kwiatkowski et al. (1992) (KPSS). The difference between the ADF, PP and the KPSS tests is that the formulation of the null hypothesis is different.[1]. The ADF (augmented Dickey-Fuller) testing procedure to test the unit root hypothesis following:

$$\Delta y_t = \theta_0 + \gamma_0 t + \gamma_1 y_{t-1} + \sum_{i=0}^{p} \theta_i \Delta y_{t-1} + \varepsilon_t \tag{1}$$

where y_t is the variable in period t; Δy_{t-1} the $y_{t-1} - y_{t-2}$; ε_t the i.i.d. disturbance with mean 0 and variance 1; t the linear time trend and p is the lag order. In order to test the null hypothesis for the presence of a unit root in y_t, we conducted the hypothesis testing that $\gamma_1 = 0$ in Equation (1). If γ_1 is significantly less than zero, the null hypothesis of a unit root is rejected. Phillips-Perron test uses similar models as the ADF tests, but lack of sensitivity to the heteroscedasticity and the autocorrelation of the residuals. Moreover, ADF tests and PP tests may be inefficient on small samples. KPSS stationarity test is more effective for small samples when it chooses a lower lag truncation parameter.

Then, to select the optimal number of lag length, k, Akaike (AIC), Hannan and Quinn (HQIC), and Schwarz's Bayesian (SIC) information criteria are used to build a decision.[2]

Once all variables are integrated of the same order, this study proceed to examine the existence of long-run relationship(s) among series. Precisely, a cointegration methodology that starts with a general approach and move to a more specific is applied to estimate the long-run relationship(s) between the variables included in vector Z_t, where Z_t includes a

[1]ADF and PP tests have non stationary series as the null hypothesis while on the other hand, KPSS assumes that the series to be investigated is stationary. In this study, all these tests are utilized to verify the variables' order of integration, I_d

[2]In the case of conflicting results of the different Information criterion, the choice done based on AIC results as suggested by Pesaran and Pesaran (1997).

 DOI: 10.1051/e3sconf/20172307001

number of integrated series at the same level. To do so, long-run relationships between the variables included in vector Z_t are estimated using Johansen Maximum Likelihood approach. Specifically, one can write Z_t as a vector autoregressive process of order k (i.e., VAR(k)):

$$Z_t = A_0 + \sum_{i=1}^{k} A_i Z_{t-i} + u_t \tag{2}$$

$$\Delta Z_t = A_0 + \Pi Z_{t-1} + \sum_{i=1}^{k} \Gamma_i \Delta Z_{t-i} + u_t \tag{3}$$

Where Z_t denotes (4×1) vector containing GDP, energy consumption, Co2 emissions, and real oil prices ($Z_t = (GDP_t, EC_t, CO2_t, OP_t)$). The four variables are measured by their natural logarithm so that their first difference approximate their growth rates. Any long-run relationship(s) are captured by the (4×4) matrix Π shown in Equation (3). To examine the long-run relationship(s) between the suggested variables, Johansen (1988) test has been established in order to test for the existence of $r \leq 3$ cointegration relationships among the four variables of the model. This is equivalent to testing the hypothesis that the rank of matrix Π in Equation (3) is at most r. Reduced-rank regression can then be used to form a likelihood ratio test of that hypothesis on the basis of the so-called trace statistic, or alternatively, the maximum eigenvalue statistic. Lüutkepohl et al. (2001) investigate the small sample properties of both tests and conclude that the trace testis slightly superior, and therefore, we favor it in our analysis. Thus, the rank of the matrix Π is imposed to estimate the un-restricted model shown in Equation (3). However, this matrix (i.e. Π) can be decomposed as shown below in Equation (4) to provide better understanding for the full system:

$$\Delta Z_t = A_0 + \alpha\beta' Z_{t-1} + \sum_{i=1}^{p} \Gamma_i \Delta Z_{t-i} + u_t, \qquad u_t \text{ is } iid \sim N(0, \Sigma) \tag{4}$$

where the cointegrating vectors are presented in a $(4 \times r)$ matrix named β and the speed of adjustments are shown in a $(4 \times r)$ matrix called α. Γ_i represents (4×4) matrices that guide short run dynamics of the model.

DOI: 10.1051/e3sconf/20172307001

4 Empirical Results and Discussion

It is clear from Table 3 that the null hypothesis of no unit roots for all the time series are rejected at their first differences since the ADF and KPSS test statistic values are less than the critical values at 1% levels of significance. Thus, the variables are stationary and integrated of same order, i.e., I (1). All variables became stationary and do not contain unit root in first difference.

Table 4 reports lag-order selection statistics using Akaike, Hannan and Quinn and Schwarz's Bayesian information criteria (i.e. AIC, HQIC and SBIC), respectively. Since the results are conflicting, we use AIC results as suggested by Pesaran and Pesaran (1997). Accordingly, this empirical work has proceeded using three lags.

Based on Π rank test developed by Johansen (1988), the results of testing for the number of cointegrating vectors are reported in Table (5), which derives two likelihood estimators named Eigen value and trace statistics test. The cointegration rank (r) can be formally tested with the trace and Eigen value statistics taking into accounts the 5% critical value of the test. Thus, results shown in Table 5 reveals that the trace statistic either rejects the null hypothesis of no co-integration among the variables or does not reject the null hypothesis that there is one co-integration relation between the variables. Start by testing a maximum rank of 0, if it is rejected, then repeat for next option where the maximum rank tested for 1. When a test is not rejected, then stop testing there and that value of r is the commonly-used estimate of the number of cointegrating relations. Here, r = 1 is not rejected at the 5% level. In other words, this trace test result does not reject the null hypothesis that these two variables are not cointegrated. The final number of cointegrated vectors with two lags is equal to one, i.e. rank (r) =1. Since, the rank is equal to 1 which is more than zero and less than the number of variables; the series are cointegrating among the variables.

Table (6) presents results of the long-run elasticities. There exists a positive relationship between energy consumption and economic growth in Bahrain, where a one percentage point increases in energy consumption significantly increases economic growth by almost 1.2%. This implies that consuming more energy - which is in fact mainly fossil

DOI: 10.1051/e3sconf/20172307001

fuels at the meantime- stimulates the economic growth significantly in Bahrain. However, it is also revealed that although this consumption stimulates economic growth, it is indeed affecting the level of emissions and contribute to the build up of carbon dioxide emissions. Thus, an increase of 1% level in Co_2 emissions in Bahrain results a significant increase economic growth by 0.49%. It is also worth to note that the relationship between oil prices and economic growth in Bahrain is positive, taking into consideration that Bahrain is in fact an oil exporting country. Looking at the lower part of table (6), the estimated short run model shows that the ECT term is negative and significant as expected in the economic growth model shown in the second column. This reflect that if the economic growth model deviates in the long-run, it could be adjusted and restored by 3.5%. However, the energy consumption model cannot be adjusted if it faces any deviation in the long-run. This may be attributed to the type of energy used in Bahrain, where the only available type used at the meantime is fossil fuels. Results suggest that energy consumption has a predictive power for economic growth in Bahrain. With regards to carbon dioxide emissions, results suggest that there is a unidirectional causality running from CO2 emissions to economic growth, where increases in CO2 emissions lead to increases in economic growth.

5 Conclusion and Policy Implications

Reducing the excess build up of CO_2 emissions has been prompted by world leaders in order to find solutions that minimize global warming problem. Kyoto Protocol represents one of global initiatives that aim to reduce the amount of global warming substances and Bahrain has ratified it and thus, being obliged to reduce it's GHG emissions. However, there has been a lack of clear and coherent policies on climate change and energy conservation within the GCC member states, including Bahrain. This might be due inter alia to concerns over the impact of energy conservation policies on the economic growth of GCC member states. This paper examined the causal relationship between economic growth, pollutant emissions, energy consumption, urban population and capital in Bahrain for the

DOI: 10.1051/e3sconf/20172307001

Acknowledging the intensive energy dependency of Bahrain's economy, one of the important challenges for the country is to reduce this dependency which will accordingly reduces the carbon emissions as required by the Kyoto Protocol. However, depending on the relationship between energy consumption, economic growth and carbon dioxide level in Bahrain, several outcomes are anticipated.

The estimation results show that there are significant relationships between GDP, energy consumption, CO2 emissions and oil prices in Bahrain. One of the significant relationships is that energy consumption causes CO2 emissions, while causality does not run from energy consumption to economic growth in the short-run estimated models. This signifies that while the implementation of energy conservation policies in Bahrain may not affect Bahrain's economic growth significantly, such policies may provide significant scope to reduce the country's CO2 emissions. The latter implication may be due to the large level of emissions attributed to conventional fuels including oil, and more significantly gas, in that country.

The main policy options shall focus on energy conservation strategies, including more efficient use of energy from both the production and user perspectives as well as the development of the renewable energy sector. The long run pursuit of high economic growth given sustained increases in energy efficiency may also reduce CO2 emissions intensity per unit of GDP. It is also important to note here that the energy conservation strategies may not exert any significant impact on the country's GDP, but may have substantial inter sectoral or equity impacts within the economy in which some sectors may gain, while some others may lose. An interesting future study would be to appraise the inter sectoral impacts of energy conservation or CO2 emissions reduction in Bahrain.

DOI: 10.1051/e3sconf/20172307001

Figure 1: Energy Consumption per capita for Bahrain and other selected neighbor countries

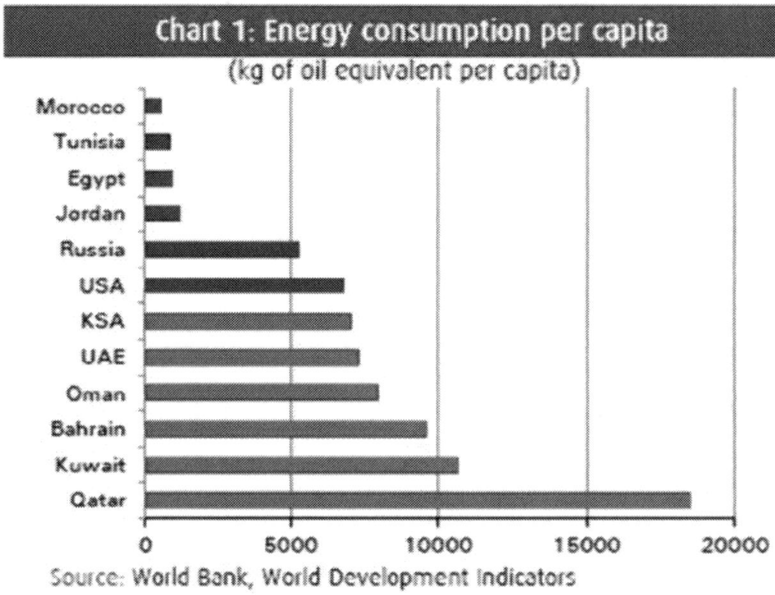

Figure 2: Share of Bahrain's GDP

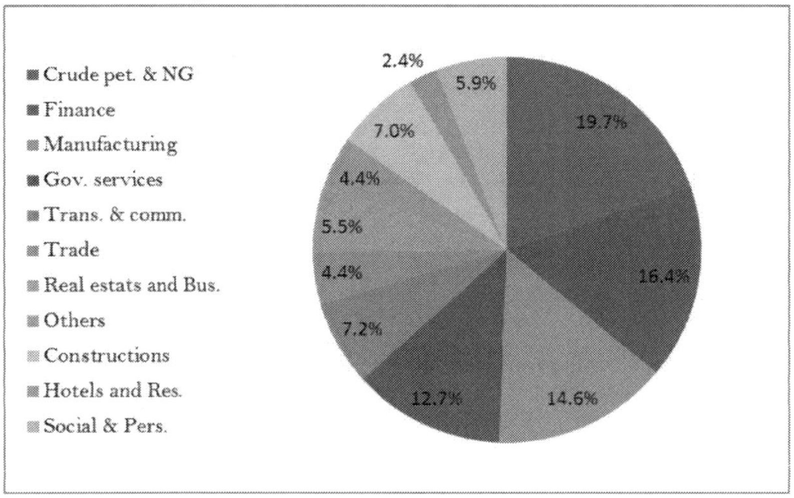

Table 1: Energy Statistics for Bahrain

Variables	1990	2013
Energy production (total million metric tones of oil equivalent)	14.3	22
Energy use (fossil fuels % of total energy use)	100	100
Energy use growth (%)	4.5	

Table 2: **Descriptive Statistics**

Variables	GDP	EC	Co2	OP
mean	9.914	9.08	2.902	1.527497
SD	0.1071	0.244	0.555	0.3154876
Skewness	-0.444	-0.98	-1.719	-0.1252766
Kurtosis	1.638	2.863	4.765	1.777607

Table 3: **Tests of Unit Root**

Unit root test	ADF	lags	PP (4)	KPSS	lags
levels					
GDP	0.79	4	-1.834	0.666***	4
EC	0.519	4	-3.141	0.274*	4
Co2	0.623	4	-3.159	0.235*	4
OP	0.286	4	-1.715	0.677	4
first difference		4			4
GDP	-2.379**	4	-6.147***	0.055	4
EC	-3.080**	4	-9.464***	0.049	4
Co2	-4.758***	4	-9.473***	0.048	4
OP	-2.242**	4	-6.862**	0.051	4

Notes: Table entries are the results obtained from unit root tests. Tests are shown in the first row: augmented Dickey and Fuller (1979) (ADF), Phillips and Perron (1988) (PP), and the stationarity test by Kwiatkowski et al. (1992) (KPSS). Regression include an intercept and trend. The variables are specified in the first column: oil price (OP), oil consumption (OC), nuclear energy consumption (NC) and real GDP (Y). All variables are in natural logarithms, while the lag length determined by Akaike Information Criteria and are in parentheses. '*' and '**' indicate significance at the 10% and 5% level, respectively. The nulls for all test except for the KPSS test are unit root.

Table 4: **lag Selection Criteria**

K	AIC	HQIC	SBIC
1	-5.886	-5.596	-5.533*
2	-6.017	-5.496	-5.128
3	-6.915*	-6.163*	-4.653
4	-6.683	-5.699	-4.945

Notes: AIC, HQIC and SBIC stand for Akaike, Hannan and Quinn and Schwarz's Bayesian information criteria, respectively. In the case of conflicting results, we use AIC results as suggested by Pesaran and Pesaran (1997). '*' indicates significant at 5% level.

 DOI: 10.1051/e3sconf/20172307001

Table 5: **Johansen's Cointegration Test**

max rank	LL	Eigenvalue	Trace statistics	95% c.v
$r = 0$	197.801	-	74.365	54.650
$r \leq 1$	222.007	0.598	25.954*	34.550
$r \leq 2$	230.719	0.280	8.528	18.170
$r \leq 3$	233.762	0.108	2.443	3.740
$r \leq 4$	234.984	0.045	-	-

Notes: The entries of the upper row show the name of the country in the first column, followed by the null hypothesis H_0, that tests for a cointegration rank of r, then H_1 shows the alternative. λ_{max} shown in the fourth column represents the maximum eigenvalue statistics, $Trace*$ shows the trace statics, $95\%c.v$ represents the critical values at 5% level, and finally $p - values$ are provided in the last column. '*', '**', and '***' indicate significance at the 10%, 5% and 1% level, respectively.

Table 6: **VECM Estimation Results**

Cointegration equations	Cointegrating Eq. (1)			
GDP (-1)	1			
EC (-1)	1.16***			
Co2 (-1)	-0.489***			
OP (-1)	0.447***			
Vector Error Correction	**Δ GDP**	**Δ EC**	**Δ Co2**	**Δ OP**
Δ GDP (-1)	0.359**	0.189	1.124*	-0.088
	(0.010)	(0.272)	(0.098)	(0.836)
Δ GDP (-2)	-0.713	0.0523	1.016	-0.047
	(0.603)	(0.831)	(0.131)	(0.912)
Δ EC (-1)	0.0365	-0.208	-0.156	0.097
	(0.611)	(0.105)	(0.657)	(0.644)
Δ EC (-2)	0.0859	-0.207*	0.169**	-0.323
	(0.218)	(0.095)	(0.620)	(0.013)
Δ Co2 (-1)	0.012	-0.067	-0.106**	-0.003
	(0.690)	(0.226)	(0.489)	(0.0497)
Δ Co2 (-2)	0.0604**	-0.0783*	-0.126	-0.052
	(0.019)	(0.087)	(0.319)	(0.514)
Δ OP (-1)	-0.055	-0.109	-0.175	0.0013
	(0.247)	(0.196)	(0.459)	(0.993)
Δ OP (-2)	-0.0543	0.077	-0.032	-0.034
	(0.262)	(0.370)	(0.892)	(0.820)
ECT_{t-1}	-0.0354**	0.189***	-0.007	0.231***
	(0.021)	(0.000)	(0.956)	(0.009)

 DOI: 10.1051/e3sconf/20172307001

References

Akbostancı, E., S. Türüt-Aşık, and G. İ. Tunç (2009). The relationship between income and environment in turkey: Is there an environmental kuznets curve? *Energy Policy 37*(3), 861–867.

Al-Iriani, M. (2006). Energy–GDP relationship revisited: an example from GCC countries using panel causality. *Energy policy 34*(17), 3342–3350.

Apergis, N. and J. Payne (2009). Energy consumption and economic growth: Evidence from the Commonwealth of Independent States. *Energy Economics 31*(5), 641–647.

Apergis, N. and J. Payne (2010). Energy consumption and growth in South America: Evidence from a panel error correction model. *Energy Economics 32*(6), 1421–1426.

Arouri, M. E. H., A. B. Youssef, H. M'henni, and C. Rault (2012). Energy consumption, economic growth and co 2 emissions in middle east and north african countries. *Energy Policy 45*, 342–349.

Dickey, D. and W. Fuller (1979). Distribution of the estimators for autoregressive time series with a unit root. *Journal of the American statistical association 74*(366a), 427–431.

Glasure, Y. (2002). Energy and national income in Korea: Further evidence on the role of omitted variables. *Energy Economics 24*(4), 355–365.

Hamdi, H., R. Sbia, and M. Shahbaz (2014). The nexus between electricity consumption and economic growth in bahrain. *Economic Modelling 38*, 227–237.

Johansen, S. (1988). Statistical analysis of cointegration vectors. *Journal of economic dynamics and control 12*(2), 231–254.

Kraft, J. and A. Kraft (1978). Relationship between energy and GNP. *J. Energy Dev.;(United States) 3*(2).

Kwiatkowski, D., P. Phillips, P. Schmidt, and Y. Shin (1992). Testing the null hypothesis of stationarity against the alternative of a unit root: How sure are we that economic time series have a unit root? *Journal of econometrics 54*(1), 159–178.

Lee, C. and Y. Chiu (2011). Oil prices, nuclear energy consumption, and economic growth: New evidence using a heterogeneous panel analysis. *Energy Policy 39*(4), 2111–2120.

Lüutkepohl, H., P. Saikkonen, and C. Trenkler (2001). Maximum eigenvalue versus trace tests for the cointegrating rank of a VAR process. *The Econometrics Journal 4*(2), 287–310.

Masih, A. and R. Masih (1997). On the temporal causal relationship between energy consumption, real income, and prices: Some new evidence from Asian-energy dependent NICs based on a multivariate cointegration/vector error-correction approach. *Journal of Policy Modeling 19*(4), 417–440.

Ozturk, I. (2010). A literature survey on energy–growth nexus. *Energy Policy 38*(1), 340–349.

Ozturk, I. and A. Acaravci (2010). Co 2 emissions, energy consumption and economic growth in turkey. *Renewable and Sustainable Energy Reviews 14*(9), 3220–3225.

Ozturk, I., A. Aslan, and H. Kalyoncu (2010). Energy consumption and economic growth relationship: Evidence from panel data for low and middle income countries. *Energy Policy 38*(8), 4422–4428.

DOI: 10.1051/e3sconf/20172307001

Patlitzianas, K. D., H. Doukas, and J. Psarras (2006). Enhancing renewable energy in the arab states of the gulf: Constraints & efforts. *Energy Policy 34*(18), 3719–3726.

Pesaran, M. and B. Pesaran (1997). *Working with Microfit 4.0: Interactive econometric analysis.* Oxford University Press.

Phillips, P. and P. Perron (1988). Testing for a unit root in time series regression. *Biometrika 75*(2), 335–346.

Reiche, D. (2010). Renewable energy policies in the gulf countries: A case study of the carbon-neutral Şmasdar cityŤ in abu dhabi. *Energy Policy 38*(1), 378–382.

Soytas, U. and R. Sari (2009). Energy consumption, economic growth, and carbon emissions: challenges faced by an eu candidate member. *Ecological economics 68*(6), 1667–1675.

Wolde-Rufael, Y. (2009). Energy consumption and economic growth: The experience of african countries revisited. *Energy Economics 31*(2), 217–224.

Yang, H.-Y. (2000). A note on the causal relationship between energy and GDP in Taiwan. *Energy Economics 22*(3), 309–317.

Energy Consumption Trends in Energy Scarce and Rich Countries: Comparative Study for Pakistan and Saudi Arabia

Uneb Gazder[a]
[a]Department of Civil Engineering, University of Bahrain, Bahrain. (Email: unebgazdar@gmail.com)

Abstract
Energy crisis is raising serious concerns throughout the world. There has been constant rise in energy consumption corresponding to the increase in global population. This sector affects the other pillars of national economy including industries and transportation. Because of these reasons, the traditional fossil-based energy sources are depleting rapidly, resulting in high and unstable energy prices. Saudi Arabia and Pakistan, although different from each other in terms of their economic stability and political systems, still rely heavily on the traditional fossil fuels. This paper presents the comparison of these two countries in terms of their energy consumption and factors affecting it. These factors include, but not limited to, economic development, and growth in population and other sectors such as; industries, transportation, etc. The comparison is also made with the regional and global energy consumption trends and these countries. Moreover, regression models were built to predict energy consumption till 2040 and compare the growth in this sector and share in global energy demand. Energy consumption in oil-rich countries (Saudi Arabia) has been driven through its economic development, while for energy insecure country (Pakistan) it is mainly because of population growth. It was also found that in the next two decades the share of Pakistan in the global energy demand will increase. This concludes that population growth will have more impact on energy consumption than economic growth. It could mean that the shift in energy sector would shift towards sustenance instead of using energy for commercial or industrial usage.

Keywords: Energy consumption; Saudi Arabia, Pakistan, Statistical comparison
Conference Track: Policy and Finance and Strategies

1. Introduction

Energy refers to "the ability to do work. It can be found in many forms including sound, vibration, electrical, mechanical, thermal, chemical, etc. The sources for production of energy can be categorized in three major categories, namely; fossil, renewable and fissile. Fossil sources include coal, petroleum and natural gas, while renewable sources include hydro, wind and solar. Lastly, fissile sources are generated through chemical processes with radioactive elements such as uranium and thorium [1]. Industrial sector has been found to be the largest consumption of energy, followed by household and transportation sector [2, 3].

Until now, the reliance for energy source has been mainly on fossil fuels with marginal investment on renewable energy sources. This trend has given rise to the crisis in this sector. It had resulted in severe impacts on economies of developing countries which are now facing issues of energy insecurity. Approximately 40% of the world energy consumption is in developing countries [4]. Economic development, urbanization and population growth have been found as the major contributors to the energy consumption sector [5, 6]. These factors have been escalating in developing countries, consequently, giving rise to energy consumption.

In spite of the current energy crisis, developing countries like Pakistan and Saudi Arabia continue to focus resources on exploration of fossil based energy sources. The differences between these countries are the facts that these countries have different scales of economies and

DOI: 10.1051/e3sconf/20172307002

indigenous resources have some impact. The reasons for selecting these countries was to investigate if there can be a common parameter affecting energy demand in these vastly different developing countries. In this paper, we have tried to study the trends of these countries in terms of factors affecting energy consumption. Moreover, comparison has also been made between them and global and regional trends for energy consumption. It is expected that the outcomes of this research will help to understand the scenario of energy consumption in developing countries as these countries represent the extremes of the developing world.

2. Methodology

The research was initiated with the study of global trends in the energy sector with reference to total energy consumption and alternative sources. It was followed by observing the trends of total energy consumption in Pakistan and Saudi Arabia. At this stage, the factors which have been found in the previous literature to affect the energy consumption were tested for their correlation in the scenario of these countries.

Regression models were also developed for predicting total energy consumption for both countries. These models were used to predict the energy parameters for future and compare the results with globally predicted energy growth rate. The data for this research was obtained from the online resources of World Bank (http://data.worldbank.org/data-catalog/world-development-indicators) for the years 1960 onwards. The factors affecting energy consumption were predicted using average annual growth rate estimated using the time series data. These predicted factors were used in the regression model for predicting energy consumption trend.

3. World Energy Consumption Trend

Bilgen [7] presented a comprehensive review of energy consumption trends for different countries and the global scenario. The trend and forecasted value for energy consumption in global scenario are given in Figure 1. It can be observed from this figure that the energy consumption trends in the regions of interest (Asia and Middle East) are same as that for global scenario.

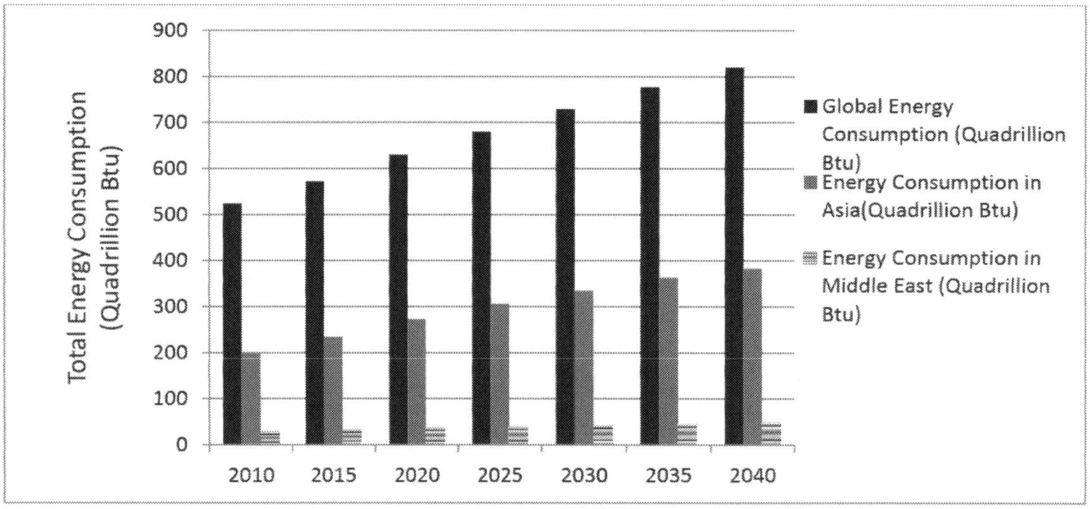

Figure 1: Global Total Energy Consumption

DOI: 10.1051/e3sconf/20172307002

Table 1 shows the trend for the use of alternative energy resources for present as well as for future predictions. It can be observed that the fossilized sources will continue to be single largest contributor to the world energy needs at least in the next two decades as well.

Table 1. Alternative Energy Global Consumption Trend

Year	2011	2012	2025	2040
Petroleum (Quadrillion btu)	36.56	35.87	36.28	35.35
Natural Gas (Quadrillion btu)	24.91	26.20	28.97	32.32
Coal (Quadrillion btu)	19.62	17.34	19.03	17.35
Nuclear (Quadrillion btu)	8.26	8.05	8.15	8.49
Hydropower (Quadrillion btu)	3.11	2.67	2.84	2.90
Biomass (Quadrillion btu)	2.60	2.53	3.74	4.26
Other renewable energy (Quadrillion btu)	1.70	1.97	3.09	3.89

4. Energy Consumption Trend in Saudi Arabia

Saudi Arabia is among the high income countries of Middle East [8]. It has the 2nd largest proven oil reserves in the world and has been involved in highest production of fossilized fuel (as per 2012). It is the largest exporter of oil and 6th largest consumer of oil. It has been reported to over-use fossilized energy sources for its own consumption [9, 10]. This fact is obvious from figure 2 which shows the growth in per-capita GDP and energy consumption for the last four (04) decades. It shows that the growth in economy has been lagging that of energy consumption for most of the time.

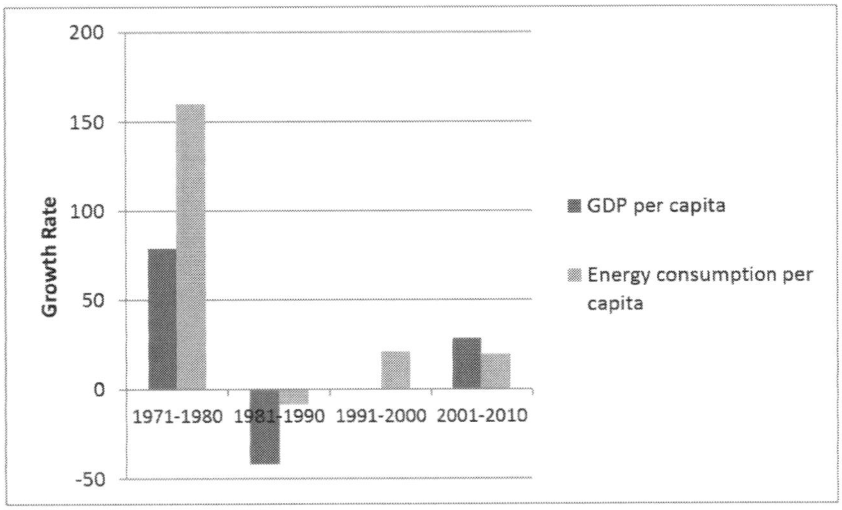

Figure 2: Growth Rate of Per Capita GDP and Energy Consumption in Saudi Arabia

Oil and liquid petroleum have been the main source of energy for Saudi Arabia. Natural gas has also become another dominant energy source for this country. The use of natural gas was started in the middle of 1970s and since then it has increased to cater for 50% of electricity production in the country. Figure 3 illustrates this fact; in addition, it also shows that the contribution of renewable energy sources in its electricity production is non-existent.

 DOI: 10.1051/e3sconf/20172307002

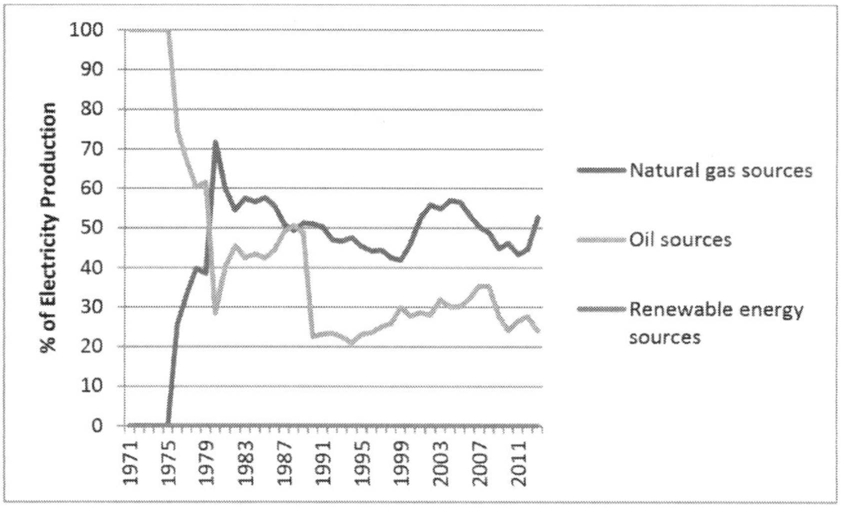

Figure 3: Alternative Energy Sources for Saudi Arabia

As population can be a major contributor to energy consumption, the energy consumption trend was compared with population growth. The comparison is shown in figure 4. It shows that the population has been growing at a consistent rate in Saudi Arabia while the energy consumption has been less robust in the previous decades. It can also be seen that the increase or decline in energy consumption does not show any cyclic behavior and shows more variability in the recent times.

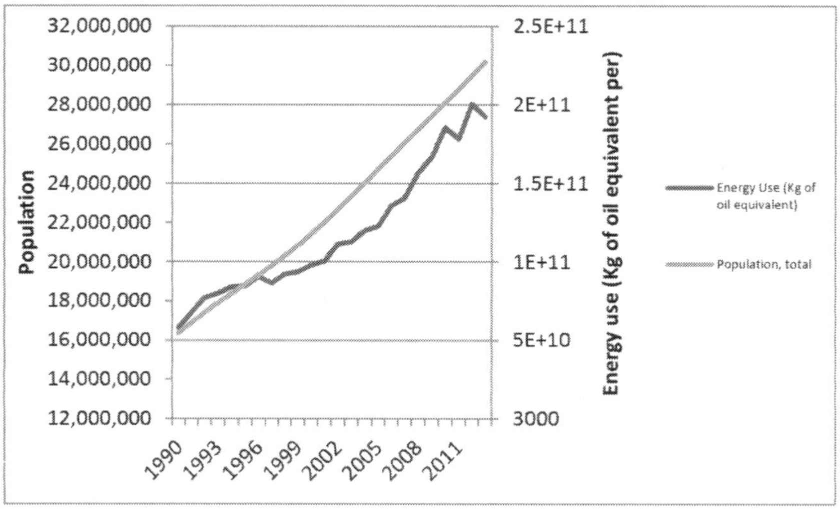

Figure 4: Trend of Growth in Population and Energy Consumption for Saudi Arabia

To further investigate the factors affecting energy consumption in Saudi Arabia, a regression model was developed. At this stage; population, GDP, industrial sector value, urban population and agricultural sector value were considered as the predictor variables. However, only GDP and

DOI: 10.1051/e3sconf/20172307002

industrial sector were found to be affecting the model statistical significance. At this point, the correlation between GDP and industrial sector value was calculated and found to be close to 1. Hence, GDP was selected for the model to predict the energy consumption as the model with this variable was giving better R-square value. The final regression model is given in table 2.

Table 2. Regression Model for Energy Consumption for Saudi Arabia

Variable	Coefficients	Standard Error	t Stat	P-value
Intercept	2.42E+10	5.28E+09	4.591795	4.13E-05
GDP (current US$)	0.268521	0.019023	14.11559	2.45E-17
R-square		0.83		

5. Energy Consumption Trend in Pakistan

Pakistan is considered among the lower middle income countries of South Asia [11]. Energy requirements are rapidly increasing in Pakistan. It is evident from the fact that energy requirements in Pakistan have increased by more than 80 percent in the last one and a half decade. Conventional sources of nonrenewable energy satisfy more than 80% of the energy requirements in Pakistan [12]. This can be observed from figure 5 which shows high contributions of nonrenewable sources.

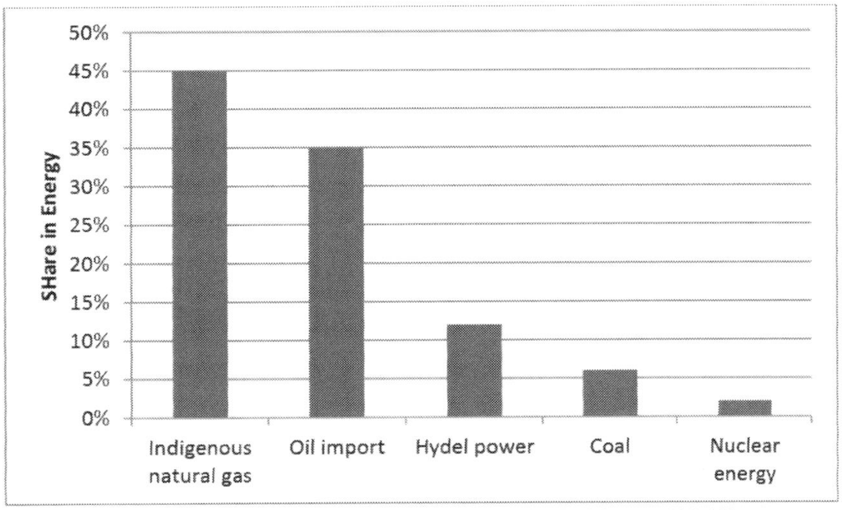

Figure 5: Share of Energy Sources in Pakistan (2010)

The energy sources used in Pakistan for electricity production mainly include hydroelectric, natural gas and nuclear power. Among which, the share of natural gas has increased in 1990s while that of hydroelectric has decreased. This situation is reversed from the perspective of sustainability as nonrenewable energy source have been more utilized. This has resulted in rapid depletion of natural gas reservoirs as shown in recent decline in its use (see figure 6). Figure 7 shows the comparison of energy consumption and population for Pakistan. The data shows that both parameters are growing with similar pattern i.e. consistent growth.

 DOI: 10.1051/e3sconf/20172307002

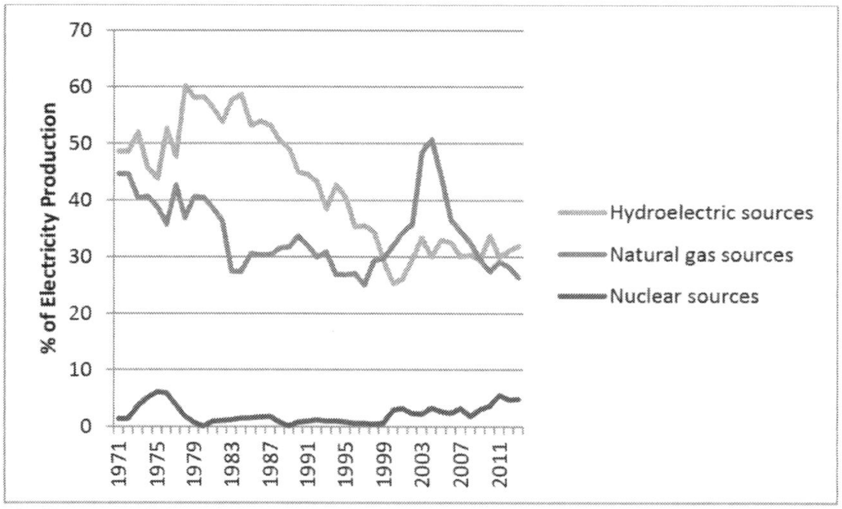

Figure 6: Share in Electricity Production of Alternative Energy Sources for Pakistan

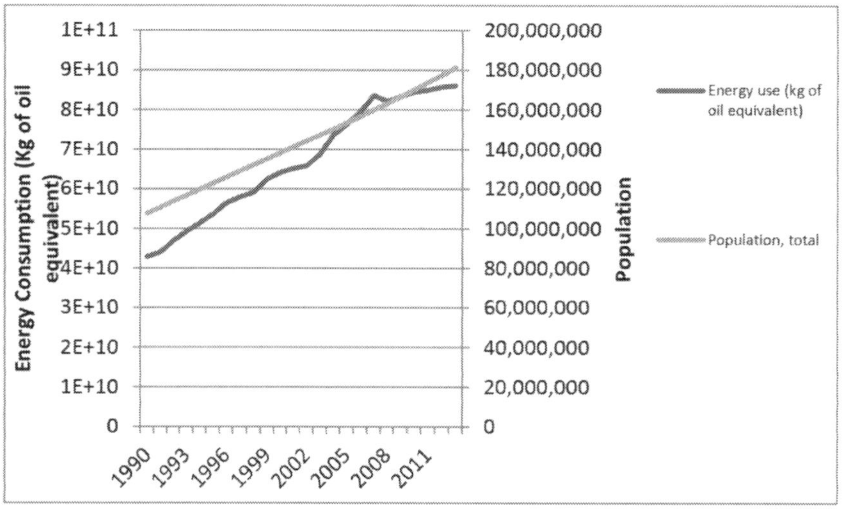

Figure 6: Trend of Growth in Population and Energy Consumption for Pakistan

The regression model was developed for Pakistan using the variables mentioned for Saudi Arabia. Population and urban population were found to be significantly affecting the energy consumption. However, the correlation for these factors was found to be close to 1, hence, population was used for predicting energy consumption as the model with population resulted in better R-square value. The final regression model is shown in table 3.

Table 3: Regression Model for Predicting Energy Consumption for Pakistan

Variable	Coefficients	Standard Error	t Stat	P-value
Intercept	-2.4E+10	1.07E+09	-22.2523	1.66E-24
Population, Pakistan	632.3549	8.822506	71.6752	1.04E-44
R-square		0.99		

6. Comparison of Energy Consumption

Table 4 presents the comparison of future energy consumption for Pakistan and Saudi Arabia. The data in this table shows that currently Pakistan is lagging behind in energy consumption with more than half of that of Saudi Arabia. However, the rate of increase in Pakistan is much higher than Saudi Arabia which could result in higher energy consumption demand in Pakistan in future.

Both countries are showing higher growth rate in energy consumption in comparison with the global trends as well as those in their specific regions. The data for the global and regional predictions was taken from Bilgen (2014). The share of Pakistan in global energy consumption is also increasing in the period 2020-2040, while that for Saudi Arabia is approximately at the same level.

Table 4: Future Energy Consumption Trend Comparison

Year	Pakistan		Saudi Arabia		Global	Asia	Middle East
	Energy Consumption (Kg of Oil Equivalent	Share in global Market	Energy Consumption (Kg of Oil Equivalent	Share in global Market			
2020	1.18E+11	7.4E-03	2.48E+11	1.6E-02	1.59E+13	6.88E+12	9.33E+11
2025	1.55E+11	9.0E-03	2.76E+11	1.6E-02	1.71E+13	7.72E+12	9.84E+11
2030	2.01E+11	1.1E-02	3.08E+11	1.7E-02	1.84E+13	8.45E+12	1.08E+12
2035	2.59E+11	1.3E-02	3.44E+11	1.8E-02	1.96E+13	9.41E+12	1.16E+12
2040	3.32E+11	1.6E-02	3.84E+11	1.9E-02	2.07E+13	9.66E+12	1.24E+12
Growth (2020-2040)	9.1		2.7		1.5	2.0	1.6

7. Conclusions

This research was focused on determining and comparing the trends in energy consumption of Pakistan and Saudi Arabia in global and regional context. The objective was to study the factors affecting energy consumption in the developing countries having different levels of economy, population and indigenous resources.

It was found that energy consumption in Saudi Arabia was mainly driven by the economic development, while that for Pakistan was contributed by the growth in population. This could also point out to the fact that energy scarce countries like Pakistan are hampered in their economic growth due to energy insecurity. In spite of this fact, Pakistan relies heavily on nonrenewable energy resources as does Saudi Arabia. The predictions for the year 2040 show that Pakistan's energy consumption growth rate will be higher than Saudi Arabia and global and regional rates. On the other hand, its share in global energy consumption will also be increasing. Hence, it can be said that population growth will become a major factor to increase the global energy consumption as the economic growth may be halted due to energy insecurities.

References

[1] Ferreira G, editor. Alternative energies: updates on progress. Berlin, Germany: Springer-Verlag; 2013.

[2] Swan, L. G., & Ugursal, V. I. (2009). Modeling of end-use energy consumption in the residential sector: A review of modeling techniques. Renewable and sustainable energy reviews, 13(8), 1819-1835.

[3] Holmberg, K., Andersson, P., & Erdemir, A. (2012). Global energy consumption due to friction in passenger cars. Tribology International, 47, 221-234.

[4] Lee, C. C., & Chang, C. P. (2007). Energy consumption and GDP revisited: a panel analysis of developed and developing countries. Energy Economics, 29(6), 1206-1223.

[5] Jones, D. W. (1989). Urbanization and energy use in economic development. The Energy Journal, 29-44.

[6] Rees, W. E. (1992). Ecological footprints and appropriated carrying capacity: what urban economics leaves out. Environment and urbanization, 4(2), 121-130.

[7] Bilgen, S. (2014). Structure and environmental impact of global energy consumption. Renewable and Sustainable Energy Reviews, 38, 890-902.

[8] Whiting, D. R., Guariguata, L., Weil, C., & Shaw, J. (2011). IDF diabetes atlas: global estimates of the prevalence of diabetes for 2011 and 2030. Diabetes research and clinical practice, 94(3), 311-321.

[9] Mehrara, M., & Oskoui, K. N. (2007). The sources of macroeconomic fluctuations in oil exporting countries: A comparative study. Economic Modelling, 24(3), 365-379.

[10] Alshehry, A. S., & Belloumi, M. (2015). Energy consumption, carbon dioxide emissions and economic growth: The case of Saudi Arabia. Renewable and Sustainable Energy Reviews, 41, 237-247.

[11] Katz, J., Lee, A. C., Kozuki, N., Lawn, J. E., Cousens, S., Blencowe, H., ... & Adair, L. (2013). Mortality risk in preterm and small-for-gestational-age infants in low-income and middle-income countries: a pooled country analysis. The Lancet, 382(9890), 417-425.

[12] Komal, R., & Abbas, F. (2015). Linking financial development, economic growth and energy consumption in Pakistan. Renewable and Sustainable Energy Reviews, 44, 211-220.

Concise Approach for Determining the Optimal Annual Capacity Shortage Percentage using Techno-Economic Feasibility Parameters of PV Power System

*M.A. Alghoul [a, b, *], Amer Ali [c], F.V. Kannanaikal [a], N. Amin [d], and K. Sopian [d]*
[a] Energy and Building Research Center, Kuwait Institute for Scientific Research, Safat 13109, Kuwait
[b] Center of Research Excellence in Renewable Energy (CoRe-RE), Research Institute, King Fahd University of Petroleum and Minerals (KFUPM), Dhahran 31261, Saudi Arabia
[c] The School of Applied Physics, Faculty of Science and Technology, Universiti Kebangsaan Malaysia
[d] Solar Energy Research Institute, Universiti Kebangsaan Malaysia, 43600 Bangi, Selangor, Malaysia
**Corresponding author e-mail: dr.alghoul@gmail.com*

ABSTRACT:
PV power systems have been commercially available and widely used for decades. The performance of a reliable PV system that fulfils the expectations requires correct input data and careful design. Inaccurate input data of the techno-economic feasibility would affect the size, cost aspects, stability and performance of PV power system on the long run. The annual capacity shortage is one of the main input data that should be selected with careful attention. The aim of this study is to reveal the effect of different annual capacity shortages on the techno-economic feasibility parameters and determining the optimal value for Baghdad city location using HOMER simulation tool. Six values of annual capacity shortage percentages (0%, 1%, 2%, 3%, 4%, and 5%), and wide daily load profile range (10 kWh - 100 kWh) are implemented. The optimal annual capacity shortage is the value that always "wins" when each techno-economic feasibility parameter is at its optimal/ reasonable criteria. The results showed that the optimal annual capacity shortage that reduces significantly the cost of PV power system while keeping the PV system with reasonable technical feasibility is 3%. This capacity shortage value can be carried as a reference value in future works for Baghdad city location. Using this approach of analysis at other locations, annual capacity shortage can be always offered as a reference value for those locations.

Keywords: PV power system, effect of different annual capacity shortages, techno-economic feasibility parameters, HOMER simulation tool, optimal annual capacity shortage

1. INTRODUCTION

Renewable energy (RE) power systems are rapidly gaining popularity as alternative energy source [1-3]. Photovoltaic has been identified as one of the promising solar technologies to replace the dependency on other conventional energy resources. The performance of a reliable PV power system that fulfils the expectations requires correct input data and careful design. Inaccurate input data when predicting the techno-economic feasibility would affect system cost, size and performance of PV power system on the long run. HOMER simulation tool was found to be one of the easiest and commonly used software [4]. It is widely used in designing solar PV power systems [5-8] and wind power systems [9-11]. HOMER simulates a diverse range of micro-power system configurations and creates a comprehensive list of all possible system designs according to cost-effectiveness. The simulation process aids in selecting the most appropriate system, which is capable of providing electrical loads judiciously by calculating the life cycle cost of the system. Allowing for a small percentage of annual capacity shortage could reduce the system cost significantly. Therefore, this study aims to predict the optimum annual capacity shortage pertaining to Baghdad city location. To accomplish this objective, six values of annual capacity shortage (0%, 1%, 2%, 3%, 4%, and 5%) are implemented and tested versus load profile range (10 kWh - 100 kWh). The operating hours of the load is assumed from 8:00am-6:00pm.

2. Materials and methods

2.1. Study Evolution Methodology

DOI: 10.1051/e3sconf/20172307003

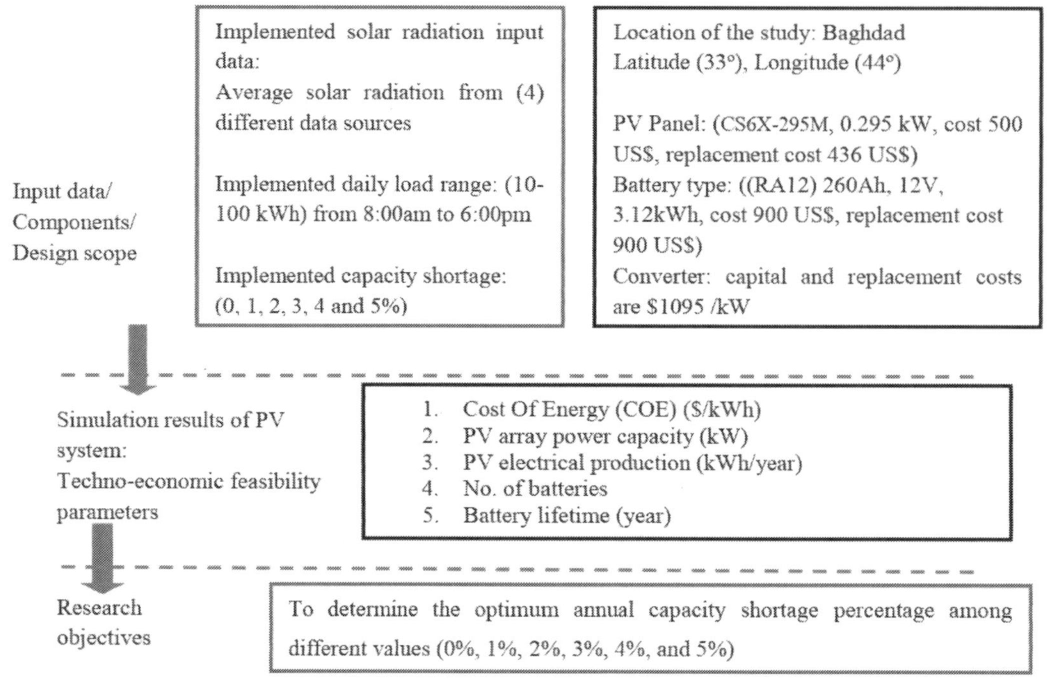

Fig. 1. Block diagram of the study evolution methodology

2.2. Location of the study and solar irradiation data sources

The present work is done for the location Baghdad, Iraq's capital located at 33° latitude and 44° longitude. The selected solar data sources are: Nasa climate site SRC1 [12], Weather base climate site SRC2 [13], meteorological department data (Iraq: 1984-2004) SRC3 [14] and meteorological department data (Iraq: 1962-1992) SRC4 [15]. Comparison between monthly solar radiations from different data sources indicates variation as illustrated in figure 2. Therefore, average monthly solar radiation from four data resources is adopted as an accurate input data in this study for Baghdad city.

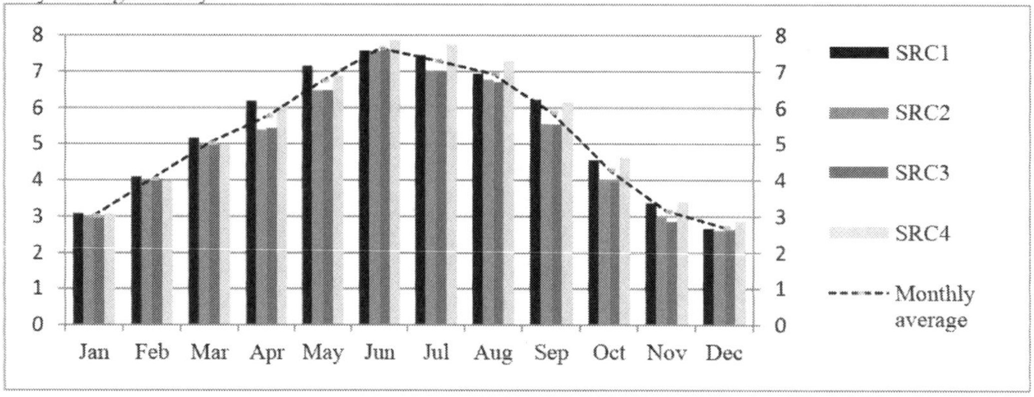

Fig. 2. Variation in solar radiation (kWh/m^2) obtained from different data sources of Baghdad city, Iraq

DOI: 10.1051/e3sconf/20172307003

3. SIMULATION RESULTS AND DISCUSSION

The aim of this section is to determine the optimum annual capacity shortage to design a feasible PV power system for Baghdad city. In this course of analysis, percentages of reduction in the values of the main feasibility parameters at different annual capacity shortages were used as an evaluation tools. The decision on optimum annual capacity shortage value compare to other implemented capacity shortage values will be made based on the below set of criteria:

1. Reduction percentage of COE (maximum or approach to maximum)
2. Reduction percentage of PV Array power capacity (maximum or approach to maximum)
3. Reduction percentage of PV electrical production (minimum or approach to minimum)
4. Reduction percentage in No. of batteries (maximum or approach to maximum)
5. Reduction percentage in battery lifetime (minimum or approach to minimum)

Figures (3-7) illustrate the percentage of reduction in the values of techno-economic feasibility parameters versus daily load profile at different annual capacity shortages. As shown in Fig 3, the minimum percentage of reduction in the values of cost of energy (COE) are observed at capacity shortages (0%), followed by (1% and 2%) where their curves are almost overlapping each other versus the daily load. So, these capacity shortages will be ignored in this round of analysis. On the other hand, the maximum percentage of reduction in the values of COE is observed at capacity shortages (3 and 5%), where their curves are mostly overlapping each other versus the studied daily load. However, to determine which of them is the most appropriate, it is essential to determine at which capacity shortage (3% or 5%) the PV power system is technically most feasible and desired.

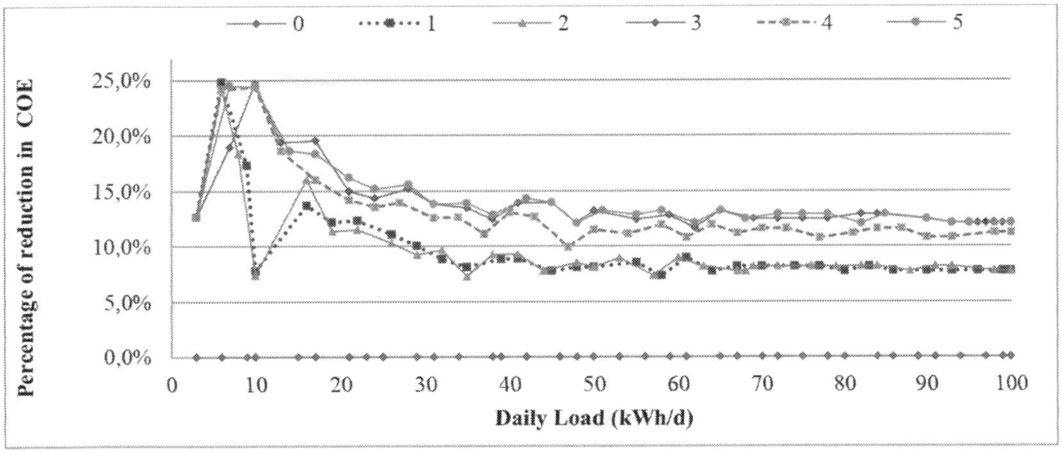

Fig. 3. Percentages of reduction in COE versus daily load profile at different annual capacity shortages

As shown in Fig 4, at the different capacity shortages, percentages of reduction in PV array power capacity are totally overlapping each other versus wide range of the studied daily load (at least 68% of the load range). However, lowest percentages of reduction are observed at capacity shortage value 4%, followed by 0%, 1%, and 5%. Therefore, these capacity shortages will be ignored, because our criterion is maximum percentages of reduction. The maximum percentages of reduction are observed at capacity shortage value 2% versus daily load range (69-100) kWh. Also, maximum percentages of reduction are observed at capacity shortage 3% versus very narrow range of

DOI: 10.1051/e3sconf/20172307003

daily load (98-100) kW/day. However, to decide which of them is the most appropriate (2% or 3%), it is essential to quantify at which capacity shortage, the percentage of reduction in PV electrical production is the minimum.

As shown in Fig 5, under the wide range of daily load, the capacity shortage values that show high percentages of reduction in PV electrical production are (2%, 4%, and 5%). So, these capacity shortages will be ignored in this round of analysis. On the other hand, the preferable low percentage of reduction is observed at capacity shortage values (0%, 1%, and 3%). At capacity shortages (0%, and 1%), the PV power system is not economically the most feasible, so capacity shortage 3% wins in this round of analysis.

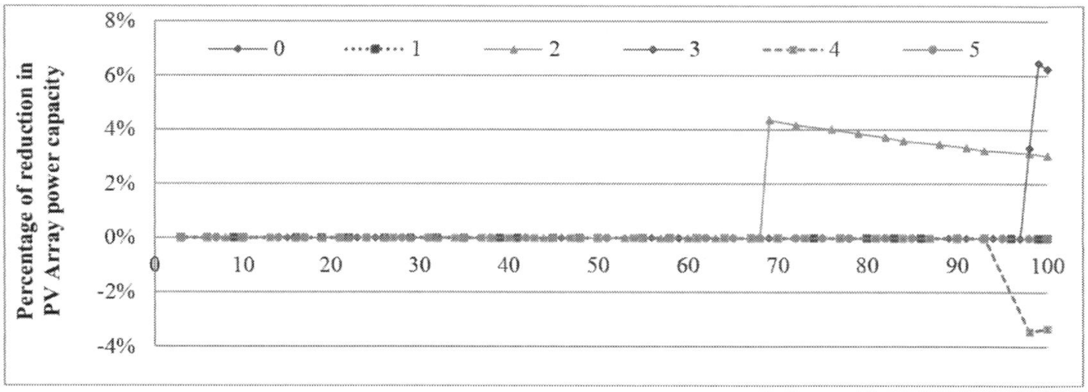

Fig. 4. Percentages of reduction in PV Array power capacity versus daily load profile at different annual capacity shortages.

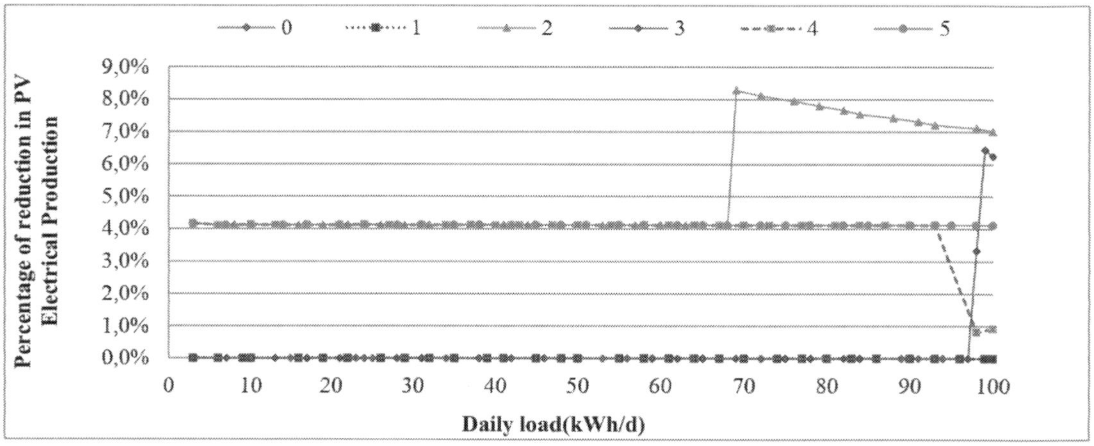

Fig. 5. Percentages of reduction in PV Electrical Production versus daily load profile at different annual capacity shortages.

As shown in Fig 6, under the wide range of daily load, the capacity shortage values that show minimum percentages of reduction in batteries number are (0%, and 1%). So, these capacity shortages are not desired and will be ignored. The maximum percentages of reduction in batteries number are found at 5% followed by (2% or 3% or 4%) depending on their trend versus the daily load. However, capacity shortages at (2% and 4%) will be ignored because the PV power system is not economically feasible. To decide which is the most appropriate capacity shortage (3% or 5%), it is essential to quantify at which capacity shortage, the percentages of reduction in battery life time is the minimum.

DOI: 10.1051/e3sconf/20172307003

As shown in Fig 7, under the wide range of daily load, the capacity shortage values that show maximum percentage of reduction in the values of battery bank life time are (5% and 4%). Since these high percentages are not desired, they will be ignored in this round of analysis. The minimum percentages of reduction in battery life time are at 0% followed by 1% and then by 3% or 2% depending on their trend versus the daily load. Since the PV power system is not economically feasible at capacity shortages 0%, 1% and 2% they will be ignored. So, the trustworthy value here is 3%.

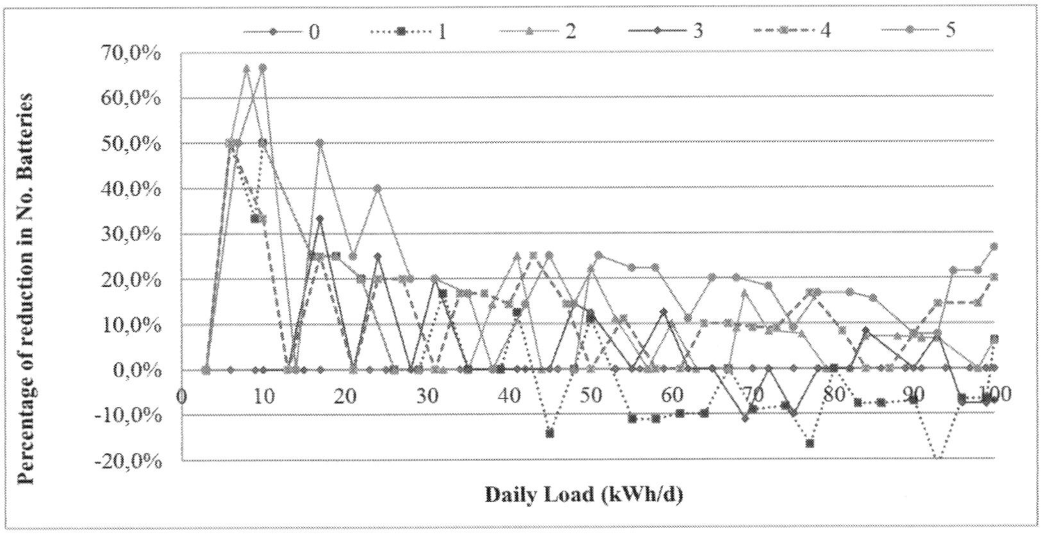

Fig. 6. Percentages of reduction in No. batteries versus daily load profile at different annual capacity shortages.

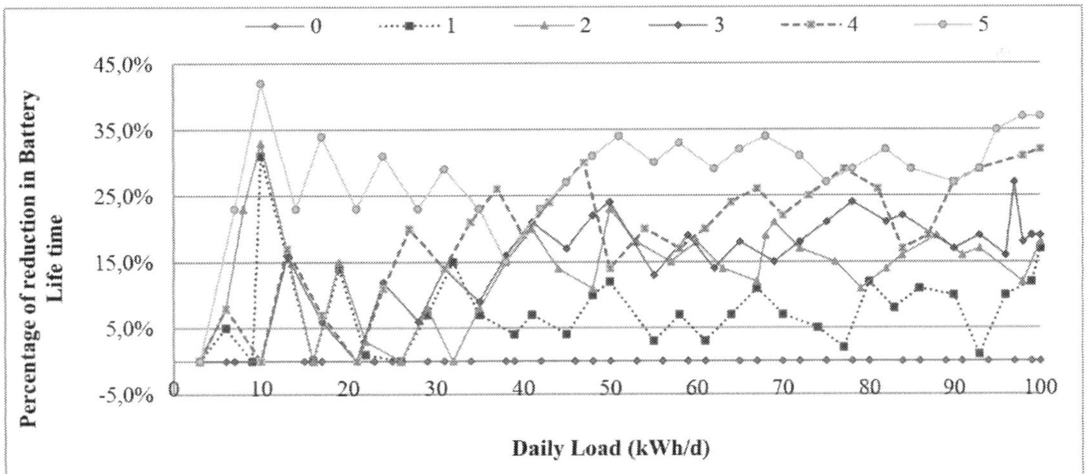

Fig. 7. Percentage of reduction in the values of battery bank life time versus daily load profile at different annual capacity shortages.

Finally, based on the results learned from Figures 3 to 7, it can be concluded that annual capacity shortage 3% is the most recommended value. Therefore, this value can be adopted as a reference annual capacity shortage in future works for designing PV power systems at Baghdad city location.

4. CONCLUSION

The outcome of this study is the followings:

1. Reaching the best economic feasibility of PV power system without affecting significantly the technical performance of the system is possible by adopting annual capacity shortage between (0 % - 5 %).
2. In this study, the results revealed that the optimum annual capacity shortage versus wide load profile range (10-100 kWh/day) for Baghdad city location is 3%. This value can be recommended as a reference capacity shortage value when designing PV power systems in Baghdad city location.
3. It is recommended to broaden the analysis further by evaluating partial values of annual capacity shortage such as (0.5%, 1.5%, 2.5%, 3.5%, and 4.5%) to reach the most accurate value of annual capacity shortage.
4. Using this approach of analysis at other locations, annual capacity shortage would be offered as reference values in future works in those locations.

REFERENCES

[1] Goodbody, C., Walsh, E., McDonnell, K. P. & Owende, P. 2012. Regional integration of renewable energy systems in Ireland - The role of hybrid energy systems for small communities. *Electrical Power and Energy Systems* **44**: 713-720.

[2] Chowdhury, S. H. & Oo, A. M. T. 2012. Study on electrical energy and prospective electricity generation from renewable sources in Australia. *Renewable and Sustainable Energy Reviews* **16**: 6879-6887.

[3] Aagreh, Y. & Al-Ghzawi, A. 2012. Feasibility of utilizing renewable energy systems for a small hotel in Ajloun city, Jordan. *Applied Energy* **103**: 25-31.

[4] National Renewable Energy Laboratory, USA. Hybrid Optimization Model for Electric Renewables (HOMER). http://www.homerenergy.com/.

[5] Shaahid, S.M & Elhadidy, M.A.2005. Economic analysis of hybrid photovoltaic–diesel–battery power systems for residential loads in hot regions—A step to clean future. *Renewable and Sustainable Energy Reviews* **11**: 635–653.

[6] Shaahid, S.M. & Elhadidy, M.A. 2006. Technical and economic assessment of grid-independent hybrid photovoltaic–diesel–battery power systems for commercial loads in desert environments. *Renewable and Sustainable Energy Reviews 11: 1794–1810.*

[7] Malika Amini .2010. Renewable Energy Systems for Rural Health Clinics in Algeria: Homer Application. *Energy* 14: 1120–1130. http://homerenergy.com/webcast-downloads/Malika_Amini/health.pdf

[8] Widatalla, A.M. and Zinko, H., 2011, November. Designing a Photovoltaic Solar Energy System for a Commercial Building. Case Study: Rosa Park Hotel in Khartoum-Sudan. In World Renewable Energy Congress-Sweden; 8-13 May; 2011; LinkÃ ping; Sweden (No. 057, pp. 2899-2906). Linköping University Electronic Press.

[9] Osorio, L., Velo, R., Fernández, M. D. & Rodríguez, M. R. 2012. Feasibility of the electrical supply by wind power in dairy cattle farms. International Conference of Agricultural Engineering, Valencia, July 2012.

[10] Adaramola, M. S. 2012. Feasibility study of off-grid hybrid energy systems for applications in Ondo State, Nigeria. *Journal of Engineering and Applied Sciences* **7**(1): 72-78.

[11] Fahmy, F. H., Ahmed, N. M. & Farghally, H. M. 2012. Optimization of Renewable Energy Power System for Small Scale Brackish Reverse Osmosis Desalination Unit and a Tourism Motel in Egypt. *Smart Grid and Renewable Energy* **3**: 43-50.

[12] https://eosweb.larc.nasa.gov/sse/RETScreen/

[13] www.weatherbase.com

[14] Tadros, M. T. Y., Mustafa, M. A. M., & Abdel-Wahab, M. 2014. Estimation of the Global Horizontal Solar Radiation in Iraq. *International Journal of Emerging Technology and Advanced Engineering* Website: www.ijetae.com (ISSN 2250-2459, ISO 9001:2008 Certified Journal, Volume 4, Issue 8, August 2014)
[15] Abdul-Wahid, S. N., & Hassan, A. M. 2010. Calculation and applications of net solar radiation in Iraq. *J Al-Qadisiyah Pure Sci*, ***15***, 1-30.

 DOI: 10.1051/e3sconf/20172308001

A Geospatial Database for Wind and Solar Energy Applications: The Kingdom of Bahrain Study Case

Khalil Al-Joburi* and Nidal Dahman*,
Email: kaljuboori@uob.edu.bh and ndahman@uob.edu.bh
Civil Engineering Department, University of Bahrain, Isa Town, Kingdom of Bahrain

ABSTRACT

This research is aimed at designing, implementing, and testing a geospatial database for wind and solar energy applications in the Kingdom of Bahrain. All decision making needed to determine economic feasibility and establish site location for wind turbines or solar panels depends primarily on geospatial feature theme information and non-spatial (attribute) data for wind, solar, rainfall, temperature and weather characteristics of a particular region. Spatial data includes, but is not limited to, digital elevation, slopes, land use, zonings, parks, population density, road utility maps, and other related information. Digital elevations for over 450,000 spot at 50 m spatial horizontal resolution plus field surveying and GPS (at selected locations) was obtained from the Surveying and Land Registration Bureau (SLRB). Road, utilities, and population density are obtained from the Central Information Organization (CIO). Land use zoning, recreational parks, and other data are obtained from the Ministry of Municipalities and Agricultural Affairs. Wind, solar, humidity, rainfall, and temperature data are obtained from the Ministry of Transportation, Civil Aviation Section. LandSat Satellite and others images are obtained from NASA and online sources respectively. The collected geospatial data was geo-referenced to Ain el-Abd UTM Zone 39 North. 3D Digital Elevation Model (DEM)-50 m spatial resolutions was created using SLRB spot elevations. Slope and aspect maps were generate based on the DEM. Supervised image classification to identify open spaces was performed utilizing satellite images. Other geospatial data was converted to raster format with the same cell resolution. Non-spatial data are entered as an attribute to spatial features. To eliminate ambiguous solution, multi-criteria GIS model is developed based on, vector (discrete point, line, and polygon representations) as well as raster model (continuous representation). The model was tested at the Al-Areen proposed project, a relatively small area (15 km^2). Optimum site spatial location for the location of wind turbines and solar panels was determined and initial results indicates that the combination of wind and solar energy would be sufficient for the project to meet the energy demand at the present per capita consummation rate..

Keywords; Bahrain, GIS, renewable energy, Multi criteria modeling,

INTRODUCTION

Providing secure and sustainable energy represents a major challenge to every country worldwide. Energy demand is continuously increasing due to population growth and urbanization. Electricity brings social and economic benefits to rural communities which represent 52% of the human population [1]. More than 2 billion rural residents in developing countries currently lack reliable electricity service [2], providing a plausible prognostication for a significant threat to

DOI: 10.1051/e3sconf/20172308001

livelihoods if the problem is not addressed. This problem and its ramification was articulated by The Millennium Goals adopted by the 2002 World Summit on Sustainable Development. [3].Coal and fossil fuel has been the mainstay of the world's energy for over 6 decades and has effected the phenomina of global warming and myriad environmental issues which weighed in on the urgency of finding an alternative. Renewable energy sources such as solar, wind, biomass, and others present the alternative sources. Nearly every country has established policies within their borders that address the need of and demand for renewable energy sources and began running pilot projects to explore such sources and ascertain ways and means of promoting its exploitation. The development of renewable energy in the Arab States of the Gulf has been relatively slow despite its many conducive elements and tremendous potential. This market has been stifled by a combination of constraints, such as the absence of relative legal and policy framework, the high initial capital costs and the lack of commercial skills and information [4].

THE KINGDOM OF BAHRAIN: OVERVIEW

The Kingdom of Bahrain is an archipelago of around 33 islands; the largest is the main island of Bahrain, all are low-lying islands between latitudes 26.10° and 25.22° North and longitudes 51.070° and 50.160° East [5]. They lay in an arm of the Arabian Gulf, known as the Gulf of Salwa, which separates the Qatar peninsula and Saudi Arabia.

The Kingdom's total land area was about 620 km^2 in 1965 [5]. Due to significant land reclamation activities since 1965, the estimated total area is 756 km^2 [5]. This represents an increase of 22% in total square area, making the Kingdom the most dynamic country in the world in terms of land reclamation.

The Climate of Bahrain

Bahrain's climate is characterized by a high average temperature and relative humidity during the summer. The minimum recorded temperature for the recorded years between 2003 and 2012 is 10.2 C, maximum is 39.7 C, the average 27.3, the mode 35.1 C, and the standard deviation is 6.8 C. The minimum relative humidity was 16%, maximum of 92%, average of 59%, mode of 59%, and standard deviation of 12.4% for the same period. Maximum rain for the same period is 34.7mm, and the numbers of days of rain are 222, the average is 0.2mm. On the other hand, the mean daily hours of sunshine range from a minimum of 7.3h in January, to a maximum of 11.3h in June [6]. The average annual wind speed is nearly 5m/s and it's mainly North to North West. Since Bahrain is an island, the sea and land breezes assure the availability of wind most of the year. June is the windiest month throughout the year and noon is the windiest time during the day [6]. The solar radiation intensity in Bahrain is very high, ranging from a minimum of 3.25 kWh/m^2 in December, to a maximum of 7.19kWh/m^2 in July [6].

Energy Consummation

Bahrain is among the highest energy consumption rates in the world. The country uses nearly 10 times more energy than Japan on a per capita consumption basis. Bahrain's estimated energy consumption is about 695 million Btu's, compared that of Japan of only 178 million [7].

Renewable Energy

Bahrain has developed some renewable energy sources and achieved some progress during the last decade. Specifically, the adoption of solar technology was very advantageous in Bahrain in the small capacity projects that were inaugurated [4]. However, high temperatures and the mixture of dust and humidity tend to reduce photovoltaic and solar cell efficiency [8]. Due to relatively low wind speed, only small and limited wind turbine may be utilized According to the study of W.E. Alnaser, a good solution that is under construction is a mobile RES (hybrid) system consisting of wind generators and solar panels to produce 1.5 kW of (alternating current) electricity for rural areas. This system utilizes solar power (which is abundant in Bahrain) and wind power and is suitable for many applications, such as mini mobile clinics, small schools, libraries, and camping sites [9].Oceans and seas hold enormous and never-ending energy resources, which could be harnessed to produce sufficient energy by using the forces generated by the waves, currents and coastal water displacement between high and low tides. This source of energy is not cost effective in Bahrain due in part to a minimal tidal range the height of which does not exceed 3m. Investment in such projects is not recommended unless a location with a tidal height larger than 6m is found, [9].

The average family doweling energy consumption in Bahrain is very high; a number of studies has been conducted regarding building energy conservation and illustrating the electricity savings potentials in the residential sector [10].

The ECC is planning to formulate policies and regulations that would enable the government to establish a long-term and comprehensive energy program, the ECC would conduct a comprehensive study on the energy efficiency of various sectors in the kingdom. It would be carried out with the assistance of a Japanese consulting firm and the Japan International Co-operation Agency (JICA). The ECC had also initiated a project in co-operation with Japanese company InterDomain to install a

DOI: 10.1051/e3sconf/20172308001

wind power measurement system for 12 months. This would enable assessment of the feasibility of utilizing the wind as a renewable energy source in Bahrain [7]. NOGA has embarked on an ambitious plan to secure more sustainable supply of energy for Bahrain.

Topography and Land Use

DEM analysis shows that Bahrain's elevation above sea level ranges from 1.00 to 134m, with 51.70% being above 5m. However, most urban areas are located between 0 to 2.0m. The Kingdom is mostly flat or near level, with 79% of Kingdom being nearly level (0–1%).

Renewable Energy Site selection

Geospatial data and Geographic Information systems (GIS) have been used in renewable energy research in three ways:

1. Decision support system [10-12]. GIS-based analyses are used to characterize spatially-related variables within a digital environment and help produce informative visualizations.
2. Model design for identifying renewable energy sources. GIS-based assessments have been used to evaluate renewable energy sources in the USA [12], developing Latin America countries, China, and Taiwan [13-15], solar radiation potential in rural India and Greece [16,17], community scale solar potential in the UK and California [18,19], and also to characterize site locations for wind farms [20] and manage community water resources [21,22].
3. Site selection [23], which is the focus of this research.

The literature suggests many different criteria for selecting the most suitable site for both solar and wind sites. The most common criteria categories are: environmental and economic. Site topography is of special importance, because of its impact of slope, aspect, and height. It also impacts directly wind speed and turbulence. Mountains result in increased wind speed; this is partly a result of altitude. Equally, topography may produce areas of reduced wind speed, such as sheltered valleys, areas in the lee of a mountain ridge or where the flow patterns result in stagnation points [24]. Wind regimes in different geographical locations can vary widely. At a given location, temporal variability on a large scale means that the amount of wind may vary from one year to the next, with even longer scale variations on a scale of decades or more [24].These long-term variations are not well understood and may make it difficult to make accurate predictions of the

DOI: 10.1051/e3sconf/20172308001

economic viability of particular wind farm projects, for instance. On timescales shorter than a year, seasonal variations are much more predictable, although there remain large variations on shorter timescales [24].

STUDY AREA

A 3.5km^2 study area is located in the south-west of the Kingdom of Bahrain. Although wind speed is moderate according to obtained data, the area has a large moderate slope terrain, which is preferable considering the slope criteria constraints of wind farm locations. Additionally, the area also has adequate highlands for air inflow. Thus, it can be stated that the city may have sufficient wind energy potential. Figure 1

DATA COLLECTION AND PREPARATION

A base map for the Kingdom of Bahrain, DEM for over 450,000 spot elevation points at 50m spatial horizontal resolution plus field surveying and GPS -at selected locations- are obtained from SLRB. Road, utilities, and population density are obtained from the CIO. Land use zoning and recreational parks are obtained from the Ministry of Municipalities and Agricultural Affairs. Wind, solar, humidity, rainfall, and temperature data are obtained from the Ministry of Transportation, Civil Aviation Section. LandSat Satellite images are obtained from NASA, others images are obtained from online sources. The collected geospatial data were geo-referenced to Ain el-Abd UTM Zone 39 North.

Data Analysis

ArcGIS 10.2 and extension are used for data processing and analysis. A 3D surface is generated based on the DEM at 50m spatial resolutions. Slope and Aspect maps are generated from the created 3D surface (figures 2 and 3). Supervised image classification to identify open space areas was performed utilizing satellite images. Other geospatial data is converted to raster format with the same cell resolution. A 100 meter buffer zone is created around all POI, roads, and national parks. A multi-criteria GIS model is developed based on vector (discrete point, line, and polygon representations) and raster model (continuous representation) to eliminate ambiguous solutions.

DOI: 10.1051/e3sconf/20172308001

Figure 24 the Study Area

DOI: 10.1051/e3sconf/20172308001

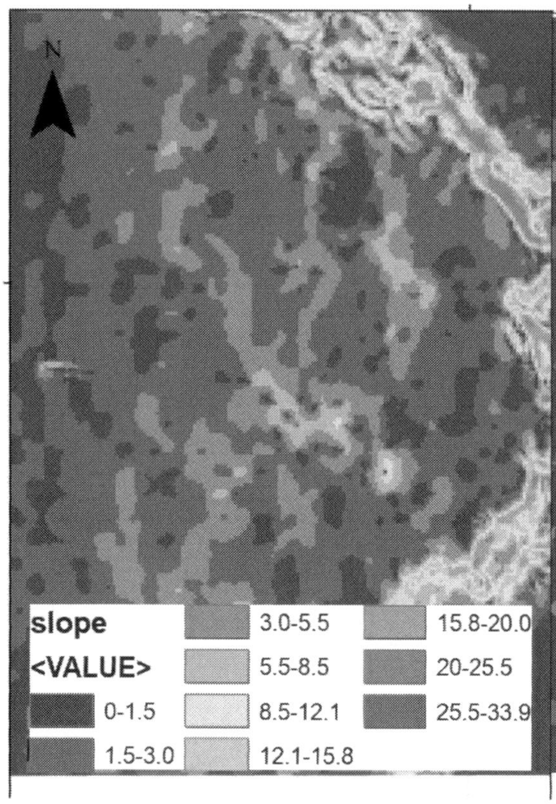

Figure 2 Slope Distribution

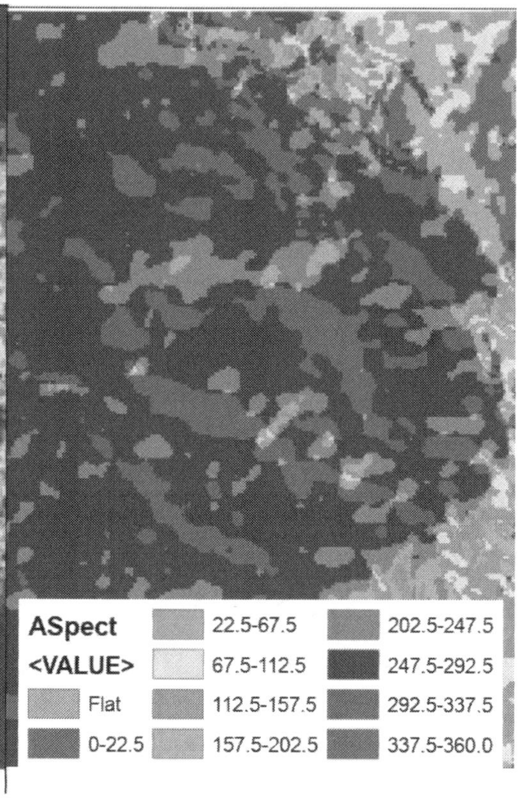

Figure 3 Aspect Distribution

RESULTS

Using the obtained DEM raster surface, slope, aspect maps, and multi-selection criteria listed in table 1, the most suitable sites map is created (figure 4). The map shows that there are broad field for suitable wind farms, however, the size of these sites differ significantly, between 500- 27,000m^2. All sites are to the west and north-west, this area is an open area with very little restrictions. The area toward the east side is mostly rough and with steep slope terrain.

DOI: 10.1051/e3sconf/20172308001

Table 1 Site selection Criteria

Class	Score	Class	Score	Class	Score
Roads		**Slope**		**Aspect**	
0-100 m	0	0-5%	100	0-180 ^{0}	0
> 100 m	100	5 -10%	50	180-270^0	50
POI		>10	0	270-300	75
0-100 m	0	**Elevation**		300- 360^0	100
> 100 m	100	0-10	0		
		> 10	100		

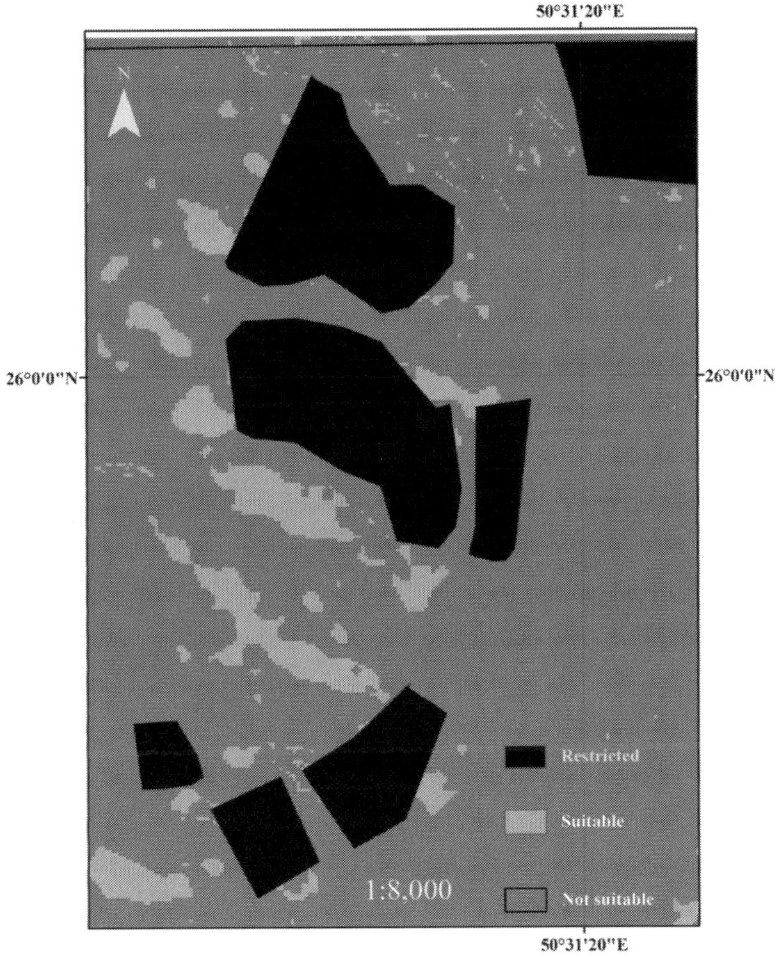

Figure 4 Obtained Suitable Sites

DOI: 10.1051/e3sconf/20172308001

CONCLUSION

Renewable energy is no longer a choice, it is a must. Traditional fossil energy sources are depleting rapidly and cause environmental degradation. Countries around the globe are searching for new energy sources and planning an effective decision making-system that can facilitate renewable energy management. Identifying, locating, and implementing wind and solar projects depends primarily on the spatial location and terrain properties. A geospatial database is at the heart of any decision making systems. It encompasses an array of multidimensional thematic spatial and non-spatial information. Examples of which are digital elevation, road centerlines, utilities, wind speed, wind direction, etc. A geospatial database for the Kingdom of Bahrain was created. It includes all necessary information needed to plan and implement a renewable energy project. The created geospatial database was tested for identifying the most suitable site for installing a wind turbine and solar farm. A 3.5km^2 area was selected in the Southern province of the Kingdom of Bahrain. A GIS based on a multi-criteria site selection was implemented to decide the most suitable wind and solar farm locations. Despite some limitations and constraints the study has portrayed both the effectiveness of GIS as a research tool and the importance's of having a geospatial database. The study established that that the Kingdom of Bahrain is capable of implementing wind power as a sustainable and renewable energy source.

DOI: 10.1051/e3sconf/20172308001

REFERENCES

[1] World Population Prospects: The 2004 Revision.

[2] HUMAN DEVELOPMENT REPORT 2004.

[3] Report of the World Summit on Sustainable Development Johannesburg, South Africa, 26 August- 4 September 2002.

[4] H. Doukas and John Psarras; Renewable energy sources and rationale use of energy development in the countries of GCC: Myth or reality? Renewable Energy 31 (2006) 755–770.

[5] Khalil Al-Joburi , Geospatial Approach for Analyzing Soil Type Impact on Bahrain Deteriorating Houses; International Journal of Geosciences and Geomatics, Vol. 3, Issue 1, 2015,ISSN:2052-5591.

[6] .W. Alnaser, R. Flanagan; The need of sustainable buildings construction in the Kingdom of Bahrain; Building and Environment 42 (2007) 495–506.

[7] Gulf Digital news, 7th APRIL 2015 - Vol.XXXVIII No.018.

[8] Alnaser WE. Renewable energy resources in the state of Bahrain. Appl Energy 1995;50(8):23–30.

[9] W.E. Alnaser, Characteristics of the available wind energy in Bahrain. Solar Energy 42(l), 3-6 (1989).

[10] Akbari H, Morsy MG, Al-Baharna NS. Electricity savings potentials in the residential sector of Bahrain. Proceedings of the 1996 ACEEE summer study on energy efficiency in buildings. American council for an energy-efficient economy 1996, Washington, DC. 1(LBL-38677) 1996 p. 11.

[10] Voivontas D, Assimacopoulos D, Mourelatos A, Corominas J. Evaluation of renewable energy potential using a GIS decision support system. Renewable Energy 1998;12:333–44.

[11] Caiaffa, E and Polliono, m; A Gis Based Methodology In Renewable Energy Sources Sustainability; International Journal of Agricultural and Environmental Information Systems, ,July-September 2014 17, 5(3), 17-36.

[12] John Van Hoesen and Steven Letendre; Evaluating potential renewable energy resources in Poultney, Vermont: A GIS-based approach to supporting rural community energy planning; Renewable Energy 35 (2010) 2114–2122.

[13] Dominguez, J. and Pinedo-Pascua, I. GIS tool for rural electrification with renewable energies in Latin America, geows, 2009 International Conference on Advanced geographic information systems and web Services 2009; 171–176.

DOI: 10.1051/e3sconf/20172308001

[14] Byrne J, Zhou A, Shen B, Hughes K. Evaluating the potential of small-scale renewable energy options to meet rural livelihoods needs: a GIS – and life cycle cost-based assessment of Western China's options. Energy Policy 2007;35:4391–401.

[15] Yue CD,Wang SS. GIS-based evaluation of multifarious local renewable energy sources: a case study of the Chingu area of southwestern Taiwan. Energy Policy 2006;34:730–42.

[16] Ramachandra TV. Solar energy potential assessment using GIS. Energy Education and Technology 2007;18(2):101–14.

[18] Gadsden S, Rylatt M, Lomas K. Putting solar energy on the urban map: a new GIS-based approach for dwellings. Solar Energy 2003;74:397–407.

[19] Johnson G, Armanino D. GIS tools for community development applications.SOLAR 2004 Conference,

[20] Sima˜o A, Densham PJ, Haklay MM. Web-based GIS for collaborative planning and public participation: an application to the strategic planning of wind farm sites. Journal of Environmental Management 2009;90:2027–40.

[21] Jankowski P. Towards participatory geographic information systems for community-based environmental decision making. Journal of Environmental Management 2009;90:1966–71.

[22] Nyerges T, Jankowski P, Tuthill D, Ramsey K. Collaborative water resource decision support: results of a field experiment. Annals of the Association of American Geographers 2006;96:699–735.

[23] A.E. Cengiz, Y. Güney and Cabuk; Determining Renewable Energy Efficiency in Eskisehir, Turkey: A GIS Based Solution; 6 International Advanced Technologies Symposium (IATS'11), 16-18 May 2011, Elazığ, Turkey.

[24]Tony Burton, 2011, Wind Energy Handbook-Second Edition, A John Wiley and Sons, Ltd., Publication, ISBN 978-0-470-69975-1.

[25] Nazli Yonca Aydin, Elcin Kentel, H. Sebnem Duzgun; GIS-based site selection methodology for hybrid renewable energy systems: A case study from western Turkey. Energy Conversion and Management 70 (2013) 90–106.

[26] Voivontas D, Tsiligiridis G, Assimacopoulus D. Solar potential for water heating bexplored by GIS. Solar Energy 1998;62:419–27.

E3S Web of Conferences 23, 08002 (2017) DOI: 10.1051/e3sconf/20172308002
World Renewable Energy Congress-17

Attracting and Maintaining Demographic Diversity in Energy: Lessons from Petroleum Engineering.

Tatyana Plaksina,
Department of Chemical and Petroleum Engineering, Schulich School of Engineering, University of Calgary, Canada
tatyana.plaksina@ucalgary.ca

Abstract

Diversity is important for any organization, its development and growth. However, frequently we observe that, if left to their own devices, many social and professional communities become self-selective, less susceptible to change, and, thus, less diverse. For example, this phenomenon can be readily observed in the academia when a professor forms a research group that is mostly comprised of people from certain origin or personal qualities. While on an individual level, this situation can be quite convenient and even beneficial, on the large scale the lack of diversity creates severe distortions and discourages many talented individuals from joining the group, organization, or industry in general.

Diversity that includes ethnic, age, and gender aspects is a serious concern in overall energy industry and academia. One of the most prominent field impacted by diversity problems is petroleum engineering which can be used as a learning opportunity for other branches of energy industry. Using petroleum engineering community, we can understand what processes have led it to become self-selective, what measures have been applied to changes the situation and introduce more diversity.

Keywords: petroleum engineering education; demographic statistical data; faculty distribution; diversity issues; gender and ethnic diversity

Introduction

Phillips in her *Scientific American* article outlined how being surrounded by people from different cultures and backgrounds promotes creativity, diligence, and hard work inside a diverse group [1]. The author emphasized that interaction between people with different focuses yields higher productivity, better problem solving, and broader perspective. Unfortunately, most energy engineering communities are still exhibiting poor diversity. In fact, most of them quickly become self-selective and discourage people with different ethnic, gender, and educational backgrounds from becoming a part of these groups. In this study, we investigate a locomotive of the

DOI: 10.1051/e3sconf/20172308002

energy field – petroleum engineering – and observe on statistical data from petroleum engineering academic programs how poor diversity and self-selection manifests itself in this most prominent energy community.

The choice to analyze data from petroleum engineering university programs in the US and Middle Eastern region is not accidental. First, unlike information about corporate workers, the data including academic rank, gender, and the country of origin are readily available on universities' official websites. Second, academic staff composition is usually representative of the actual diversity in the field as the same PhD graduates work in both industry and academic institutions (here we consider only positions that require PhD qualification, MSc and BS are not considered as there are no equivalent academic positions for holders of these degrees). Final, petroleum engineering is the biggest and most influential field in the energy technology industry, thus, all conclusions drawn from this analysis would be relevant to most of other smaller energy fields.

Methodology

In the previous section, we highlighted the importance of maintaing diversity inside then energy community for better problem solving and overall health of the industry. To assess the situation in petroleum engineering field as a flagman of the energy technology industry, we collect and analyze the statistical data from 14 US petroleum engineering university program and 4 similar programs in the Middle East. While looking at the data, we specifically focus on rank distribution of faculty members (meaning, only tenure-track professors) to establish approximate age diversity of the academic staff, faculty ethnic background (determined from the country where he or she obtained bachelor degree, as majority of people tend to go for the first degree in their home country), and finally gender diversity (that could only be established for those representatives who provided profile pictures if their name cannot be used for identification purposes). Because many Middle Eastern programs do not provide pictures of their faculty members on their offical websites, a very small sample of programs (four of them) that do provide this means of identification was analyzed here. Thus, in this study, we consider 14 US programs including Texas A&M University, TX, University of Texas in Austin, TX, Penn State University, PA, Texas Tech University, TX, University of Houston, TX, Louisiana State University at Baton Rouge, LA, University of Louisiana at Lafayette, LA, University of Tulsa, OK, University of Oklahoma, OK, New Mexico Institute of M&T, NM, West Virginia University, VA, University of North Dakota, ND, and University of Alaska, AK. From the Middle Eastern universities we selected four including Abu Dhabi Petroleum Institute, UAE, Texas A&M University in Qatar, Amirkabir University, Iran and University of Tehran, Iran.

As shown in **Fig. 1**, the academic staff in the US universities was conservatively growingover last 15 years even though the number of PhD students was rapdly doubling and tripling from the number of students in 2000. This low sensitivity of the faculty count to the increased number of students can be explained by relative unattractiveness of academic careers in petroleum engineering in comparison to those in oil and gas operating companies. While there is a great mobility of highly qualified experts from the industry to academic institutes, most people who choose to become a university professor as a first option usually make a lifetime career choice that unfortunately they cannot not change later on. There are almost no examples of a petorleum engineering university professor who would assume a position in a major corporation as a full time employee. As a result, most PhD graduates try to find industrial positions first and then consider academic career.

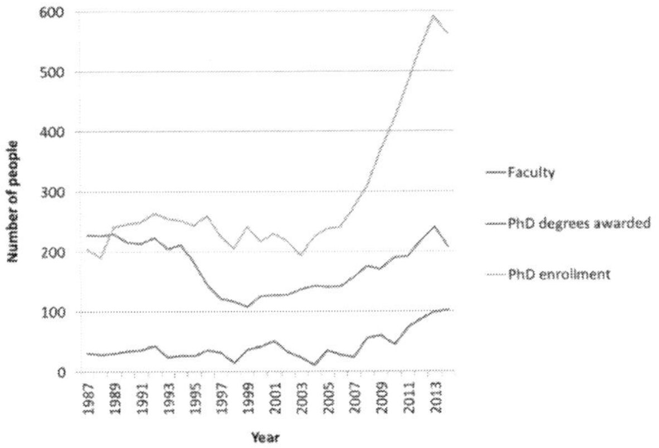

Figure 1. Faculty hiring and PhD enrollment and graduation trends from 1987 to 2014 in the US.

Results

Now, let us observe how these distortions affect the diversity of people in petroleum engineering faculty. Because of the low mobility between the academia and industry, people with certain backgrounds tend to join the university faculty. **Fig. 2(a)** demonstrates how the academic community becomes less diverse and self-selective due to certain restrictions to a certain population of PhD holders. Specifically, second largest percent (42 faculty members) of petroleum engineering academic staff in the US comes from the Middle East. This might look surprising because the number of PhD's from the Middle East to other nationals in petroleum engineering in the US is not that high. However, it is important to keep in mind that for a long time nationals from Iran were not eligible for any industrial positions inside the US.

 DOI: 10.1051/e3sconf/20172308002

As a result, many of PhD graduated would gladly accept an academic position as means of obtaining legal status in the country.

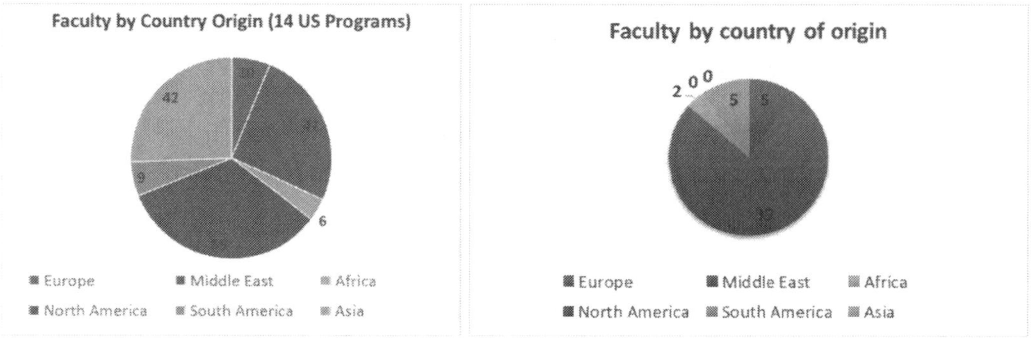

(a) **(b)**

Figure 2. Faculty distibution in (a) the US and (b) in the Middle East based on their country of origin (or the country where they obtained their bachelor degree).

While this situation is not too detrimental to diversity on a scale of one school, taken to the national level, it might create tremendous problem. Specifically, professors of certain background (let us say, Iran) are much more comfortable with students from the same ethnic and cultural background, thus, the number of people from the same geographic region rapidly increases making the community self-selective and discouraging students from other ethnic and gender backgrounds to enter programs. **Fig. 2(b)** illustrates similar distribution for the four Middle Eastern universities. As clear from the chart, the diversity situation in these programs are much worse than in the US: some regions (Europe and North America) are not represented at all. While it might be a good idea to attract local talent to educate local students, these programs and, thus, their graduates would never be able to compete with those who were educated in a diverse environment, saw many different cultures and ways of solving problems.

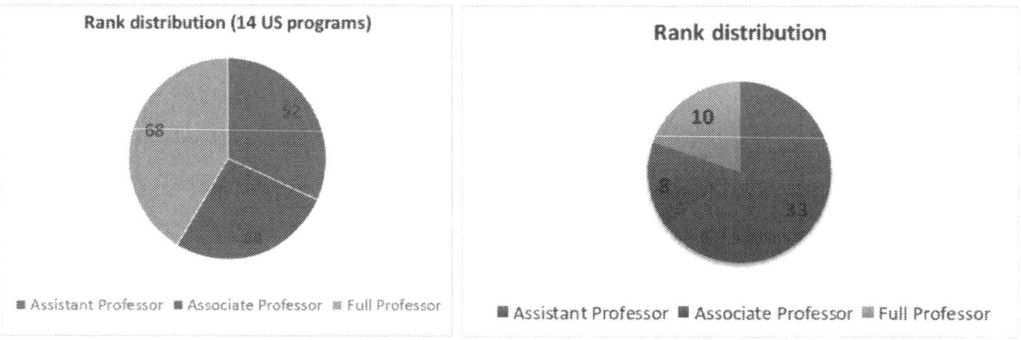

DOI: 10.1051/e3sconf/20172308002

(a) **(b)**

Figure 3. Faculty rank distibution in (a) the US and (b) in the Middle East based on tenure-track faculty.

Another important indication of diversity is age of university faculty. **Fig. 3(a)** shows rank distribution of faculty in the US for 14 chosen schools. Typically, assistant professors are relatively young faculty members from late 20s to 40s, associated professors are from 40s to 50s and then they become full professors. As the chart shows, in the US the ratio of the three categories is almost ideal and is roughly 1:1:1 with a slight distortion toward older faculty members (full professors). **Fig. 3(b)** shows a completely different situation in the Middle East with about 65% of the faculty being assistant professors. It is hard to explain why would the programs have so many junior faculty members without promoting them to associate or full rank. Either the Middle Easter universities do not retain their faculty well and, thus, have very few senior faculty members, or they have a completely different definition of academic ranks that has nothing to do with description of these ranks in the US. Perhaps, in these universities tenure is not even an option for academic staff and award of associate and full professorship has completely different criteria.

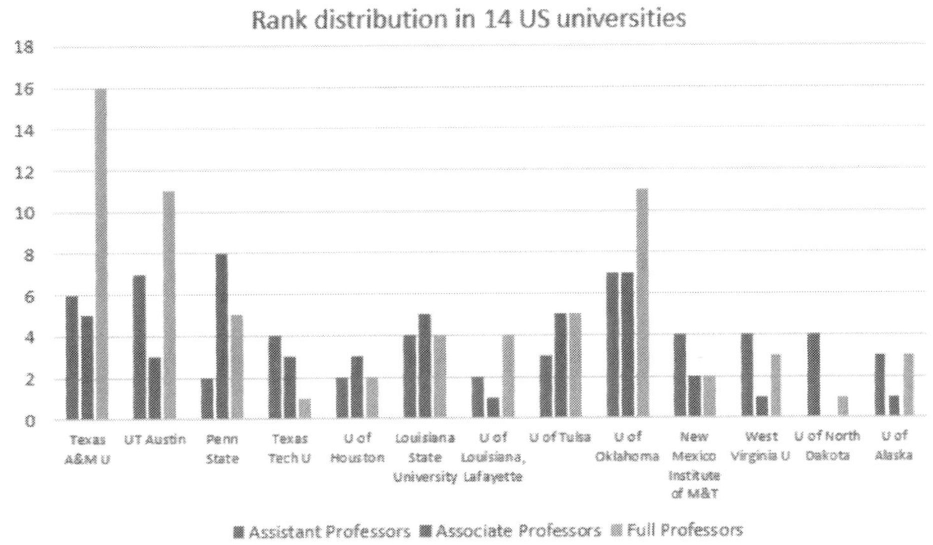

Figure 4. Faculty rank distibution in the US based on tenure-track faculty plotted by university.

DOI: 10.1051/e3sconf/20172308002

Fig. 4 provide a more detailed distribution of faculty by rank by each US university. On this plot, the reader can observe that in almost each school there is some distortion toward either rank. For example, Texas A&M has the highest number of full professors and incomparably low numbers of associate and assistant professors. University of North Dakota, however, has no associate professor at all. These features can be explained by relative age of these programs and their school rankings. While Texas A&M has a relatively old and well-established program that can attract by high pay many prominent scholars, University of North Dakota is a new program that cannot afford to pay for a well-known name and, thus, primarily chooses assistant professors from their own graduates.

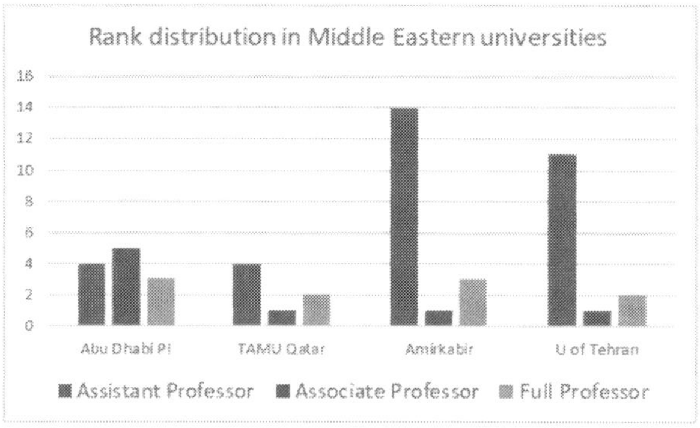

Figure 5. Faculty rank distibution in the Middle East based on tenure-track faculty plotted by university.

Fig. 5 provides similar rank distribution for the four universities in the Middle East. From this bar chart, the reader can observe that UAE and Qatar programs, being influenced by the western system of education and frequently visited by western scholars, are very similar to the majority of the US based programs. Iranian programs, on the other hand, are drastically different and more alike University of North Dakota (which is mostly staffed with the academic members of Iranian heritage). This is an interesting observation how communities with certain traits become self-selective and start looking similarly even though they are in different geographic regions.

DOI: 10.1051/e3sconf/20172308002

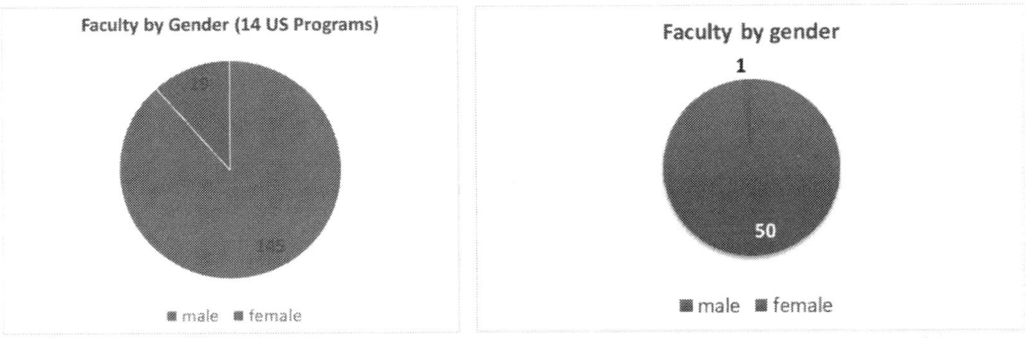

(a) **(b)**

Figure 6. Faculty gender distibution in (a) the US and (b) in the Middle East based on tenure-track faculty.

The final aspect of diversity to discuss in this section is gender diversity. Petroleum engineering has been traditionally a male-dominated community. Recently, big academic programs in the US have stated their commitment to great gender diversity and made a continuous effort to attract more qualified female PhD graduate to their academic staff. While the ratio of men to women among PhD petroleum engineering students is about 3:1, the percent of female professors is still only 13% in the US (**Fig. 6(a)**). The situation with gender diversity in the Middle East is still very alarming. As shown in **Fig. 6(b)** in the four petroleum engineering programs selected in this study, women comprise only 2%. This distortion can be easily explained by unattractiveness of these institutions (due to their regional culture) for most western female PhD holders and scarcity of domestic female PhD students.

Discussion

Demographic, ethnic, and gender diversity is key to any organization, its health and prosperity. Nonetheless, frequently we observe that, if left to their own devices, many social and professional groups become self-selective and less diverse. For example, this phenomenon can be readily observed in the academia when a professor forms a research group that is mostly comprised of people from certain origin or personal qualities. While on an individual level, this situation can be quite convenient and even beneficial, on the large scale the lack of diversity creates severe distortions and discourages many talented individuals from joining the group, organization, or industry in general.

Conclusions

In this study, we analyzed statistical data from various petroleum engineering departments at universities from the US and the Middle East and observed some alarming trends. Our analysis revealed an unexpected picture inside the petroleum engineering discipline. For example, there is obvious inefficiency in diversity staffing in universities worldwide. We carefully investigated this phenomenon and uncovered the true reasons behind staggering demographic gaps. The statistical assessment shows a particularly alarming situation among junior faculty members coming from preferential national group which manifests in scarcity of female representatives (some schools such as University of Tulsa, Texas Tech, ULL and University of Wyoming do not have any female faculty members in their petroleum division, while the rest of the programs have barely 5-15% of the entire faculty counts with rare exceptions such as Texas A&M and UT Austin who try to maintain greater demographic diversity). This phenomenon observed even though most programs try to keep their graduate student male to female ratio close to 3:1 with substantial number of women continuing their education toward PhD in recent years. Clearly, there are some discouraging factors for women to become university educators that we tried to elaborate on.

Moreover, our investigation of mobility of faculty members revealed that most schools organize conclaves of their own graduates by retaining them as faculty members which also impacts diversity in a negative way. Poor mobility leads to a rapid decline in research and demographic diversity because graduates of other schools and those with other research interests have difficulties joining these self-selective programs.

References

[1] Phillips, K. 2014. How Diversity Makes Us Smarter. *Scientific American* Oct. 1, 2014. Accessed at https://www.scientificamerican.com/article/how-diversity-makes-us-smarter/.

 DOI: 10.1051/e3sconf/20172308003

The Effect of Solar Reflective Cover on Soak Air Temperature and Thermal Comfort of Car Parked under the Sun

A.A. Lahimer [a], M.A. Alghoul, [b, c], K. Sopian*, [a], N. G. Khrit [a]*
[a] *Solar Energy Research Institute, Universiti Kebangsaan Malaysia, 43600 Bangi, Selangor, Malaysia*
[b] *Energy and Building Research Center, Kuwait Institute for Scientific Research, P.O. Box 24885, Safat 13109, Kuwait*
[c] *Center of Research Excellence in Renewable Energy (CoRe-RE), Research Institute, King Fahd University of Petroleum and Minerals (KFUPM), Dhahran 31261, Saudi Arabia*
**Corresponding authors: dr.alghoul@gmail.com (M.A. Alghoul), ksopian@ukm.edu.my (K. Sopian)*

ABSTRACT

Parking a vehicle under the sun for a short period of time can rapidly increase the interior air cabin temperature no matter in clear sky days or even in partially cloudy days. These circumstances can be anxieties to car occupants upon entry. The aim of this paper is to evaluate experimentally the effect of solar reflective cover (SRC) on vehicle air temperature and cabin thermal comfort. Experimental measurements of parked cars were conducted in UKM, Bangi city, Malaysia (latitude of 2.9° N and longitude of 101.78° E) under partially cloudy day where average ambient temperature is 33°C. The experimental measurements cover the following cases: case (I): car with/ without SRC (at different measurement time); Case (II): using two identical cars concurrently (SRC versus baseline); Case (III): using two identical cars concurrently (solar reflective film (SRF) versus baseline) and Case (IV): using two identical cars concurrently (SRF versus SRC). Experimental results dedicated to case (I) revealed that the maximum cabin air temperature with SRC (39.6°C) is significantly lower than that of baseline case (57.3°C). This leads to temperature reduction improvement of 31% and the difference between the cabin and the ambient air temperature was minimized by approximately 73%. In addition, the results revealed that the air temperature at breath level of car with SRC dropped to comfort temperature (27°C) after 7 min while baseline car reached comfort temperature after 14 min. Results of the other cases are discussed inside the paper. Overall, it is learned that SRC is found superior as an efficient thermal insulation system limits solar radiation transmission into the cabin through the glass; keeps cabin air temperature close to the ambient temperature; and provide acceptable thermal environment to the occupants as they settle into their parked car.

Keywords: parked vehicle under the sun, cabin soak air temperature, solar reflective aluminum cover, cabin thermal comfort.

1. INTRODUCTION

Most vehicles have large-sized glazed surfaces area with respect to the total cabin surface area. Large percentage of glass is inclined with respect to the horizontal which maximizes the entry of the incident solar radiation. (Rugh and Farrington 2008) claimed that typically 50% to 75% of the thermal energy entering the car cabin is due to transmitted and absorbed solar energy from the glazing. Windshield alone accounts for more than 40% of heat transmitted into the cabin (Rugh, Chaney et al. 2013). Therefore, the incident solar radiation transmitted through the glass of the car is the main cause of high cabin temperature, it can increase the interiors temperature for the first 15 min at a rate of 1°C/min (Parker 1988; McLaren, Null et al. 2005). As a result, the extreme maximum cabin temperature which can be as much as two times of the ambient temperature or even more can be reached at about 45 min. The unpleasant environment will take long time (usually 15 min) to cool down by A/C system to reach human thermal comfort level. In addition, this peak cooling load has a negative impact on A/C size, fuel consumption and emission. Therefore, blocking the sun rays is the only approach that can have a significant cabin temperature reduction. For example, (Mezrhab and Bouzidi 2006) achieved a considerable reduction of the temperature inside the compartment up to 10°C by using reflecting glazing (low transmissivity and absorptivity). (John P. Rugh, Lawrence Chaney et al. 2007) achieved 9.7°C temperature reduction by using solar-reflective glass and solar-reflective paint. (Rugh and Farrington 2008) recommended that, exterior mounted window shades would be more suitable than interior window shades. (Hoke and Greiner 2005; Rugh and Farrington 2008; Levinson, Pan et al. 2011) studied the effect of solar reflectance

DOI: 10.1051/e3sconf/20172308003

on the cabin air soak temperature (low solar reflectance for dark color and high solar reflectance for light color). Based on their studies, the cabin air soak temperature was reduced by approximately 1°C for each 0.1 increase in the solar reflectance of the cabin coatings.

2. Materials and Methods

The objective of this study is to study the effect of SRC (a commercially available half top cover which made of aluminum laminated with non-woven; low solar absorptance and low thermal emittance) on cabin soak air temperature, thermal comfort and AC compressor of car parked under the sun during cloudy days. In addition to compare SRC to other thermal load reduction strategies such as solar reflective film SRF.

2.1. Experimental Setup

Two identical cars with same interior and exterior were parked, side by side, in an open grassy space under sun and oriented facing south without shadow interference during the experimental measurements. The seats were adjusted to the same position and all windows remained closed throughout the test. For the evaluation of cabin soak temperature and thermal comfort, each vehicle was instrumented with HD32.3A (fig.1) to measure the interior parameters (globe thermometer temperature T_g, natural wet bulb temperature T_n, environment temperature T, relative humidity RH and air speed V_a) in 60 sec intervals at breath-level height midway between the front seat headrests. Weather data such as ambient temperatures were measured by TECMAN TM820M (fig.1) at a nearby building and global horizontal solar irradiance was obtained by SP-110: self-powered silicon-cell pyranometers (fig.1). The average values of the ambient temperature and global horizontal solar irradiance were considered.

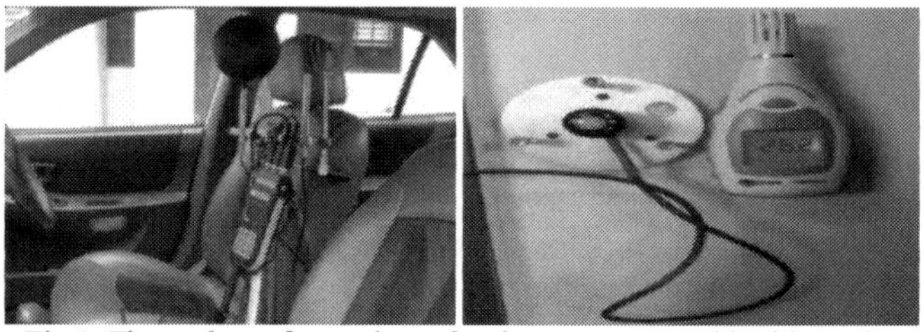

Fig 1: Thermal comfort and weather instruments used in this study

2.2. Experimental Procedures

Sets of experimental measurements on parked cars have been conducted for the following cases, including:

- *Case (I) using one car with/ without SRC (measurement at different time)*

In May 23, 2016, a thermal soak test from12:00-1:00 PM was performed on one car covered with SRC followed immediately by 18min cooling. The initial cabin and average ambient temperatures were 22°C and 32°C, respectively. After 5min from the end of the previous cooling trial, another soaking experiment from1:25-2:25 PM was performed on the same car but without SRC followed immediately by 18min cooling trial. The initial cabin temperature in this test was 27.6°C and the average ambient temperature was 33°C.

- *Case (II) using two identical cars concurrently (SRC versus baseline)*

Soaking trial from 12:55-2:10 PM (75min) was performed on two identical cars in October 15, 2016. One car was covered with SRC and the other one served as a baseline and remained unmodified as shown in fig 2. The aim was to study the effect of SRC at equally initial cabin and ambient temperatures for both cars. The average ambient temperature and the solar irradiance were 35°C and 696 W/m^2, respectively.

DOI: 10.1051/e3sconf/20172308003

Fig 2: Two identical cars with and without solar reflective cover at the experimental field

- *Case (III) using two identical cars concurrently (SRF versus baseline)*

In November 13, 2016, a thermal soaking test from 12:30-1:04 PM (34min) was performed on two identical cars followed immediately and simultaneously by 21min cooling trial. One car served as a baseline without integrating any temperature reducing techniques; while the other one was fitted with SRF on every window (about 65% infrared rejection) except the front windshield remained unmodified. Prior to testing, maximum fan speed and recirculation mode were used in each vehicle to monitor and adjust the AC vent temperatures. Accordingly, the vent temperatures set at the same values for both cars to assure similar AC cooling performance. The average initial cabin temperatures for both cars were 36.7°C. The aim was to study the effect of SRF on cabin soak air temperature and occupants thermal comfort.

- *Case (IV) using two identical cars concurrently (SRC versus SRF)*

Immediately, after the end of the previous cooling trial, another soaking experiment from1:27-2:33 PM was performed on the same cars but the baseline was covered with SRC followed immediately and simultaneously by 16min cooling trial. The average initial cabin temperatures for both cars were 37.6°C. The average ambient temperature and the solar irradiance were 33.8°C and 779 W/m^2, respectively for both trials.

3. RESULTS AND DISCUSSION

3.1 Cabin Soaks Air Temperature Comparison

- **Case (I)**

The cabin car measurements started after cooling the cabin at comfort level for 20 min using the AC system then it is turned off. As the AC turned off in the parking lot, the results showed that the cabin air temperature rapidly reached the outside ambient temperature within the first 5 minutes at a rate of 2°C/min. This result was also observed in (McLaren, Null et al. 2005).

Fig. 3 shows one hour of car soak air temperature under direct sun. It can be seen that the maximum cabin air temperature observed for both cases were reached at the same period of time at about 45 min regardless of any approach used, which was expected when the same car was used for the study. Also, it can be seen in Fig. 3 that the maximum cabin air temperature for case I with SRC (39.6 °C) is significantly lower than that of baseline case (57.3 °C). The breath level air temperature in case I with SRC is approximately 17.7°C cooler than the baseline of case I without SRC. These results could be attributed to the low solar low thermal emittance and absorptance of the cover, which lead to maintain an average temperature difference of 5°C between the cabin and the ambient. However, the soaking test results showed that the SRC demonstrated a temperature reduction improvement of 31%, and consequently reducing the heat gained by the cabin. This means that the difference between the cabin and the ambient air temperature was reduced by approximately 73% by using SRC.

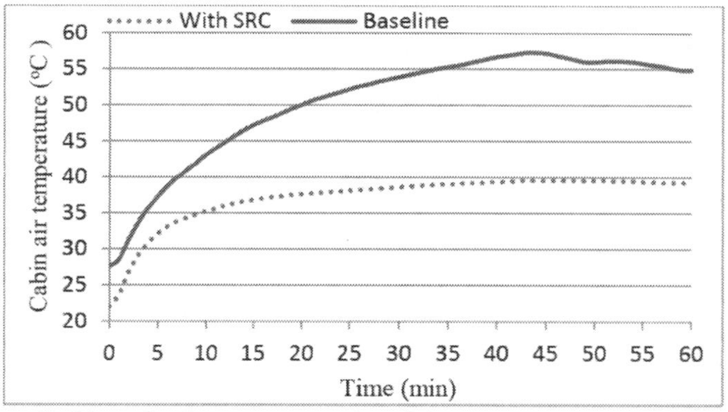

Fig 3: Comparison of cabin soaks air temperature at breath level (SRC versus baseline) using one car with measurements at different time, **case I**

- **Case (II)**

In this trial, both identical cars were subjected to 75min thermal soak trial under direct sun and started the soaking trial with an initial cabin air temperature as same as the average ambient air temperature. Fig. 4 shows that, for the first 15min the cabin air temperature in the car with SRC increased sharply till reached the maximum value of 42°C and then almost continued steady until the end of 1h. SRC was able to maintain an average temperature difference of 6°C between the cabin and the ambient. However, the cabin air temperature in the car without SRC also increased sharply for the first 15min and then continued to increase till it reached the maximum value of 55.4°C at the end of 1h of the soaking test. This resulted in breath level air temperature difference approximately 13.4°C between the SRC and baseline case. Although for the car without SRC it should be also noted that the cabin air temperature have to increase or at last remains steady during the peak solar irradiance (12:00-3:00 PM), however the temperature drop is contributed to the weather condition as the sky become very cloudy.

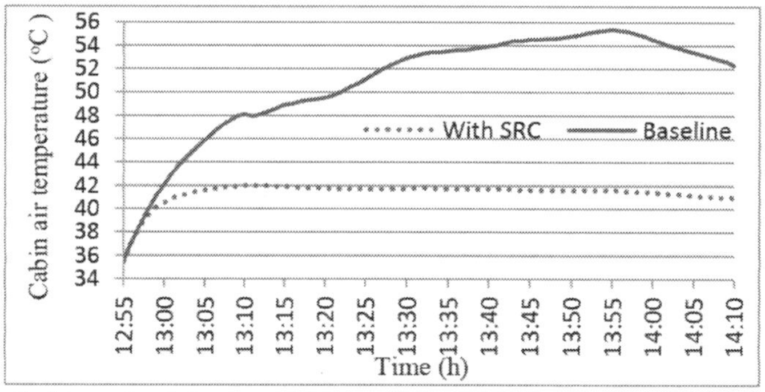

Fig 4: Comparison of cabin soaks air temperature at breath level (SRC versus baseline) using two identical cars concurrently

- **Case (III)**

Fig. 5 shows the soak cabin air temperatures of the baseline and SRF cases. At the beginning of the thermal soak trial, the air temperatures inside SRF car were hotter than the baseline temperatures. In fact, the breath-level air temperatures in the baseline case were approximately 0.83°C cooler than in the case with the SRF. After 18min the soak cabin air temperatures for the baseline case and the SRF experiments are almost identical, where the average difference in cabin air temperature between the two cars was 0.39°C and the maximum difference was 0.1°C. This

DOI: 10.1051/e3sconf/20172308003

shows inability of SRF in reducing the interior cabin air temperature and does not have any significant difference from the baseline.

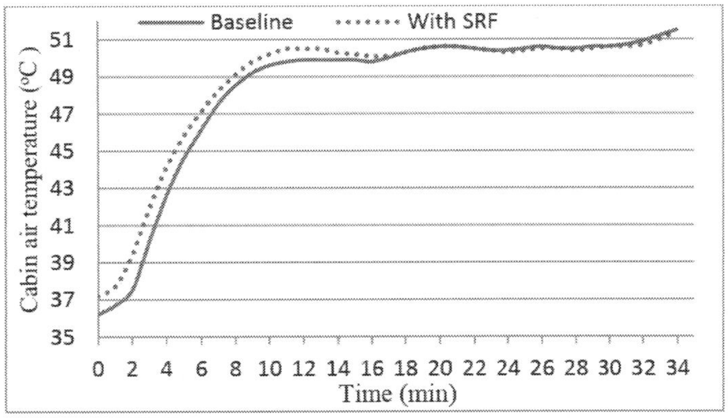

Fig 5: Comparison of cabin soaks air temperature at breath level (SRF versus baseline) using two identical cars concurrently

- **Case (IV)**

The soak cabin air temperatures of SRC and SRF cases were compared and shown in fig. 6. The air temperatures increased sharply in SRC case until 7min and then become almost steady with maximum value of 45.9°C. On the other hand, in SRF case the air temperatures increased sharply until 14min and then continue to increase to reach the maximum value of 60.1°C. Moreover, the breath-level air temperatures in the case with the SRC were approximately 14.2°C cooler than the SRF case, demonstrating a significant impact from SRC for this thermal soak test. However, as expected, the SRF case has the highest interior temperatures during thermal soak due to the direct solar radiation transmitted through the front windshield and also windows tint allows some. In addition, windows tint leads to increase the absorption due to its color and become an extra source of heat.

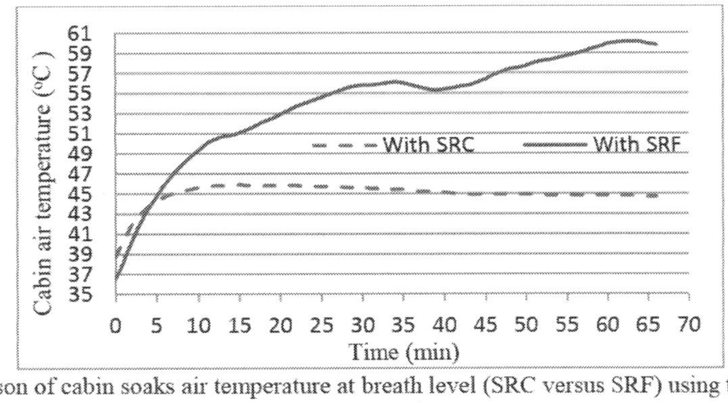

Fig 6: Comparison of cabin soaks air temperature at breath level (SRC versus SRF) using two identical cars concurrently

3.2 Thermal Comfort Evaluation

In this round of analysis, the comfort level is considered with PMV between (-0.50 and 5) and the PPD is less than 10%; therefore comfort temperature (Tc) is 27°C. The cooling down measurements of cases (I, III and IV) are plotted

 DOI: 10.1051/e3sconf/20172308003

in figures 7, 8 and 9 respectively. Since, the cooling down behavior of the other cases is similar to Case I, however, the discussion will be dedicated to case I only.

After 1 h of soak under direct sun, 18 min transient temperature measurements using AC system to cool-down the cabin for both trials in case I are shown in Fig 7. In general, it is clearly seen that the cabin air temperature in case I with SRC cooled faster than baseline case (without SRC) since the cooling down trail started. The cabin breath level air temperature dropped to comfort temperature (27°C) after 7 min in case I with SRC while baseline case reached comfort temperature after 14 min. Thus, it can be concluded that occupants used the cover will reach the comfort level 7 min faster than baseline case during the cool-down trials. The time difference is attributed to the lower initial air temperature of the cabin when it started the cool-down for case I with SRC, whereas the cabin for baseline case started the cool-down with a higher air temperature. However, at the end of the cooling down trials, the air temperature dropped to 21.5 °C with SRC and approximately 24.5 °C in baseline case. A temperature difference of approximately 3°C is still existed between the modified and baseline cases.

The lessons learned from figures (3, 4 and 7) that in both situations (with/without SRC) at the end of the soaking test, the comfort indices values were out of the comfort zone. But the occupants will still feel a significant temperature reduction in case with SRC upon entry. Therefore, reducing the thermal solar loads using SRC could reduce the exposure time of occupants for high temperature, thereby reducing their thermal stress and anxiety and improving vehicle fuel consumption.

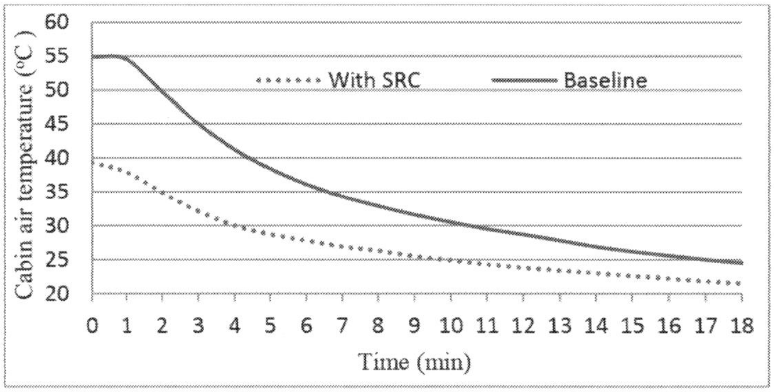

Fig 7: Car cooling down (SRC versus baseline) for case (I)

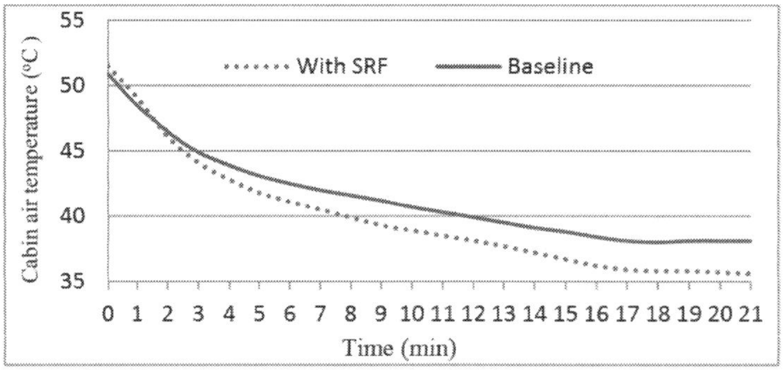

Fig 8: Cooling down of two identical cars (SRF versus baseline) for case (III)

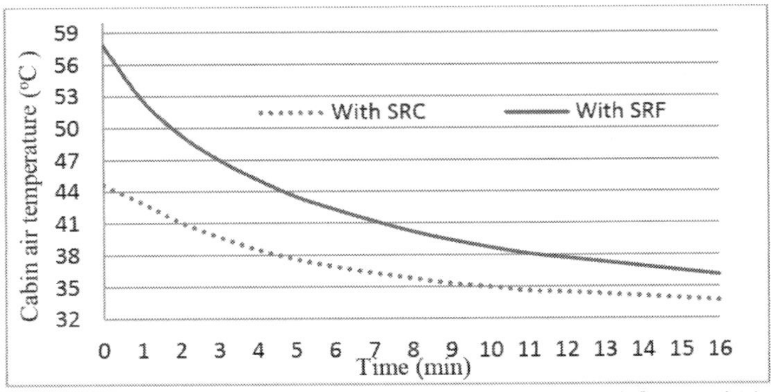

Fig 9: Cooling down of two identical cars (SRC versus SRF) for case (IV)

4. CONCLUSION

In the work, experimental approach is adopted to evaluate the effect of SRC on vehicle soak air temperature, thermal comfort under partial cloudy sky conditions of UKM, Bangi city, Malaysia. The results confirmed the significant effects of SRC over other approaches such as SRF in terms of cabin soak air temperature, reducing time required to achieve thermal comfort and preventing cracking and fading of cabin's interior.

References

1. Fanger, P. O. (1970). Thermal comfort: Analysis and applications in environmental engineering, Danish Technical Press.
2. Hoke, P. B. and C. M. Greiner (2005). Vehicle Paint Radiation Properties and Affect on Vehicle Soak Temperature, Climate Control System Load, and Fuel Economy, SAE Technical Paper 2005-01-1880, doi:10.4271/2005-01-1880. Detroit, Michigan, USA, Warrendale, PA: SAE International Publication.
3. John P. Rugh, Lawrence Chaney, et al. (2007). Reduction in Vehicle Temperatures and Fuel Use from Cabin Ventilation, Solar-Reflective Paint, and a New Solar-Reflective Glazing. SAE World Congress. : 1-8. Detroit, Michiga, USA, Warrendale, PA : SAE International Publication
4. Levinson, R., H. Pan, et al. (2011). "Potential benefits of solar reflective car shells: Cooler cabins, fuel savings and emission reductions." Applied Energy **88**(12): 4343-4357.
5. McLaren, C., J. Null, et al. (2005). "Heat Stress From Enclosed Vehicles: Moderate Ambient Temperatures Cause Significant Temperature Rise in Enclosed Vehicles." Pediatrics **116**(1): e109-e112.
6. Mezrhab, A. and M. Bouzidi (2006). "Computation of thermal comfort inside a passenger car compartment." Applied Thermal Engineering **26**(14–15): 1697-1704.
7. Rugh, J. (2002). Integrated Numerical Modeling Process for Evaluating Automobile Climate Control Systems Paper presented at Future Car Congress, Crystal City, VA, USA, Paper # 2002-01-1956,Society of Automotive Engineers.
8. Rugh, J., L. Chaney, et al. (2013). Impact of Solar Control PVB Glass on Vehicle Interior Temperatures, Air-Conditioning Capacity, Fuel Consumption, and Vehicle Range. SAE 2013 World Congress & Exhibition. 2013-01-0553. U.S, National Renewable Energy Laboratory (U.S.)) ; 5400-57489.
9. Warrendale, Penn.: Society of Automotive Engineers
10. Rugh, J. and R. Farrington (2008). Vehicle Ancillary Load Reduction Project Close-Out Report: An Overview of the Task and a Compilation of the Research Results, National Renewable Energy Laboratory. USA, Golden Publication, CO.

DOI: 10.1051/e3sconf/20172308003

11. Rugh, J. P., T. J. Hendricks, et al. (2001). Effect of Solar Reflective Glazing on Ford Explorer Climate Control, Fuel Economy, and Emissions. International Body Engineering Conference. Detroit, MI, USA, Society of Automotive Engineers Publication.

 DOI: 10.1051/e3sconf/20172309001

The Feasibility of Utilizing Wind Energy in Commercial Buildings With Special Reference to the Kingdom of Bahrain

Saeed Abdulrahim Saeed*
Department of Architecture and Interior Design
College of Engineering, University of Bahrain
sabdulrahim@uob.edu.bh; saeedibnouf@gmail.com

Abstract

This article shall investigate the feasibility of utilizing wind energy for commercial buildings with special reference to the Kingdom of Bahrain. Bahrain World Trade Center which was built in 2008, is located in the city of Manama. The fifty-story complex contains identical twin towers that rise over 240 meters in height. The towers are connected by three bridges which hold three turbines each 29 meters long. The three turbines were originally design to provide electric energy required for lighting. The Bahrain World Trade Center was selected as a case study to investigate the feasibility of utilizing wind energy technologies in skyscrapers with special reference to the Kingdom of Bahrain. It is hoped that the findings and conclusion of the study shall be of some value for future utilization of wind energy in the GCC countries and the world at large.

Keywords: wind energy, renewable energy in buildings, wind turbines and skyscrapers

Introduction

Buildings in the Gulf Countries; Kingdom of Saudi Arabia, United Arab Emirate, Kuwait, Oman and the Kingdom of Bahrain are consuming the largest proportion of the generated fossil energy production. In fact the electric energy consumption is increasing at a very high rate due to the increase in population and the vast expansion of urban settlements. Bahrain produces 2800 MW of electricity and 140 million gallons per day of desalinated water using fossil fuel as the main energy source. In 2009, a consultation committee was established with a view to establishing a solar and wind hybrid pilot project with a capacity of between 3 and 5 MW. Bahrain ratified the Kyoto Protocol in May 2006. However, the country remains heavily dependent on the oil sector. Petroleum revenues make up about two-thirds of government revenue and export earnings. Hydrocarbons also provide the foundation for Bahrain's two major industries; refining and aluminum smelter. In addition, in Bahrain as with other countries across the GCC, electricity production is heavily subsidized and supplied at a price which is far less than the cost of generation[1]. Cheap electric energy cost may not encourage people to reduce their energy consumption. Recently the Water and Electricity Authority in the Kingdom of Bahrain, is in the process to increase the cost of both water and electricity, this might help in decreasing electricity and water consumption. Approximately, 60% of the generated electric energy in Bahrain is consumed in buildings, and the remaining percentage is consumed by all other sectors, i.e. industry, transportation, communication, agriculture etc. Furthermore, buildings emits 30% of carbon dioxide and 25% of nitrogen oxide found in the air [2]. Currently the vast majority of this energy is produced from non-renewable fossil fuel sources, i.e. oil and gas. With uncertainty over the availability of oil and gas productions into the future, rising demands for fossil energy and increasing the emission of CO2 which is causing considerable climatic changes. Due to the global warming, there is a vital need for reducing the emission of CO2. The proposed strategy could summarized in the Figure 1.

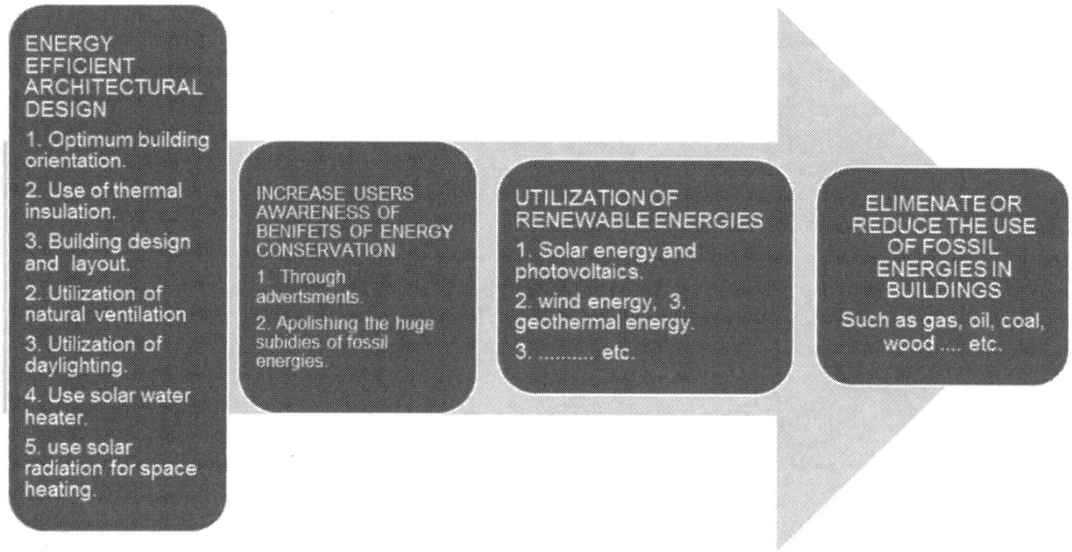

Figure 1. Energy Conservation Strategies.

a). Energy Efficient Architectural Design:

Architects and designers should adopt energy efficient architectural design strategies with the objective of utilizing the natural means to reduce energy needs. Energy conservation measures, and passive solar architectural design techniques which should be implemented, include; optimum building orientation in relation to sun movement and prevailing wind direction, adequate thermal insulation, use of reflected glass for openings, sun protection and proper shading devices ... etc. When energy efficient architectural design or the so called passive solar energy fails to provide optimum thermal comfort requirement renewable energy technologies could be adopted. Renewable energy is generally defined as energy that is collected from sources which are naturally replenished on a human time scale, such as sunlight, wind, rain, tides, waves and geothermal heat. Renewable energy often provides energy in four important areas: electricity generation, air and water heating and cooling, transportation and rural energy services. Based on REN21's 2016 report, renewable energies contributed 19.2% and 23.7% to humans' global energy consumption and the generation of electricity in 2014 and 2015, respectively. This energy consumption is divided as 8.9% coming from traditional biomass, 4.2% as heat energy (modern biomass, geothermal and solar heat), 3.9% hydro-electricity and 2.2% is electricity from wind, solar, geothermal, and biomass. REN21's also indicated that worldwide investments in renewable technologies amounted to more than US$286 billion in 2015, with countries like China and the United States heavily investing in wind, hydro, solar and biofuels. According to the International Energy Agency 2011, solar energy, radiant light and heat from the sun, is harnessed using a range of ever-evolving technologies such as solar heating, photovoltaic's, concentrated solar power (CSP), concentrator photovoltaic's (CPV), solar architecture and artificial photosynthesis[3].

The main renewable energy technologies which are being widely used in buildings, are photovoltaic's (active solar energy) and wind energy. Solar cells when properly located on roofs and walls may provide a good proportion of electric energy which could be used for different indoor activities. The solar radiation falling on the solar cells shall be converted to a D.C. electric current which could be transformed to A.C electric current ready to be used by any electrical appliances. The production of electricity by photovoltaic's is getting a considerable attention by scientists and engineers, and a lot of developments is in progress with the objective of increasing the efficiency of electric energy production and reducing the capital and running cost.

b). Wind Energy:

Wind moves horizontally due to difference in pressure, from high pressure zone to a lower pressure zone, and moves vertically due to difference in temperature. In fact wind exists because the sun unevenly heats the surface of the earth. The temperature of air molecules which are in touch of the heated surface of the earth will increase, their density will decrease and become lighter and will move upward being replaced by cooler molecules. This air movement will continue as long as the sun shines. Wind power is the conversion of wind energy into alternate forms of energy, including wind turbines for the production of electrical wind power, windmills for mechanical power, and wind pumps for water pumping or drainage. Wind energy was used by mariners to move boats and ships, and by farmers to grind their grains and pump water. Recently more people are using wind turbines to generate electricity. Wind is a clean energy that

 DOI: 10.1051/e3sconf/20172309001

produces no air or water pollution. Since the wind is free, operational costs are nearly zero once a turbine is erected. Mass production and technology advances are making turbines cheaper, which will spread the utilization of wind energy in buildings[1].

The term wind energy or wind power describes the process by which the wind is used to generate mechanical power or electricity. Wind turbine convert the kinetic energy into mechanical power. A wind turbine is a machine that transforms the kinetic energy of the wind into mechanical or electrical energy. Generally, wind turbines consist of a foundation, a tower, a nacelle and a rotor. The foundation is for supporting the turbine, while the tower holds up the rotor and a nacelle (or box). The nacelle contains large primary components such as the main axle, gearbox, generator, transformer and control system. The rotor is made of 2 or 3 blades and the hub which holds them in position as they turn[1]. Wind turbines could be basically classified into two groups; horizontal-axis turbines and vertical axis turbines. Mathias Aarre Maehlum[4], listed the most important facts about the wind power that should be included on any balanced wind energy, these are as horizontal and vertical wind turbines[4]. Horizontal Axis Wind Turbines require a "yaw" mechanism to turn the turbine against the wind direction in order to harvest the wind energy. They are usually installed high up above ground level on a tower in order to harvest the ample wind energy up in the sky. There are 2 types of blades that are used on these turbines[4]. The turbines are usually installed on or close to the ground level. However, the wind speed is always lower than that in the upper sky. There are also 2 types of blades that are used on these turbines[5].

Horizontal Axis Wind Turbines require a "yaw" mechanism to turn the turbine against the wind direction in order to harvest the wind energy. They are usually installed high up above ground level on a tower in order to harvest the ample wind energy up in the sky. There are 2 types of blades that are used on these turbines[4].

HAWT

VAWT

While the Vertical Axis Wind Turbines do not require a "yaw" mechanism to harvest the wind energy. Since the blades rotate 360 degree on the vertical shaft of the wind turbines, wind of any direction can turn the turbine.

The continuous development and technological advancement in the wind technology resulted in a wide variation in the design of wind turbines. This

DOI: 10.1051/e3sconf/20172309001

variations in the design of wind turbines reflects ongoing commercial, technological, and inventive interests in harvesting wind for the generation of energy. A quantitative measure of the wind energy available at any location is called the Wind Power Density (WPD). It is a calculation of the mean annual power available per square meter of swept area of a turbine., and is tabulated for different heights above ground. Calculation of wind power density includes the effect of wind velocity and air density[6].

Advantages and disadvantages of wind energy technology could be summarized as follows[4];

i).). Advantages:

- Wind energy is a green energy source. Harnessing wind energy does not pollute the environment nearly as much as fossil, coal and nuclear power do.
- The potential of wind power is absolutely incredible. Harnessing wind energy can be done almost anywhere.
- Wind energy is a renewable source of energy. Wind is naturally occurring and there is no way we can empty the energy resources.
- i). Advantages:
- Wind energy is a green energy source. Harnessing wind energy does not pollute the environment nearly as much as fossil, coal and nuclear power do.
- The potential of wind power is absolutely incredible. Harnessing wind energy can be done almost anywhere.

ii). Disadvantages:

- Wind is unpredictable and the availability of wind energy is not constant. Wind energy is therefore not well suited as a base load energy source if we had a cost-effective ways of storing wind energy the situation would be different.
- We can hope for breakthroughs in energy storage technologies in future, but right no turbines have to be used in tandem with other energy sources.
- Birds, bats and other flying creatures have slim chances to of surviving when taking a direct hit from a rotating wind turbine.
- Noise is a problem for some people that live in the proximity of wind turbines building wind turbines in urban environments should be avoided

Alex Wilson[7], advocated that, building-integrated wind often doesn't live up to its promise. The turbines must overcome several challenges to meet performance expectations and be cost effective. Noise and vibration from wind turbines are among the greatest obstacles to integrating them into buildings. Based on the recent surge in building-integrated wind, one might think that engineers had beaten this problem. In truth, some wind turbines are a lot quieter than others- vertical-axis machines among them, but managing noise and vibration remains a huge challenge. Wind speed typically increases with height, as it is less affected by trees and surrounding topography. Putting wind turbines on top of buildings—especially tall buildings—should allow them to take advantage of height without an expensive, full-size tower. In some cases, building geometry can enhance wind turbine performance. Several manufacturers of building-integrated wind turbines are taking advantage of the increased wind velocities at building parapets—where the wind rises up the façade of a large building and curls over the edge. Some architects are designing wind scoops right into the structures of buildings or situating building towers to funnel wind into turbines. Most of our electricity is used in buildings, and generating the electricity on site reduces the need for transmission. This in turn reduces transmission losses as well as the materials needed for wiring and poles. In addition to this practical benefit, wind turbines spinning on a building provide a visible testament to a building owner's commitment to the environment[8].

In the wind turbines, the wind passes over the blades creating mechanical movement, which causes the rotor to turn. The blades turn a low-speed shaft inside the nacelle gears connect the low speed shaft of the rotor with a high speed shaft that drives a generator. Some of wind turbines do not contain gearbox and instead use a direct drive mechanism to produce power the generator. The rapidly spinning shaft drives the generator to produce electric energy. Electricity from the generator goes to a transformer which converts it to the right voltage for electricity grid. The electricity is then transmitted via the electricity network. The optimum number of blades for a wind turbine depends on the job the turbine has to do. Turbines for generating electricity need to be operate at high speeds, but do not need turning force. These machines generally have three or two blades. Two bladed machines are cheaper and lighter, with higher running speeds which reduces the costs of the gearbox, and they are easier to install. Their performance is somewhat similar to the three blades turbines, but according to experts, they produce more noise when compared with the three blades turbines, and they are not visually attractive when they are running[8]. Wind energy technologies can be used as stand-alone applications, connected to a utility grid, or even combined with photovoltaic system. Single small turbines, below 100 kilowatts, are used for homes, telecommunications dishes, or water pumping. Small turbines are sometimes used in connection with diesel generators, batteries, and photovoltaic systems. These systems are called hybrid wind systems and are typically used in remote areas where the connection to the grid is not available. Turbines sometimes have to be stopped for maintenance for repairing components or if there is a failure that needs to be checked. Also if the wind is too little or too much

 DOI: 10.1051/e3sconf/20172309001

the turbines should be stopped, and high wind velocity might damage the turbine. Big environmental and nature conservation groups like Birdlife, Green Peace, Friends of the earth … etc, support wind energy. Birdlife recently stated that climate change was the single largest threat to birds, and wind and renewable energy strategies were a clear solution to climate change. Wind energy emits no toxic substance such as mercury and air pollutants like smog-creating nitrogen oxides, acid rain-forming sulfur dioxide and particulate deposits which can cause serious respiratory diseases[8]. There are three options locating wind turbines on top of buildings, the first option is installing the wind turbines on top of the building as an independent object to utilize the high wind speed. The second option is the integration of wind turbine in the construction of buildings, where the building itself mount the turbine. The third option is the installation of wind turbines between two towers and well integrated with the structure of the two towers, such as Bahrain World Trade Center.

c). Bahrain World Trade Center (BWTC).

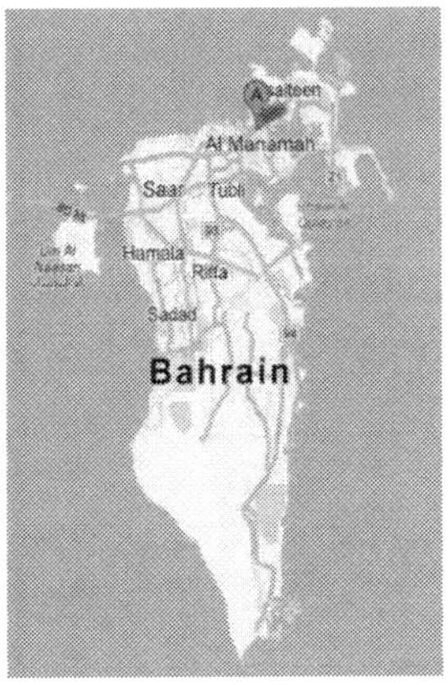

Left :

Map of Bahrain showing the location of the Bahrain World Trade Center

Right:

World trade Center, Sheraton Hotel, Diplomatic area

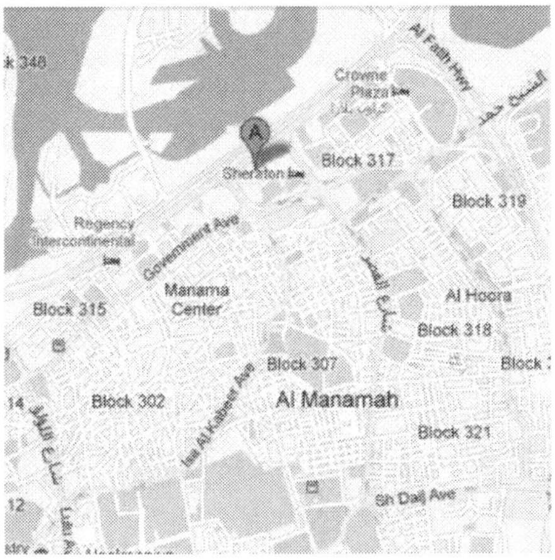

The Bahrain World Trade Center (BWTC) is composed of two towers located in Manama Bahrain. The height of the two towers is approximately 240-meter-high (787 ft.) each containing fifty floors. The towers were designed and built in 2008 by the multi-national architectural firm, Atkins. It is regarded as the first skyscraper in the world to integrate wind turbines into its structural design. The wind turbines were developed, built and installed by Danish company Norwin A/S. The Bahrain World Trade Center (BWTC) is built close to the King Faisal Highway, near popular landmarks such as the towers of Bahrain Financial Harbor

 DOI: 10.1051/e3sconf/20172309001

(BFH), National Bahrain Bank (NBB) and Abraj Al-Lulu. The Bahrain World Trade Center is the second tallest building in Bahrain after the twin towers of the Bahrain Financial Harbor (BFH). The twin towers of BWTC are linked with three bridges, each is holding a 225 kW wind turbine, totaling to 675 kW of wind power capacity. Each turbine is about 29 meters (95ft.) in diameter. The three turbines are facing the north direction, i.e. the prevailing wind direction in Bahrain is north and north-north west. On the other-hand the sail-shaped twin towers are designed to funnel the wind through the gap to provide accelerated wind passing through the turbines. This was confirmed by wind tunnel tests, which showed that the building create an S shaped flow, insuring that any wind coming within 45° angle to either side of the central axis will create a wind stream that remains perpendicular to the turbines. This significantly increases their potential to generate electricity[5]. The three 29m-diameter turbine blades on Bahrain's iconic landmark are the first in the world to be integrated on such a scale into a commercial development. According to the studies, the three turbines are expected to provide 11-15% (11000-13000 megawatt-hours per year), enough to provide light in 300 homes for three years. The successful rotation of the blades involved collaboration between Atkins architects and engineers and turbine specialists Norwin of Denmark. The noise of wind turbines has been reduced significantly. Improved design has drastically reduced the noise of mechanical components so that most audible sound is that of the wind interacting with the rotor blades[9]. Victor T. Barrera[10], in his article, "Ten of the Greenest Skyscrapers in the World" listed the BWTC the second of the ten greenest skyscrapers in the world, describing them as "*Breathtaking design and practical, clean technology are merged to give us the greenest skyscrapers the world has seen*". He stated that "clean technology and green design have both resulted in numerous innovations that continue to push the boundaries of energy generation and efficient resource consumption. Today, urbanization and the increasingly dense populations of the world's largest cities are pushing architects to reach for new heights in green skyscraper design. The Chief Architect of Atkins, Shaun Killa, inspired by nautical expression of a sail to the consistent onshore breeze generating electrical energy. The plans of the two towers were adequately design to help funneling the North and North-west wind to increase the wind speed passing through the three wind turbines. The positioning of the three wind turbines and the unique aerodynamic design of the two towers, the prevailing wind, which is generally from the North and North-west, is funneled into the bath of the turbines producing clean energy. According to Atkins engineers, the idea of the turbines was not only to supply power to the building, but to serve as a landmark in the Kingdom of Bahrain, and be a show case to the entire world as the first commercial building where wind turbines directly integrate into the structure producing non-polluting renewable energy. Three anemometers installed on each

DOI: 10.1051/e3sconf/20172309001

of the bridges are meant to monitor the wind speed. Different readings from the three anemometers are used to establish the extent of turbulence experienced by the blades at any point of time. The wind turbine communicates with the central unit (UCC) and the BMS through fiber optic cables. Also the controller is required with online connectivity enabling programing from remote locations. For safety considerations, the turbines and its support bridges are integrated to the building structure, any failure of its parts and components could be catastrophic. In view of this, extensive modeling involving around 2000 load cases were carried out to validate and theoretically demonstrate that the turbines and the bridges could survive without excessive fatigue. The main objectives was to incorporate the value of substantially harnessing wind energy by direct integration into a building structure. The knowledge and experience gained could then be disseminated to design teams globally. The outline design of the bridge and turbines was prepared by Atkins. Feasibility study carried out together with the project's unique location facing the Arabian Gulf in the direction of prevailing wind, proves techno-commercial validity. The architectural detailed design and engineering was then carried out by wind turbine specialist, M/s Ramboll and their partners M/s Norwin with both companies being headquartered in Denmark.

 DOI: 10.1051/e3sconf/20172309001

The topmost turbine is on the 35th floor.
The second one is on the 25th floor.
while the lowest one is on the 16th floor.

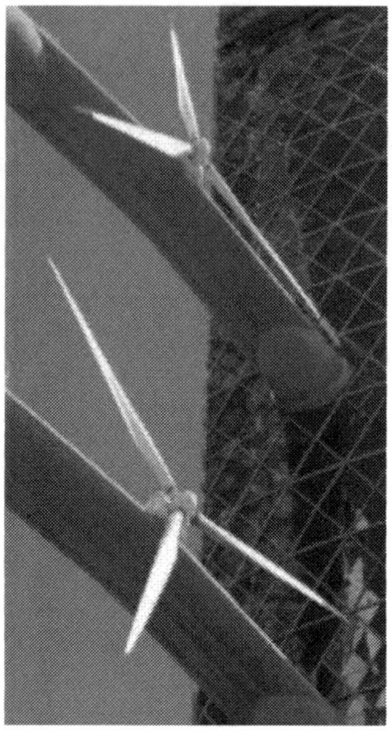

The technical features of the configuration of the wind turbine are as follows [2]:-

Nominal generating power	225kW at 15-20 m/s wind speed.
B/lade diameter	29.200 meters
Sweep Area	670 Sq. meters.
Rotor speed	27 rpm at full load.
Generator speed	1500 rpm-5Hz.
Survival wind speed	80 meters/sec.
Design life	20 years.

d). Architectural and Technical Design

The main design criteria's considered for incorporating the wind turbines can be classified as follows[3]:-

i. Aerodynamic Building Shape:-
ii. The elliptical plan forms and sail-like profiles acts like aero foil, funneling the on-shore breeze between the twin towers creating a positive pressure in the north elevation and a negative pressure in the south elevation, which increases the wind velocity between the two towers. The unique location of the building facing directly opposite the Arabian Gulf insured free availability of the onshore wind.
iii. Operational Limits:-
iv. A number of design parameters have been built into the system, in consideration of the turbines operational efficiency and safety features

 DOI: 10.1051/e3sconf/20172309001

(website 2). The operational wind allowed into the system can be broadly defined as follows:-

- Original wind direction +45 deg. to -45 deg.
- Current wind direction +60 deg.to -60 deg.
- Cut-in wind speed 4meters/sec. Wind speed.
- Wait time for star up from onset of ideal conditions 30 minutes.
- Cut-out wind speed 21 meters/sec.
- Reverse power protection level 70% for 5 min, 80% for 30 min. 90% for 3 sec.

According to the consultant engineer, the achievements can be broadly summarized as follows:-

i. The turbines have been successfully commissioned and have been demonstrated to generate in a manner as intended.
ii. Generated power have been successfully utilized with no detrimental impact on any of the electrical systems.
iii. All safety features have been put to real time test and have been passed.
iv. All automatic controls have been demonstrated to be working in a logical manner in synchronization with the operation of the turbines.
v. Double Glazing (30 mm in thickness) is used on surfaces exposed to direct solar radiation. This will reduce the surface absorptivity by 90-95% . It allows the light to come in but it cuts the heat. The indoor seems cold even if the outdoor is hot and the air conditioners are off.
vi. The building uses all the glass overall so that to allow the natural lighting into the building and to use less electricity and to save energy and the lights come on when it gets dark or when its night.
vii. Demonstration to local electricity authority EDD have been successfully carried out.
viii. Cost of implementing the scheme have been established to a fraction of the total cost of the project.

Table (1): Wind Turbine Power Production Vs. Diversified Load Connected*

YEAR	Total Production (KWh) (A)	Total Running Hours (B)	YEAR	Total Production (KWh) (A)	Total Running Hours (B)
2008-2009	96,909.69	239:14:54	2013	91,187.10	3288:37:00
2010	122,209.10	3612:48:03	2014	80,572.90	2232:43:00

2011	138,699.97	4234:24:01	2015	109,411.70	3412:49:00
2012	91.686,40	3397:57:02	2016	28,258.94	974:27:00

Total Production = 758935.80 KWh, while the Total Running Hours = 2139:00:00

Table (2): Turbine Production Data from 2008-2016*

Turbine location	Total Operating Hours (hrs.)	Total Electricity Produced During Operation (kwhr.)
Turbine No. 1 (lowest)	8735	352,054.5
Turbine No. 2 (middle)	7434	238,358.0
Turbine No. 3 (top)	5224	168,523.3
Total Running Hours	21393	-
Total KWhrs Produced	-	758,935.80

Source: Mr. Surajit Biswas; Senior Operation Manager, Bahrain World Trade Center, Oct. 2016.

In the light of the above, and the fact that this design solution was the first of its kind in the world, defines the success of the project. The World Trade Center project offers designers, architects and engineers around the world a huge knowledge on how large wind turbines could be integrated with the building structure as a source of generating renewable energy. The Bahrain World Trade Center project establishes a technological precedent which is set to raise the awareness of environmental design and its importance in the built environment and pave the way for designers and clients to incorporate renewable and energy efficient measures into their future developments to reduce carbon emissions which is causing climate changes. The Bahrain World Trade Center (BWTC) has received international acclaim for its commitment to sustainable development, being awarded LEAF Awards 2006 for "Best Use of Technology within a Large Scheme and recently the Arab Construction World Sustainable Design Award. Wind energy is one of the cleanest, most environmentally friendly energy sources. It emits no greenhouse gases or air pollutants.

e). Concluding Remarks:-

1. The Bahrain World Trade Center with its magnificent dynamic form is becoming one of the dominant iconic skyscrapers emphasizing the tremendous possibilities of utilizing wind energy in buildings. Its huge

three wind turbines supported and integrated with the structure of the twin towers will stand there for a long time inspiring the new generation, architects and designers to go for renewable energy technologies.

2. The strong integration of the wind turbines with the structure of the towing towers was a great challenge in itself. There are a number of technological and structural problems which should have been carefully studied and analyzed through rigorous wind tunnel experiments. Of course the findings of the simulation experiments, i.e., wind tunnel experiments will not reflect the exact performance of a full scale building, since such experiments are based on a number of assumptions. Therefore the BWTC could be taken as a full scale model for the utilization of wind energy needed by skyscrapers.
3. The overall design process for designing, constructing and running the project is a real demonstration of how architects, engineers from different specializations and scientists could join together as one team to produce a magnificent building project.
4. The dynamic twin towers and their architectural building form, with its three huge wind turbines pointing toward the outer space gives the feeling that the building is about to take-off like a rocket.

References:

1. http://www.nortonrosefulbright.com/knowledge/publications/33679/renewable-energy-in-bahrain.
2. https://google.com.bh/webhp?source=search_appHp=epa;US Environmental Protection.
3. https:/en.wikipedia.org/wiki/Reneable_energy.
4. http://energyinformative.org/wind-energy-pros-and-cons/, Mathias Aarre Maehlum.
5. Bahrain World Trade Center Bahrain" Design Build-network, 2007.
6. "NREL" Dynamic Maps, GIS Data, and Analysis Tools-Wind Maps". Nrel.gov.2013-09-03. Retrieved 2013-11-06.
7. https://www.buildinggreen.com/feature/folly-build-integrated-wind.
8. www.ewea.org/wind-energy-basis/faq/Wind Energy frequently asked questions (FAQ).
9. Justin Thomas (March 2007), "World's First Building Integrated Wind Turbines", Renewable Energy.
10. http://www.energytrendsinsider.com/2009/05/12/ten-of-the-greenest-skyscrapers-in-the-world/

Model for the techno-economic analysis of common work of wind power and CCGT power plant to offer constant level of power in the electricity market

Z. Tomsic[*], I. Rajsl, M. Filipovic

e-mail: zeljko.tomsic@fer.hr; ivan.rajsl@fer.hr, matea.filipovic@fer.hr

Faculty of Electrical Engineering and Computing, University of Zagreb, Croatia

ABSTRACT

Wind power varies over time, mainly under the influence of meteorological fluctuations. The variations occur on all time scales. Understanding these variations and their predictability is of key importance for the integration and optimal utilization of wind in the power system. There are two major attributes of variable generation that notably impact the participation on power exchanges: Variability (the output of variable generation changes and resulting in fluctuations in the plant output on all time scales) and Uncertainty (the magnitude and timing of variable generation output is less predictable, wind power output has low levels of predictability). Because of these variability and uncertainty wind plants cannot participate to electricity market, especially to power exchanges. For this purpose, the paper presents techno-economic analysis of work of wind plants together with combined cycle gas turbine (CCGT) plant as support for offering continues power to electricity market. A model of wind farms and CCGT plant was developed in program PLEXOS based on real hourly input data and all characteristics of CCGT with especial analysis of techno-economic characteristics of different types of starts and stops of the plant.

The Model analyzes the followings: costs of different start-stop characteristics (hot, warm, cold start-ups and shutdowns) and part load performance of CCGT. Besides the costs, the technical restrictions were considered such as start-up time depending on outage duration, minimum operation time, and minimum load or peaking capability.

For calculation purposes, the following parameters are necessary to know in order to be able to economically evaluate changes in the start-up process: ramp up and down rate, time of start time reduction, fuel mass flow during start, electricity production during start, variable cost of start-up process, cost and charges for life time consumption for each start and start type, remuneration during start up time regarding expected or unexpected starts, the cost and revenues for balancing energy (important when participating in electricity market), and the cost or revenues for CO_2-certificates. Main motivation for this analysis is to investigate possibilities to participate on power exchanges by offering continues guarantied power from wind plants by backing-up them with CCGT power plant.

KEYWORDS

Wind plant, CCGT plant, electricity market, PLEXOS model

[*] Corresponding author

 DOI: 10.1051/e3sconf/20172309002

INTRODUCTION

Today energy market is very uncertain and represents a great challenge for new investments. In each moment in time production and consumption need to be in balance and load is covered by the production from various power sources. But different energy sources have different issues. Renewables represent a clear energy and they don't produce GHG emissions, but they have intermittent nature (wind and solar), their geographical distribution is typically uneven and the variability of renewables requires more operational flexibility to compensate fluctuations. This problem limits their participation in the energy market. In today market with huge share of renewable, some thermal power plants struggle to create high enough profit to cover their investment cost because they are working at very low efficiency or they have a big number of starting up and shutting down what increases operational costs or they don't work at all. Frequent changes of the load cause changes of fuel consumption and emission levels. These changes in the operating modes have a significant effect on the O&M costs of conventional technologies which in any cause should not be ignored in long term planning and investment costs. According to previous work [1, 2], wind power reduces the price of electricity and the attractiveness for the investment in other power stations as gas fired power plants. Feed-in guarantees the purchase of the entire produced electricity, is opposed to the full exposure to market conditions - in other words, exposure to the market price of the electricity and entire responsibility for the deviation of the planned and actual production of electricity. On the other hand, having huge share of stochastic renewables as wind and solar creates need in the system for balancing, especially when the produced electricity is lower than we expected. This working regime forms peaks and gaps and solutions are energy storages, pumped hydro, OCGT and CCGT plants. The option considered in this paper is balancing the wind production with production from CCGT power plants taking into account production cost of three different models of CCGT – model with one, two and three gas turbines and one steam turbine. In this regime CCGT has a possibility to earn additional revenues in the market where one conventional power plant can difficult break through and cover the initial investment and earn profit. Previous works related to this theme [3, 4, 5] recognised great opportunities in this area by finding a set of optimal solutions in moments when the load is variable which makes CCGT one of the most flexible ways of balancing the unpredictable wind power.

 DOI: 10.1051/e3sconf/20172309002

THE OUTLINE OF THE PROBLEM AND METHODOLOGY

Objective of this resarch is to examine, on the basis of the model, technical and economic indicators of the CCGT power plant in function of balancing or stabilizing variable electrical production from wind farms according to the different number of gas turbines.

Technical Aspects of CCGT power plant

Analysis is related to the technical ability, i.e. flexibility of the CCGT power plant for balancing variable production of the wind farms. Whereas this type of drive CCGT depends on the nature of the wind in the area, frequent switching of CCGT plant and the associated high costs are expected [6]. Therefore, three different models are analyzed: 1+1 model (one steam and one gas turbine), 2+1 model (one steam and two gas turbines) and 3+1 model (one steam and three gas turbines). All three models have the same major characteristics: total installed capacity, as the other related features such as part load efficiency etc.

DESCRIPTION OF THE MODEL

Model is made in PLEXOS, a software tool for modelling and optimization of relations on the electricity market with prominent comprehensive range of features delivered through a simple interface developed by Energy Exemplar. This is a general simulation tool based on object modelling which defines a hierarchical set of classes, while the user of simulator creates parts or the entire system designing the instances of objects. The definition of the class describes which collections may belong to objects of a particular class and how they act according to the objects of the same or different types. More details regarding modeling approach can be found in [7].

CCGT model

Analysis is related to the technical ability, i.e. flexibility of the CCGT power plants for following variable production of the wind,whereas this type of drive CCGT depends on the nature of the wind in the area, frequent switching of CCGT plant and the associated high costs are expected. Table 7 and Table 8 are CCGT data taken from elsewhere [8, 9, 10, 11, 12, 13, 14]. Following characteristics are used for the purpose of the detailed model of the CCGT power plant according to the given input data: Number of the units of the gas and steam turbines; Installed capacity - maximum; minimum stable power; detailed heat rate curves; FO&M fee; VO&M fee; Start-up costs for multiple start profiles; Start-up time for multiple start profiles; Lifting speed of power at the start for multiple start profiles; Lifting

DOI: 10.1051/e3sconf/20172309002

speed of power in operating mode; Lowering speed of power in operating mode; Motor fuel; Fuel consumption at the start for multiple start profiles; Fuel prices; The limitation of the production; The limitation of the start; auxiliary consumption of the power plant and others.

Table 7. Input data for Gas and Steam Turbine - 1+1 Model

1+1 Model	Max Cap. (MW)	Min Stable Lvl. (MW)	VO&M (€/MWh)	FO&M (€/kW/year)
GT	307	61.4	3.22	20
ST	138	27.6	3.22	20

	Aux (MW)	Ramp rate (MW/min)	Eff. 100% (%)	Eff. 80% (%)	Eff. 20% (%)
GT	10	15.35	40	38	23
ST	4	6.9	31.17	31.17	31.17

Table 8. CCGT start profile data, 1+1 Model

Start	Idle Time (hrs)	Fuel Offtake (GJ)	Start Cost (€)	Start Time (min)
Hot	8	89.2	10235	30
Warm	48	93.9	14685	90
Cold	96	112.67	20025	190

Table 9. Input data for Gas and Steam Turbine - 2+1 Model

2+1 Model	Max Cap. (MW)	Min Stable Lvl. (MW)	VO&M (€/MWh)	FO&M (€/kW/year)
GT1=GT2	153.5	30.7	3.22	20
ST	138	27.6	3.22	20

	Aux	Ramp rate	Eff. 100%	Eff. 80%	Eff. 20%

DOI: 10.1051/e3sconf/20172309002

	(MW)	(MW/min)	(%)	(%)	(%)
GT1=GT2	5	7.7	40	38	23
ST	4	6.9	31.17	31.17	31.17

Table 10. Input data for Gas and Steam Turbine - 3+1 Model

3+1 Model	Max Cap. (MW)	Min Stable Lvl. (MW)	VO&M (€/MWh)	FO&M (€/kW/year)	
GT1=GT2=GT3	102.33	20.47	3.22	20	
ST	138	27.6	3.22	20	
	Aux (MW)	Ramp rate (MW/min)	Eff. 100% (%)	Eff. 80% (%)	Eff. 20% (%)
GT1=GT2=GT3	3.33	5.117	40	38	23
ST	4	6.9	31.17	31.17	31.17

Electricity market model

Prices are taken from the stock exchange EPEX for the year 2014.

Wind model

In order to take into account interdependence of wind speeds in neighboring hours ARIMA (Auto Regressive Integrated Moving Average) modelling approach basic statistical parameters are used. The ARIMA models are traditionally very well suited to capture short range correlations, and hence have been used extensively in a variety of forecasting applications. ARIMA model predicts future values of a time series by a linear combination of its past values and a series of errors. Three different patterns of wind have provided three different scenarios with capacity factors: 33%, 24% and 61%. First two patterns have the same stochastic parameters (Table 11) but different capacity factors. It could be considered as a wind with same stochastic parameters but in two different weeks in year. Mean values and standard deviation in third scenario are higher and represent a strong wind, which can occur in some occasions, used to show impact of the higher wind capacity on the CCGT production price. Each scenario has five cases with different maximum value, mean value and deviation. Case one is wind with highest

DOI: 10.1051/e3sconf/20172309002

maximum value (400 MW) and Case 5 with lowest maximum value (100 MW). The data used for wind modelling can be found in Table 11 and Table 12.

Table 11. Wind power output stochastic parameters for wind capacity factor 33% and 24%

	Min Value (MW)	Max Value (MW)	Mean Value (MW)	Std. Dev. (MW)	ARIMA a & b
Case 1	0	400	104	92	0.5
Case 2	0	300	78	69	0.5
Case 3	0	200	52	46	0.5
Case 4	0	150	39	34,5	0.5
Case 5	0	100	26	23	0.5

Table 12 Wind power output stochastic parameters for wind capacity factor 61%

	Min Value (MW)	Max Value (MW)	Mean Value (MW)	Std. Dev. (MW)	ARIMA a & b
Case 1	0	400	200	200	0.5
Case 2	0	300	150	150	0.5
Case 3	0	200	100	100	0.5
Case 4	0	150	75	75	0.5
Case 5	0	100	50	50	0.5

Dummy generator model

Dummy power plant is super flexible power plant which marginal cost and cost of launching are significantly larger than the marginal cost of CCGT units, but significantly lower than in the model defined VoLL. Dummy generator is activated only in cases when CCGT power plant achieves technical limits. Modelled Dummy has same characteristics as in [7] - max capacity of 500 MW and VO&M cost of 5000 €/MWh. amount of VoLL in the model is fixed on 100000 €/MWh to

DOI: 10.1051/e3sconf/20172309002

ensure activation of Dummy generator only in case of an emergency and generator activation price is €1,000,000.

Electricity consumption model

Consumption is fixed amount which is equal throughout the entire optimization period, but different in each scenario. Consumption is changing depending on the modelled wind power level.

RESULTS AND DISCUSSION

The presented results were obtained on 3.6 GHz based processor with 32 GB RAM using CPLEX solver under PLEXOS® 7. Results are presented for the three different patterns of wind, with three different wind capacity factors. For each wind capacity factor there are examined three different models of CCGT at different wind levels (as is shown in Table 11 and Table 12) in order to examine the impact of gas turbine number on CCGT production cost.

Fig. 25 represents wind with capacity factor 33% and maximum power output 100 MW. Fig. 26 is wind with capacity factor 24% and maximum output 100 MW. Wind with capacity factor 33% and wind with capacity factor 24% have the same average power and standard deviation. Fig. 27 is wind with capacity factor 61%. Profile and deviation of wind on Fig. 27 is higher and it can be characterized as a strong wind. Each figure represents the data for the one week in hourly resolution.

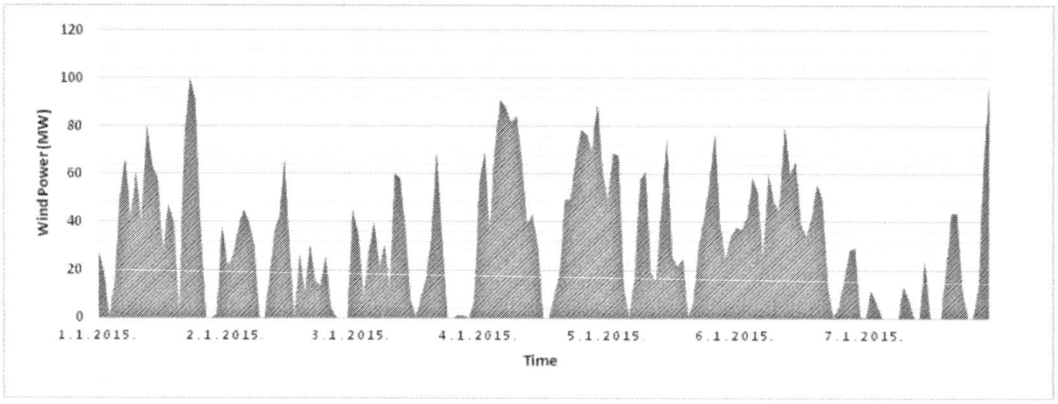

Fig. 25 Profile of wind, capacity factor = 33%, case 5 (max wind power 100 MW)

 DOI: 10.1051/e3sconf/20172309002

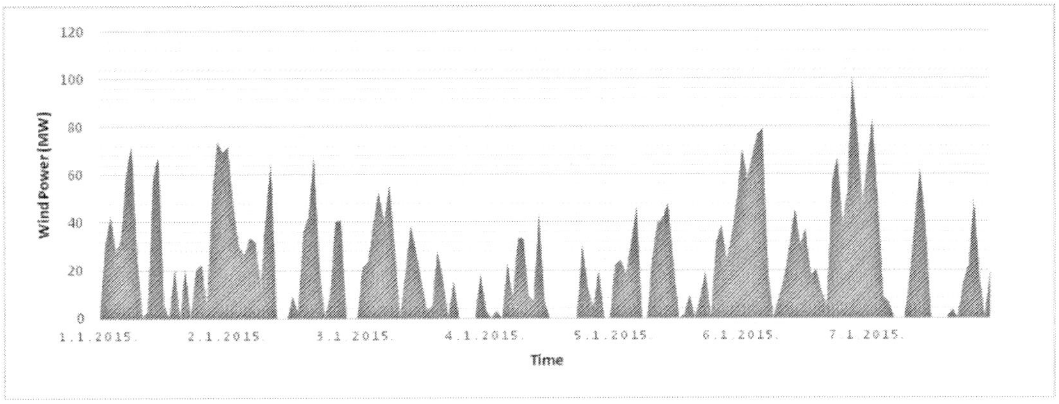

Fig. 26 Profile of wind, capacity factor = 24%, case 5 (max wind power 100 MW)

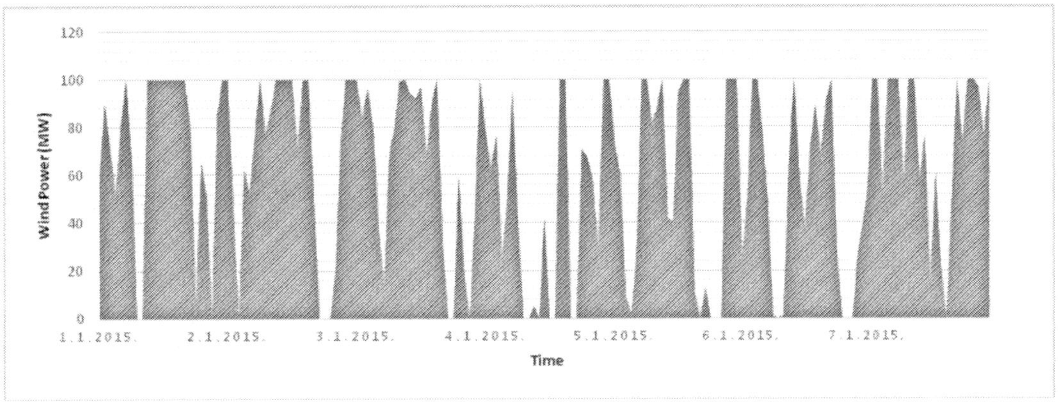

Fig. 27 Profile of wind, capacity factor = 61%, case 5 (max wind power 100 MW)

In this research cost of production and unserved energy are analysed for three models of CCGT power plant: 1+1 model (CCGT with one gas turbine and one steam turbine), 2+1 model (CCGT with two gas turbines and one steam turbine) and 3+1 model (CCGT with three gas turbines and one steam turbine). Results form optimization tool PLEXOS can be found on following figs. 4 and 5. Fig 4 illustrates cost of production and and Fig.5 is for unserved energy for the case 5 (maximum wind power 100 MW) in three different scenarios (wind capacity factor 33%, 24% and 61%).

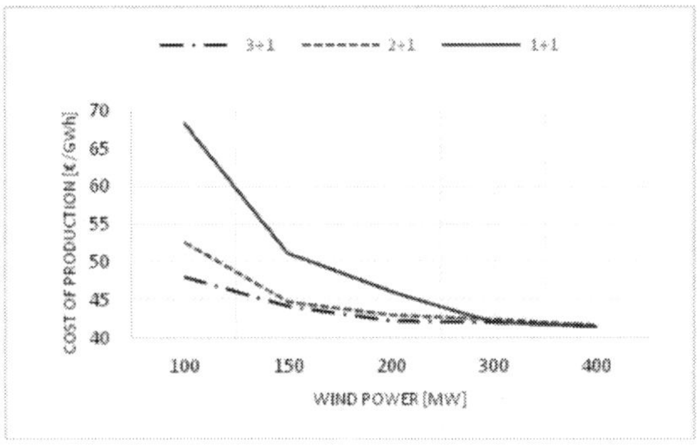

Fig. 28 Cost of production for 1+1, 2+1 and 3+1 model for scenario 1 (wind capacity factor = 33%) and case 5 (wind production 5.54 GWh)

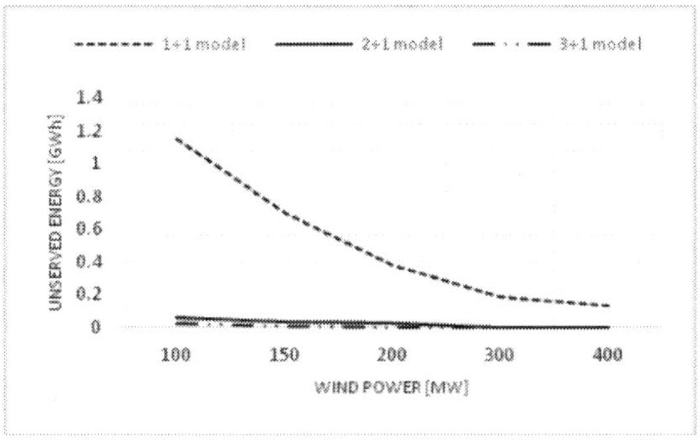

Fig. 29 Unserved energy for 1+1, 2+1 and 3+1 model for scenario 1 (wind capacity factor = 33%) and case 5 (wind production 5.54 GWh)

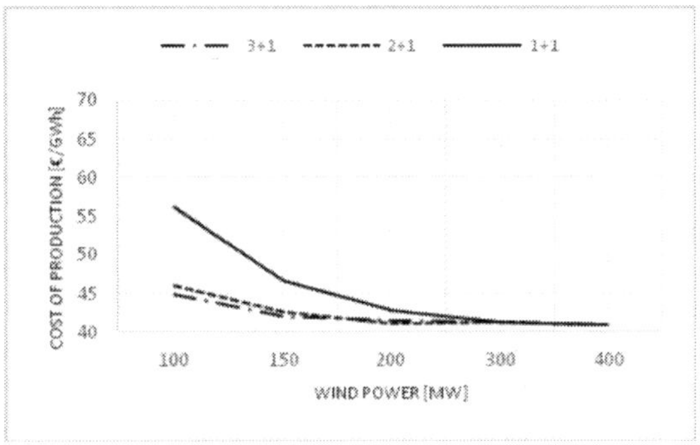

Fig. 30 Cost of production for 1+1, 2+1 and 3+1 model scenario 2 (wind capacity factor = 24%) and case 5 (wind production 5.54 GWh)

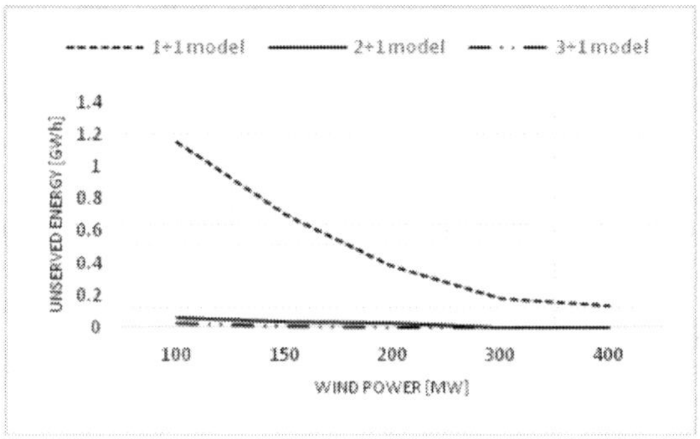

Fig. 31 Unserved energy for 1+1, 2+1 and 3+1 model for scenario 2 (wind capacity factor = 24%) and case 5 (wind production 5.54 GWh)

Fig. 30 and Fig. 32 shows that wind nature, which CCGT is balancing, has a significant impact on decision on number of gas turbines. On the lower wind power levels, it is very important if CCGT has one, two or three gas turbines. CCGT with one gas turbine has a lower efficiency at low power outputs (100-200 MW) than CCGT with two or three gas turbines. Due to lower flexibility, CCGT with one gas turbine has to shut down and start up more often resulting in increased costs (start up cost, consumption of the fuel). That difference is significant: in scenario with wind capacity factor of 61% difference in cost of production between CCGT model with one and three gas turbines is slightly less than 40 €/GWh. If the wind power level in the system is higher (300-400 MW),

DOI: 10.1051/e3sconf/20172309002

difference in these costs becomes smaller, especially difference between CCGT models with two and three gas turbines. If the wind power is higher than 150 MW, the both, two turbines and three turbines model, have very high efficiency and flexibility regarding balancingw the wind production. For the power levels higher than 300 MW number of gas turbines it is no more important because costs of production are almost the same – all three models have high efficiency at these power levels.

It is also very important to notice the influence of the wind capacity factor on CCGT production costs. In the scenario with largest wind capacity factor (capacity factor 61%) CCGT cost of production is significantly larger in comparison with scenarios with lower wind capacity factors (capacity factors 33 and 24%), regardless of wind power level and CCGT model. As can be seen from Fig. 31 and Fig. 33, unserved energy in each scenario depends on number of gas turbines; for the CCGT power plant with one gas turbine unserved energy is significatly higher than unserved energy for the CCGT plant with two or three turbines due to higher min stable level value.

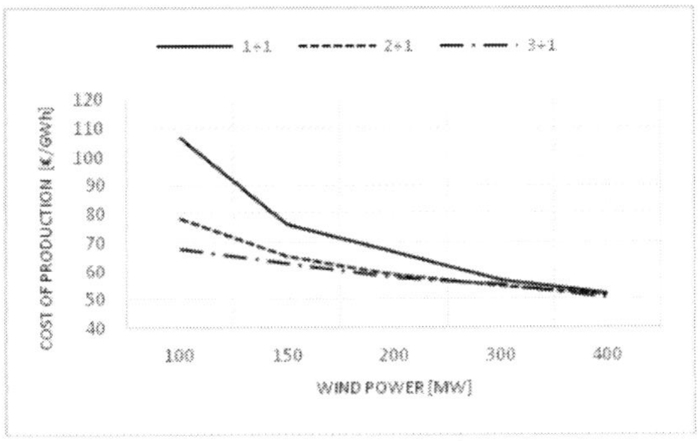

Fig. 32 Cost of production for 1+1, 2+1 and 3+1 model for scenario 3 (wind capacity factor = 61%) case 5 (wind production 5.54 GWh)

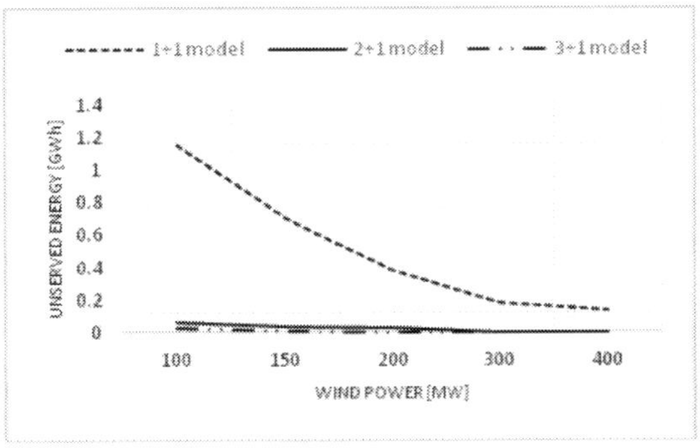

Fig. 33 Unserved energy for 1+1, 2+1 and 3+1 model for scenario 3 (wind capacity factor = 61%) and case 5 (wind production 5.54 GWh)

Detailed data regarding production of each power plant in the modeled system for each scenario and case can be found in Appendix.

DOI: 10.1051/e3sconf/20172309002

CONCLUSION

This paper presents techno-economic analysis for the balancing of volatile nature of renewables, in this case wind farms, with different CCGT power plant models having different number of gas turbine as basis for results comparison.

Investments in the CCGT power plant with one gas turbine is lower than investment in CCGT with two or three gas turbines. The CCGT with one gas turbine has a low efficiency and flexibility at lower load and consequently higher costs because of limitations as the limited speed of raising power and limited speed of the start and those limitations have a huge impact on the production costs.

In order to avoid restrictions associated with the start of the CCGT power plant it is desirable that CCGT constantly has load higher than minimum stable level, but in the system with a huge share of renewables it is not often possible. If CCGT often works at lower capacity it is better to have more gas turbines which will result in decrease of production cost while if the load is higher, all three CCGT models will come in the area with good efficiency and all three CCGT models will have almost the same production cost.

BIBLIOGRAPHY

[1] T. Traber and C. Kemfert, "Gone with the Wind?: Electricity Market Prices and Incentives to Invest in Thermal Power Plants under Increasing Wind Energy Supply," 2009.

[2] W. I. Horowitz, J. Moore and A. Pacheco, "The impact of wind generation on the electricity spot-market price level and variance: The Texas experience," *Energy Policy,* no.7, pp. 3939-3944, July 2011.

[3] M. Valdés, M. D. Duran and A. Rovira, "Thermoeconomic optimization of combined cycle gas turbine power plants using genetic algorithms," *Applied Thermal Engineering,* no. 17, pp. 2169-2182, December 2003.

[4] A. Rovira, C. Sánchez, M. Muñoz and M. D. Duran, "Thermoeconomic optimisation of heat recovery steam generators of combined cycle gas turbine power plants considering off-design operation," *Energy Conversion and Management,* no. 4, pp. 1840-1849, April 2011.

[5] E. Godoy, J. S. N. and J. B. S., "Families of optimal thermodynamic solutions for combined cycle gas turbine," *Applied Thermal Engineering,* pp. 569-576, 2010.

[6] E. Mearns, "The balancing capacity issue: A ticking time-bomb under the UK's Energiewende," Energy Matters, 30 June 2014. [Online]. Available: http://euanmearns.com/the-balancing-capacity-issue-a-ticking-time-bomb-

 DOI: 10.1051/e3sconf/20172309002

under-the-uks-energiewende/. [Accessed April 2016].

[7] Ž. Tomšić, I. Rajšl and M. Filipović, "Techno-Economic analysis of common work of Wind and CCGT plant by offering continuous level of power to electricity market," in *11th Conference on Sustainable Development of Energy, Water and Environment Systems – SDEWES* , Lisabon, Portugal, 2016.

[8] Siemens AG Energy Sector, *Flexible performance, convincing quality – economical and future-proof packages SGT5-PAC 4000F,* 2014.

[9] WBCSD Cement Sustainability Initiative (CSI) / ECRA GmbH, "List of Constants and Default CO2 emission factors," 8 December 2014. [Online]. Available: http://www.cement-co2-protocol.org/v3/Content/Internet_Manual/constants.htm. [Accessed March 2016].

[10] C. L. Hart and J. G. Wright, "Gas power plant fuel requirements and uncertainty considering increasing renewables penetration," in *African Utility Week*, Cape Town, 2015.

[11] D. Chase and P. Kehoe, *GE Combined-Cycle Product Line and Performance,* GE Power Systems.

[12] N. Kumar, P. M. Besuner, S. A. Lefton, D. D. Agan and D. D. Hilleman, "Power Plant Cycling Costs," Intertek APTECH, 2012.

[13] A. Schröder, F. Kunz, J. Meiss, R. Mendelevitch and C. v. Hirschhausen, "Current and Prospective Costs of Electricity Generation until 2050," DIW BERLIN - Deutsches Institut für Wirtschaftsforschung, Berlin, 2013.

[14] S. Simoes, W. Nijs, P. Ruiz, A. Sgobbi, D. Radu, P. Bolat, C. Thiel and S. Peteves, "The JRC-EU-TIMES model - Assessing the long-term role of the SET PLan Energy technologies," European Commission, Joint Research Centre, Institute for Energy and Transport, 2013.

Appendix

Table 13 Production of CCGT, Dummy and Wind power plant for scenario 1

Model	Load [MW]		100	150	200	300	400
1+1	Production [GWh]	CCGT	10.11	16.2	22.15	33.61	44.93
		Dummy	1.15	0.7	0.38	0.18	0.13

 DOI: 10.1051/e3sconf/20172309002

		Wind	5.54	8.3	11.07	16.61	22.14
2+1		CCGT	11.2	16.87	22.51	33.79	45.06
		Dummy	0.06	0.03	0.02	0	0
		Wind	5.54	8.3	11.07	16.61	22.14
3+1		CCGT	11.24	16.89	22.53	33.79	45.06
		Dummy	0.02	0.01	0	0	0
		Wind	5.54	8.3	11.07	16.61	22.14

Table 14 Production of CCGT, Dummy and Wind power plant for scenario 2

Model	Load [MW]		100	150	200	300	400
1+1	Production [GWh]	CCGT	11.88	18.57	25.32	38.07	50.83
		Dummy	0.83	0.49	0.09	0.05	0
		Wind	4.09	6.14	8.19	12.28	16.37
2+1		CCGT	12.7	19.05	25.41	38.12	50.83
		Dummy	0.01	0.01	0	0	0
		Wind	4.09	6.14	8.19	12.28	16.37
3+1		CCGT	12.71	19.06	25.41	38.12	50.83
		Dummy	0	0	0	0	0
		Wind	4.09	6.14	8.19	12.28	16.37

Table 15 Production of CCGT, Dummy and Wind power plant for scenario 3

Model	Load [MW]	100	150	200	300	400

DOI: 10.1051/e3sconf/20172309002

1+1	Production [GWh]	CCGT	5.28	3.85	17.59	19.17	25.68
		Dummy	1.23	0.76	0.57	0.35	0.34
		Wind	10.29	20.59	15.44	30.88	41.18
2+1		CCGT	6.29	4.45	18.05	19.46	25.96
		Dummy	0.22	0.16	0.11	0.06	0.06
		Wind	10.29	20.59	15.44	30.88	41.18
3+1		CCGT	6.44	4.58	18.14	19.51	26.01
		Dummy	0.07	0.03	0.02	0.01	0.01
		Wind	10.29	20.59	15.44	30.88	41.18

Online Bayesian Learning with Natural Sequential Prior Distribution Used for Wind Speed Prediction

Nawal Cheggaga
Department of electronics, Faculty of Technology, Saad Dahlab University, Blida, Algeria, Laboratory of Electrical Systems and Remote control. P O Box 270 route de Soumaa ,Blida 09000, Algeria
cheggaga@yahoo.fr

Abstract

Predicting wind speed is one of the most important and critic tasks in a wind farm. All approaches, which directly describe the stochastic dynamics of the meteorological data are facing problems related to the nature of its non-Gaussian statistics and the presence of seasonal effects .In this paper, Online Bayesian learning has been successfully applied to online learning for three-layer perceptron's used for wind speed prediction. First a conventional transition model based on the squared norm of the difference between the current parameter vector and the previous parameter vector has been used. We noticed that the transition model does not adequately consider the difference between the current and the previous wind speed measurement. To adequately consider this difference, we use a natural sequential prior. The proposed transition model uses a Fisher information matrix to consider the difference between the observation models more naturally. The obtained results showed a good agreement between both series, measured and predicted. The mean relative error over the whole data set is not exceeding 5 %.

Keywords: artificial neural network, Bayesian learning, Fisher information, online learning , wind speed forecasting.

Introduction

Due to the high penetration of wind power in the electricity system, the forecast accuracy of wind power prediction systems becomes increasingly important. In recent years, many scholars have done a lot of research on wind power prediction. The forecast accuracy has improved constantly, and it can be expected that intense research and development efforts are already on track.
Many researches were achieved in order to predict wind behaviour. However, it is still one of the most difficult quantities to forecast [1], namely due its stochastic nature. The actual state of the art includes five main families of methods: persistence Method [2], physical Methods[3], spatial Correlation Models[5], artificial Intelligence Methods[6] and hybrid Methods[7].However, there will always be an inherent and irreducible uncertainty in every prediction.

Bossanyi [8] used a Kalman Filter with the last 6 values as input and got up to 10% improvement in the RMS error over persistence for 1-min averaged data for the prediction of the next time step. This improvement decreased for longer averages, and disappeared completely for 1-hourly averages.

Dambrosio and Fortunato [9] used a one-step-ahead adaptive control by means of a recursive
least squares algorithm for the electrical part of the turbine. They show a fast and reliable response to a step in the wind.

Nogaret et al [10] reported that for the control system of a medium size island system, persistent forecasting is best with an average of the last 2 or 3 values, ie 20-30 minutes. Tantareanu [11] found that ARMA models can perform up to 30% better than persistence for 3-10 steps ahead in 4-sec averages of 2.5Hz-sampled data.

Dutton et al [12] used a linear autoregressive model and an adaptive fuzzy logic based model for the cases of Crete and Shetland. They found minor improvements over persistence for a forecasting horizon of 2 hours, but up to 20% in RMS error improvement for 8 hours horizon. However, for longer horizons, the 95% confidence band contained most of the likely wind speed values, and therefore a meteorological-based approach was deemed more promising on this time scale.

In the same team, Kariniotakis et al [13,14] were testing various methods of forecasting for the Greek island of Crete. These included adaptive linear models, adaptive fuzzy logic models and wavelet based models. Adaptive fuzzy logic based models were installed for on-line operation in the frame of the Joule II project CARE (JOR3-CT96-0119).

Fukuda et al [15] worked on an Auto Regressive model for blade angle optimisation. Using data mining, they found that the use of additional variables was helpful only in December, but not in June. Hunt and Nason [16] used an analysis of principal components of wavelets derived from wind speed time series for a measure-correlate-predict technique. In [17] comprehensive comparison study on the application of different artificial neural networks in 1-h-ahead wind speed forecasting is presented. Three types of typical neural networks, namely, adaptive linear element, back propagation, and radial basis function, are investigated. The wind data used are the hourly mean wind speed. The results show that even for the same wind dataset, no single neural network model outperforms others universally in terms of all evaluation metrics. Moreover, the selection of the type of neural networks for best performance is also dependent upon the data sources. Different network structures, learning rates, and inputs are believed to result in different forecast accuracies. Among the optimal models obtained, the relative difference in terms of one particular evaluation metric can be as much as 20%. This indicates the need of generating a single robust and reliable forecast by applying a post-processing method.

For this purpose in this paper, the variation in wind data is formulated as a system identification problem, where the input of the system is the past values (z(t - 1), z (t - 2), z(t - 3), …) of a time series and its desired output zt is the future of one value. So data is generated sequentially from unknown time-varying systems, the problem of approximating the target system by a parameterized probabilistic model via the data sequence is called an online learning problem. A reasonable approach to such problems is online Bayesian approach considered with an observation model P(y(t) |θt) parameterized by time-varying parameter θt (∈ IRk) for the behavior of the observation data zt.

Used Data

The dataset of these parameters were recorded in (Algeria), the collected experimental data is the hourly average wind speed. Data have been collected during 2013 at a height of 60m.

Figure 1 shows the evolution of the different wind speed time series during April 2013. Therefore, the objective of this work is to predict these data based on the past observations.

Results

There is a lot of scope in meteorological time series prediction and what is needed is the accuracy of forecast. First the performances of a three-layer perceptron in combination with an online Bayesian learning, was comprehensively investigated for forecast value of next hourly average wind speed.

The three-layer perceptron defined as follows is used as the function f (·)

$$f(x_t;\theta_t) = a_{0,t} + \sum_{i=1}^{n_h} a_{i,t}\sigma(x_t; b_{i,t})$$

where the integer nh (> 0) denotes the number of hidden units, and the integer nx (> 0) is the dimension of the input variable xt . The parameter a0,t (∈ IR) is the output bias, the parameter ai,t (∈ IR)(i = 1,..., nh) describes a weight parameter of the i th hidden unit, and the parameter vector bi,t(∈ IRnx+1)(i = 1,..., nh) denotes a weight parameter vector connecting from the inputs xt and the input bias to the i th hidden unit.

The observation model use $P(z_t|\theta_t)$ parameterized by time-varying parameter θt (∈ IRk) for the behavior of the observation data zt at time t , and a transition model (prior distribution) P(θt |θt−1) to describe the behavior of the time-varying parameter θt .

DOI: 10.1051/e3sconf/20172309003

Transition models P(θt |θt−1) based on the squared norm of differences between the current parameter θt and the previous parameter θt−1, which can be written as

$$P(\theta_t|\theta_{t-1}) \propto \exp\left(-\frac{\gamma_t}{2}||\theta_t - \theta_{t-1}||^2\right) \qquad (1)$$

have been considered [18]–[20]. Here, the variable γt is often called the hyperparameter, which controls the scale of the transition. However, the squared norm ||θt−θt−1||2 does not adequately relate to the difference between the current observation model P(zt |θt) and the previous observation model P(zt−1|θt−1), and the changes between the observation models P(zt |θt) and P(zt−1|θt−1) are highly dependent on the values of θt−1.

In figure 1 The neural network cannot explore or predicts the speed below 1.5m / s or up to 11m / s because the change between the observation models is extremely small at θt−1 = θa, whereas the change between the observation models is large at θt−1 = θb with the transition model (1). We may consider that this phenomenon can happen when the eigenvalues of the Fisher information matrix are small at θt−1 = θa and when the eigenvalues of the Fisher information matrix are large at θt−1 = θb. This phenomenon of the transition model (1) has affect the tracking of the system or data zt .

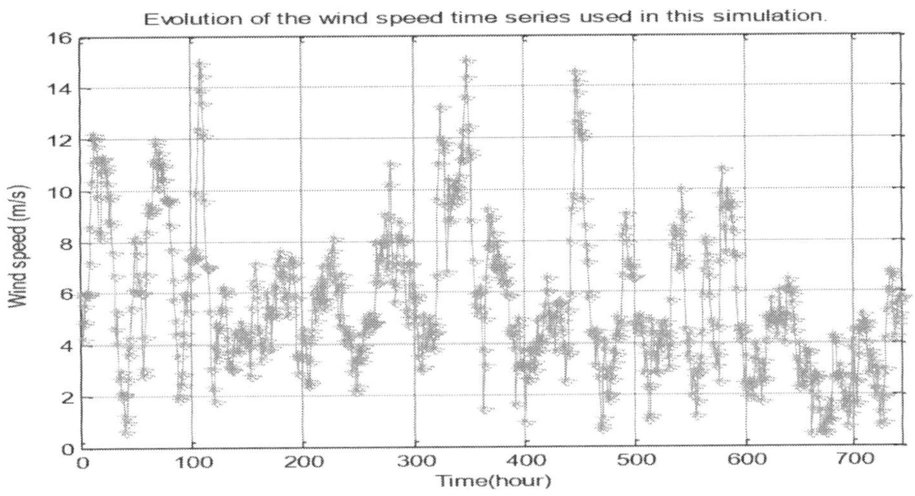

 DOI: 10.1051/e3sconf/20172309003

Fig. 1: Evolution of the wind speed time series used in this simulation.

To break this bridge we will use the same structure of artificial neural network with an online Bayesian learning. the observation data zt generated by an input–output system. The observation model P(zt |θt ,βt)describing the observation data zt will be:

$$P(y_t | x_t, \theta_t, \beta_t) = \frac{1}{Z(\beta_t)} \exp\left(-\frac{\beta_t}{2}(y_t - f(x_t; \theta_t))^2\right) \quad (20)$$

where the variables yt (∈ IR) and xt (∈ IRnx) stand for the output variable and the input variable at time t , respectively, and the observation data zt is considered as the set of these variables xt and yt , i.e., zt = (yt , xt). The function P(xt) describes a probability distribution of the input variable xt. And the online Bayesian approach use natural sequential prior transition model [21]:
to adequately consider the difference between the observation models P(zt |θt) and P(zt−1|θt−1), and to avoid influences on the learning and/or tracking of the target system. Here,θt= θt − θt−1,and T stands for the transpose operator of vectors and matrices. The natural sequential prior uses a Fisherinformation matrix Ft−1

P(z|θt−1)(∂ log P(z|θt−1)/∂θt−1)(∂ log P(z|θt−1)/∂θt−1)T dz derived from the observation model P(zt−1|θt−1), in order to consider the difference between the observation models more naturally. In information geometry approaches, this Fisher information matrix Ft−1(∈ IRk×k) is defined as a metric matrix of a model manifold corresponding to the observation model P(zt−1|θt−1) [11]. The natural sequential prior is based on the norm θtT Ft−1θt, which considers such a metric of the model manifold. Given the observation model in (19), the Fisher information
matrix Ft can be written as

$$F_t = \iint P(x_t) P(y_t | x_t, \theta_t, \beta_t) \times \frac{\partial \log P(y_t | x_t, \theta_t, \beta_t)}{\partial \theta_t} \frac{\partial \log P(y_t | x_t, \theta_t, \beta_t)}{\partial \theta_t}^T dx_t\, dy_t. \quad (24)$$

by assuming that the variable yt contains Gaussian noise. The variable βt ∈ (0,∞) stands for a hyperparameter describing the inverted variance of the Gaussian noise at time t .

DOI: 10.1051/e3sconf/20172309003

Figures 2 and 3 present the simulation results for generating a single-step (one hour) prediction with natural sequential prior transition model.

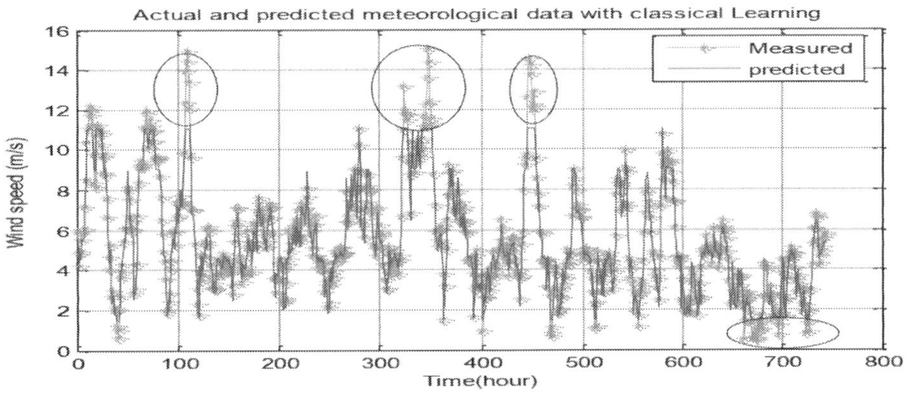

Fig. 2. Actual and predicted wind speed data with classical learning.

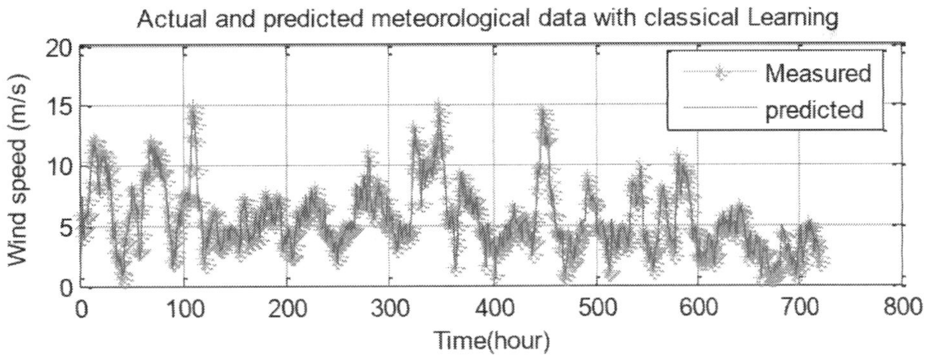

Fig.3. Actual and predicted wind speed data with our approach.

 DOI: 10.1051/e3sconf/20172309003

In this work we performed experiments with a step time of one hour. The input network is a sequence of hourly mean values prior. The target output is made of future values.

Root Mean Square the Error (RMS) differences between observed and estimated values, by chosen models, were used to evaluate the performance of models RMSE were computed by:

$$RMSE = \sqrt{\frac{1}{N}\sum_{i=1}^{N}(y_t - y\hat{}_t)^2} \quad (9)$$

Where y_t *the original time series,* $y\hat{}t$ *is* is the computed.

The RMS error and Mean relative error has been calculated for the different estimation of the wind speeds and compared at real measured data wind speed. The resultants are given in table1. According to this table it should be noted that the MAE is improved from 21% to 3% and the RMSE is improved from 1.6102 to 0.1892, the coefficient of correlation (r) is improved at 0.996.

Table 1: Comparison between observed and predicted meteorological data

a ***models***	***E (%)***	***RMSE***	***r***
Our algorithm predictor	***03%***	***0.1892***	***0.9986***
classical learning	***21%***	***1.6102***	***0.9433***

CONCLUSION

In this paper, a natural sequential prior which uses a Fisher information matrix was applied to the online learning of three-layer perceptron used for wind speed prediction purpose. As a result, the model consider the difference between the observation models more naturally and confirmed the superiority of the proposed model in terms of errors. One of the most important aspects of future work will be to generalize and/or extend the proposed model to apply it to other renewable energy sources such as solar irradiation and air temperature.

ACKNOWLEDGMENT

This work was supported by the Ministry of Higher Education & Scientific Research (Algiers) under project number: F00420120007

REFERENCES

[1] Lazić L, Pejanović G, Živković M, Wind forecasts for wind power generation using the Eta model, Renewable Energy, Volume 35, Issue 6, June 2010, Pages 1236-1243, ISSN 0960-1481
[2] Zhao, X., Wang, S.X. and Li, T. (2011) Review of Evaluation Criteria and Main Methods of Wind Power Forecasting.
[3] Ma, L., Luan, S.Y., Jiang, C.W., Liu, H L. and Zhang, Y. (2009) A Review on the Forecasting of Wind Speed and Generated Power. Renewable and Sustainable Energy Reviews, 13, 915-920. http://dx.doi.org/10.1016/j.rser.2008.02.002
[4] Saima, H.; Jaafar, J.; Belhaouari, S.; Jillani, T. A., "Intelligent methods for weather forecasting: A review," National Postgraduate Conference (NPC), 2011 , vol., no., pp.1,6, 19-20 Sept. 2011 Energy Procedia, 12, 761-769. http://dx.doi.org/10.1016/j.egypro.2011.10.102
[5] Ma, L., Luan, S.Y., Jiang, C.W., Liu, H L. and Zhang, Y. (2009) A Review on the Forecasting of Wind Speed and Generated Power. Renewable and Sustainable Energy Reviews, 13, 915-920. http://dx.doi.org/10.1016/j.rser.2008.02.002
[6] Ma Lei, Luan Shiyan, Jiang Chuanwen, Liu Hongling, Zhang Yan, A review on the forecasting of wind speed and generated power, Renewable and Sustainable Energy Reviews, Volume 13, Issue 4, May 2009, Pages 915-920, ISSN 1364-0321
[7] G. Jianming Hu, Jianzhou Wang, Guowei Zeng, A hybrid forecasting approach applied to wind speed time series, Renewable Energy, Volume 60, December 2013, Pages 185-194, ISSN 0960-1481
[8] Bossanyi, E.A.: Short-Term Wind Prediction Using Kalman Filters. Wind Engineering 9(1), pp. 1-8 (1985)
[9] Dambrosio, L, and D. Fortunato: One-step-ahead control of a wind-driven, synchronous generator system. Energy 24, pp. 9-20 (1999), doi:10.1016/S0360-5442(98)00067-X
[10] Nogaret, E., G. Stavrakakis, J.C. Bonin, G. Kariniotakis, B. Papadias, G. Contaxis, M. Papadopoulos, N. Hatziargyriou, S. Papathanassiou, J. Garopoulos, E. Karagounis, J. Halliday, G. Dutton, J. Pedas-Lopes, A. Androutsos, and P. Pligoropoulos: Development and Implementation of an Advanced Control System for Medium Size Wind-Diesel Systems. Proceedings of the EWEC '94 in Thessaloniki, 10.-14. Okt, pp. 599-604

DOI: 10.1051/e3sconf/20172309003

[11] Tantareanu, C.: Wind Prediction in Short Term: A first step for a better wind turbine control. Nordvestjysk Folkecenter for Vedvarende Energi, October 1992, ISBN 87-7778-005-1

[12] Dutton, A.G., G. Kariniotakis, J.A. Halliday, and E. Nogaret: Load and Wind Power Forecasting Methods for the Optimal Management of Isolated Power Systems with High Wind Penetration. Wind Engineering 23(2), pp. 69-87 (1999) .

[13] Kariniotakis, G., E. Nogaret, and G. Stavrakis: Advanced Short-Term Forecasting of Wind Power Production. Proceedings of the European Wind Energy Conference held in Dublin, Ireland, October 1997, pp. 751-754, ISBN 0 9533922 0 1 .

[14] Kariniotakis, G.N., E. Nogaret, A.G. Dutton, J.A. Halliday, and A. Androutsos: Evaluation of Advanced Wind Power and Load Forecasting Meghods for the Optimal Management of IsolatedPower Systems. Proceedings of the European Wind Energy Conference, Nice, France, 1-5 March 1999, pp. 1082-1085, ISBN 1 902916 00 X .

[15] Fukuda, H., S. Tamaki, M. Nakamura, H. Nagai, F. Shijo, S. Asato, K. Onaga: The Development of a Wind Velocity Prediction Method Based on a Data-Mining Type Auto-Regressive Model. Proceedings of the European Wind Energy Conference, Copenhagen, Denmark, 2-6 June 2001, pp. 741-744, ISBN 3-936338-09-4 .

[16] Hunt, K., and G.P. Nason: Wind speed modelling and short-term prediction using wavelets. Wind Engineering 25 (1), pp. 55-61 (2001).

[17] Gong Li and Jing Shi. On comparing three artificial neural networks for wind speed forecasting. Applied Energy Volume 87, Issue 7, July 2010, pp. 2313-2320.

[18] J. F. G. de Freitas, M. Niranjan, A. Gee, and A. Doucet, "Sequential Monte Carlo methods to train neural network models," Neural Comput.,vol. 12, no. 4, pp. 955–993, Apr. 2000.

[19] N. de Freitas, C. Andrieu, P. Højen-Sørensen, M. Niranjan, and A. Gee,"Sequential Monte Carlo methods for neural networks," in SequentialMonte Carlo Methods in Practice, A. Doucet, N. de Freitas, andN. Gordon, Eds. New York, USA: Springer-Verlag, 2001, pp. 359–380.

[20] T. Kurihara, Y. Nakada, K. Yosui, and T. Matsumoto, "Bayesian online learning: A sequential Monte Carlo with importance resampling," inProc. IEEE Signal Process. Soc. Workshop Neural Netw. Signal Process.,Sep. 2001, pp. 163–172.

[21] Yohei Nakada and Takashi Matsumoto, Online Bayesian Learning WithNatural Sequential Prior Distribution, IEEE transactions on neural networks and learning systems, vol. 25, no. 1, january 2014.

E3S Web of Conferences **23**, 09004 (2017)
World Renewable Energy Congress-17
DOI: 10.1051/e3sconf/20172309004

RENEWABLES CANNOT BE STORED ECONOMICALLY ON A WELL-RUN POWER SYSTEM

Donald Swift-Hook,
Secretary, WREN and Visiting Prof, Kingston University,
Woking, Surrey GU21 4XX
United Kingdom
email: donald@swift-hook.com

ABSTRACT

Economic storage on a power system must rely on arbitrage, buying electrical power when it is cheap and selling when it is dear. In practice, this means a store must buy power at night and sell it during the day.
There is no solar power at night [by definition], so solar power cannot be stored economically on a well-run power system. Also renewables [and nuclear] are installed commercially to save fuel but fuel costs the same at night as it does during the day, so there is no arbitrage on fuel-saving to justify storage.
Pumped water storage has always been widely used on power systems and is still the only method that is economic today, although many others have been tried, including fuels cells, compressed air and batteries. Devices for power correction and balancing [e.g. capacitor banks and batteries] may physically involve the storage of energy [just as a mobile phone does] but it is misleading to describe them as methods of power system storage, [just as it would be misleading to call a School bus a fuel transportation system, even though it does transport fuel].
When a power system has different sorts of plant generating - coal, gas, nuclear, wind etc - any power being put into storage is from the plant that would need to be switched off [because less power was needed] if storage ceased [e.g. because the store became full or failed]. On a well-run power system, that always has the highest fuel/running cost, but the wind blows free and has zero fuel/running cost, so wind is never [normally] stored unless there is no other plant on line i.e. wind power is the last to be stored.

1 INTRODUCTION

Storing electricity is part of normal power system operation and pumped water is the only system of bulk storage of electricity that has been in use since power systems were first installed (1) and it is still the only system in practical use today. Figure 1 shows the upper storage lake and dam of a 400 MW pumped storage scheme at Cruachan that has been in operation in Scotland since 1965 and Figure 2 shows diagram-matically how such a system operates. 5% of all EU generation is associated with pumped water storage systems.

2 ARBITRAGE.

 DOI: 10.1051/e3sconf/20172309004

The economics of such a store rely on arbitrage: electricity is put into store (e.g. by pumping water) when it is cheap and re-generated when it is dear.

Figure 1. The upper lake and dam at Cruachan, above Loch Awe in Scotland, a 400 MW reversible-pumped water storage scheme commissioned in 1965.

If the extra value of the electricity is sufficient to pay for the cost of the storage plant (pumps, etc) as well as the round-trip loss of energy in pumping and then regenerating (typically 10% or 20% being lost in each direction), the system can be economic.

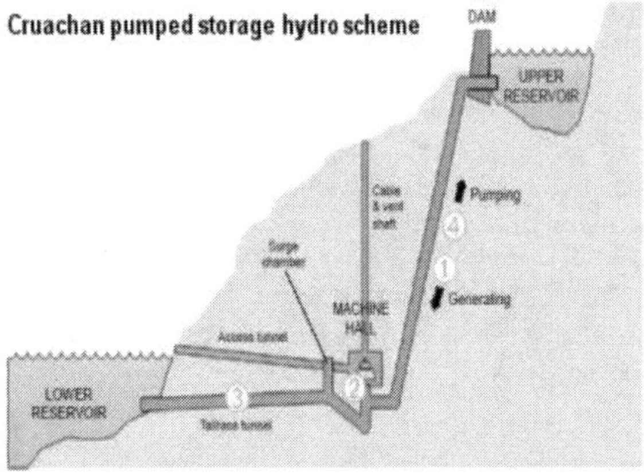

Figure 2. Typical features of a modern pumped water storage facility

On most power systems, this arbitrage between cheap and expensive electricity is between night-time and day-time generation.

3 SOLAR POWER

An immediate consequence is that solar power cannot be stored economically on a power system because there is no sunshine (by definition) at night and therefore no solar power when power needs to be stored.

When solar photovoltaic power plant stands alone, not grid-connected, storage is needed to provide power when the sun does not shine. A typical example is shown in Fig 3 of a solar

 DOI: 10.1051/e3sconf/20172309004

powered garden light, which includes a battery that is charged during the day by a solar cell while the sun shines and then can provide illumination when it discharges through a small light bulb at night. It is cheaper to provide a battery to store as much as 12 hours of electricity than it is to extend the power system for a mains-connection to the electricity network.

Figure 3. With no mains supply, a garden light has a solar cell on top to charge a battery during the day which discharges through a small light bulb to provide illumination at night.

So it is not to be said that you cannot store solar power, only that you cannot do so economically when you are mains connected. Recommendations to the contrary are often made and, although they are clearly physically possible, they are simply not economic.

4 FUEL SAVING

Avoiding emissions to limit global warming does not have any direct financial advantage for a generator and nor is the commercial justification for renewables normally to provide generating capacity: it is to save fuel.

Those countries with the greatest interest in saving fuel are those with the greatest fuel production /consumption, so it is no co-incidence that China, the USA and India, who are amongst the greatest fuel producers, are also installing the most wind power, more than half of the worlds capacity between them (3). The EU too relies heavily on fuel imports, so it too has a great interest in installing wind power, to save the need for fuel imports.

Having saved a certain amount of fuel by using wind energy, to put the electricity into a store and then regenerate it involves losses, so storage must always waste fuel and less fuel will have been saved at the end of the storage process than without it. There is no variation in the cost of fuel between when the wind blows and does not blow to counter-balance this loss and allow arbitrage. So storing wind energy must always expect to waste money as well as fuel. The same is true for other renewables. This means that storage cannot be economic for a fuel saver and the suggestion that storage should be installed with renewables is unrealistic.

5 OTHER TYPES OF STORAGE

DOI: 10.1051/e3sconf/20172309004

In 1994, Ter-Gazarian (2) described many other methods of storage based on many different physical principles, including various types of thermal energy storage, fly wheels, compressed air, various electro-chemical systems, capacitor banks and superconducting coils.
Few methods have been added to this list since then and, despite continuing research activities and development projects, no new practical systems have emerged for large-scale use. Pumped water is still the only realistic method to be installed for electrical energy storage on power systems today.
Pumped air has been advocated instead of pumped water and two commercial scale demonstration projects were put into operation in the last century, a 290 MW scheme at Huntorf in Germany in 1978 and a smaller 110 MW plant 13 years later at McIntosh, Alabama. Neither of these proved economic but studies have continued, including recent experimental installations at Gaines in Texas in 2012 generating 2 MW and at Seabrook, New Hampshire in 2013 generating 1.65 MW. Nothing on a commercial scale has materialised.
Fuel cells, too, have been widely advocated for use on power systems. The first practical fuel cell systems were used in the first satellites and then on a much larger scale in the Apollo space missions, Figure 4, but they have not proved commercially viable for Earth-bound applications.
Large banks of lead-acid batteries have been used to power submarines silently for nearly 100 years. Today, huge efforts are going into the development of lithium-ion batteries for electronic applications (because they have much higher energy density, much lower loss of charge and no memory effect) and they are making some inroads for vehicle applications. On power systems, batteries are used for control and balancing purposes but they are far too expensive to consider for bulk energy storage.

Figure 4. Alkaline fuel cells have been used by NASA in the Apollo missions (shown here) and in the Space Shuttles

6 MISLEADING TERMINOLOGY

A great deal of confusion is being created by battery enthusiasts who assert that the variability of wind makes energy storage essential and suggest that new developments are leading to batteries being widely used for energy storage on power systems. This is misleading terminology.
Batteries are being introduced on a widespread basis to enhance the frequency response of power systems but that is control and balancing. The physical processes certainly involve the storage of energy, just as charging a mobile telephone does, but energy storage [putting energy in and taking it out later] is not what they are for in either case.
UK's National Grid Plc recently tendered for Enhanced Frequency-Response equipment and, of the 64 sites that bid, 61 were for batteries. Successful companies included EdF Energy

DOI: 10.1051/e3sconf/20172309004

Renewables Ltd., Vattenfall AB, EON SE, Low Carbon, Element Power Ltd, RES and Belectric Solar Ltd. This is the first time that batteries will have been introduced onto the UK power system on such a scale.

The amount of energy stored is not significant in bulk energy terms - they will not be called upon to supply electricity for more than 15 minutes - but it is their ability to do so at 1 second's notice that is the attraction, to enhance the power system's frequency response. It is misleading to claim that such units are for power system energy storage when they are only for power system control or balancing.

7 STORAGE *VIS À VIS* WIND POWER.

It has been pointed out that storage cannot be economically associated with solar power but nor can it be with wind power. There are several obvious mistakes that are frequently made.

For instance, it is often suggested that storage would be valuable to transform intermittent wind into base-load generation. However, during periods of base-load, capacity must be standing idle and freely available for generation. The value of yet more capacity at such a time is clearly negligible, so the installation of storage or, indeed, any other type of power plant, cannot be justified economically then and the idea of transforming intermittent wind into base-load generation is misconceived. [The times when capacity is needed are at peak periods but these are so brief that, again, storage cannot be justified then, unless it is competitive with cheap gas turbine peaking plant and that is totally unrealistic].

Storage and wind power are not compatible contractually (4). The wind only blows for a fraction of the time, typically around one third on average in the UK, although there are rarely periods when there is no wind at all. Any associated storage would therefore find wind taking three times as long and therefore three times the cost to be stored as nuclear or any other continuously available type of power would. [This argument applies to any other type of power plant on a power system. If the storage were associated with just one plant, it too would be out of service whenever the plant was. The store should be associated with the whole power system.]

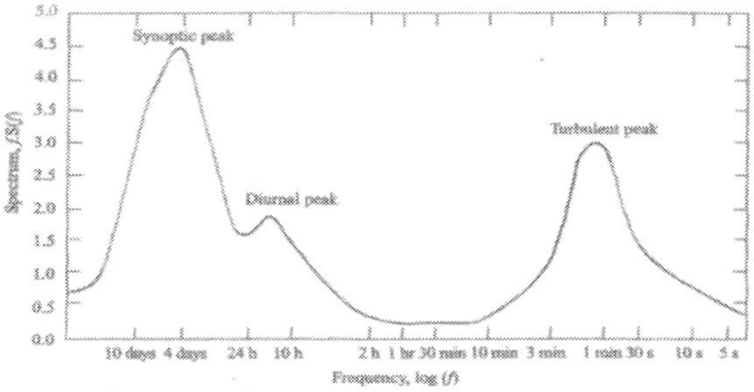

Figure 5. The classical frequency spectrum measured by van der Hoven at Brookhaven National Labs on Long Is.

The classic work of van der Hoven (Figure 4) showed three peaks in the wind frequency spectrum: one due to the local turbulence (1 to 2 min), a small diurnal (day to night) peak and a broad synoptic peak from 2 to 10 days when the wind blows for up to 10 days (as a weather pattern passes) and ceases for up to 10 days. If a store cannot cope with 10 days of wind it is not effectively tackling the wind storage problem. A store coping with 10 days of wind must be up to

 DOI: 10.1051/e3sconf/20172309004

20 times larger and more costly than a typical night-to-day store making effective storage for wind power prohibitively expensive.
On a more practical note, enthusiasts who claim that wind power needs storage to cope with it's intermittency manage to ignore the fact that, with more than 300 GW of wind power generating capacity installed - and there is now more wind capacity operating than nuclear, in the UK and the EU and world-wide - no wind-farm has ever had to install a storage plant alongside it.

8 WIND IS THE LAST TO BE STORED

Having shown that to store wind power is very costly, let us consider when it is likely to happen on a power system. It has been pointed out that power systems use storage very widely (1), with or without wind power on them, and it might then be thought that, when electricity is being put into a store, typically during base-load periods at night, and the wind happens to be blowing at the same time, wind is as likely as any other source of power to be stored or perhaps even more likely. More careful consideration shows that this is not the case.
To discover which source of energy is being stored at any particular time when many different types of power plant are on line, we need to look at the state of the power system with and without storage. The difference between the two tells us which power is being stored. The power plant, which would need to be switched off if storage ceased (because the store failed or became full) would be the power plant that was filling the store.
The plant to be switched off on a well run power system, which always uses the cheapest power available, would be the *dearest* generation on the system at the time. The "wind blows free" and has zero marginal cost so wind will always be amongst the *cheapest* plant and never the dearest, unless the cheapest generation is also the dearest, that is to say wind is the only generation left on the system.
A proviso is that some plant may have less than zero marginal cost and be prepared to pay to stay on line. Combined Heat and Power may have periods [e.g. in the early hours of a cold winter night] when heat is needed but no electric lights and it can't do one without the other. Nuclear and other plant may find thermal cycling very damaging and be prepared to pay to avoid it. With these provisos, it can be said that wind is the last to be stored.

9 CONCLUSIONS

1. Despite continuing research activities and other development projects, **pumped water** is still the only system of bulk energy storage being installed on power systems today.
2. Energy storage is economically justified by **arbitrage** between cheap [night-time] and more expensive [day-time] electricity.
3. Pumped waster storage plant can also provide many **control and balancing** services such as spinning reserve but these are quite distinct from storage; batteries, capacitor banks and other

plant can also offer such control and balancing functions but it is misleading and confusing to claim that they are "for power system energy storage".

4. With no solar power at night, of pv **solar power cannot be stored** economically on a power system.
5. Contrary to frequent misunderstandings, the installation of storage cannot be justified on a power system to store wind energy either. Storage and wind power are **contractually not compatible**.
6. **Wind power is to save fuel** and to store wind power would waste fuel on round-trip losses. **Storage cannot be economic for a fuel saver**
7. Wind only blows for one third of the time on average, so it will take 3 time as long and **cost 3 time as much to store only wind** compared with nuclear or any other continuously available type of power.
8. The wind frequently blows for up to 10 days at a time and a store of such capacity would **cost 20 times as much as current day/night energy storage** costs.
9. If storage is already installed on a well-run power system with various types of generation, normally **wind is the last to be stored**.

10 REFERENCES

(1) Swift-Hook, D.T., Grid-connected intermittent renewables are the last to be stored, Renewable Energy, Vol. 35, p.1967, 2010

(2) Ter-Gazarian, A., Energy storage for power systems Vol. 6 in Swift hook DT editor IEE Energy Series. London: Peter Peregrinus, 1994

(3) Swift-Hook, D.T., The case for renewables apart from global warming, Renewable Energy, Vol. 49, p. 147-150, 2013

(4) Swift-Hook, D.T., Wind energy really is the last to be stored and solar energy cannot be stored economically, Renewable Energy, Vol. 50, p. 971-976, 2013

DOI: 10.1051/e3sconf/20172310001

Impact of nanoparticles and butanol on properties and spray characteristics of waste cooking oil biodiesel and pure rapeseed oil

K. H. Ahmad and A. K. Hossain*

Sustainable Environment Research Group, Aston University, Birmingham B4 7ET, UK
**Corresponding author email: a.k.hossain@aston.ac.uk*

Abstract

Renewable biofuels can offset greenhouse gases by replacing fossil fuels destined for internal combustion engines. However, biofuels have their own setbacks and may lead to poor combustion inside the engine cylinder. In this study, nanoparticles and butanol were blended either separately or together with waste cooking oil biodiesel and neat rape seed oil to investigate the impact of these additives on the properties and spray characteristics. The investigation comprised of three stages, with each having an effect on how the next stage of the investigation was conducted. Initially, the physicochemical characteristics of 25ppm, 50ppm, 75ppm and 100ppm concentrations of aluminium oxide and copper oxide nanoparticle blends with fossil diesel, waste cooking oil biodiesel and rapeseed oil were investigated. The results from first stage investigation showed that, in general, blends containing aluminium oxide nanoparticles gave better results for almost all the concentrations when compared with copper oxide nanoparticle blends with the same nanoparticle concentrations. Overall, waste cooking oil biodiesel blended with 100ppm aluminium oxide nanoparticle showed most promising results like the flash point of 159.3°C, kinematic viscosity @40°C of 4.66 cSt, and gross calorific value of 44.43 MJ/kg. These values were 61.6% higher, 51.3% higher and 3.2% lower than that of corresponding fossil diesel values. Subsequently, in the second stage of the study, the addition of butanol was investigated to assess its ability to enhance the emulsion of biofuel-nanoparticles blends. Four blends containing 90% biodiesel & 10% butanol, and 90% rapeseed oil & 10% butanol, with and without 100ppm Al_2O_3 were prepared. Results showed that the kinematic viscosity of the fuel blends containing 100ppm aluminium oxide nanoparticles were decreased by 0.4% and 3.3%, for 90% biodiesel & 10% butanol and 90% rapeseed oil & 10% butanol blends respectively, when compared to without the nanoparticles. The results obtained from the second stage of investigation proved that butanol acted as a surfactant and thus addition of butanol helped to improve the properties of the biofuel-nanoparticle blends. In the third stage of the study, the spray characteristics of fossil diesel, biodiesel, biodiesel + 100ppm aluminium oxide nanoparticles, rapeseed oil, rapeseed oil + 100ppm aluminium oxide nanoparticles, 90% biodiesel & 10% butanol, 90% biodiesel & 10% butanol + 100ppm aluminium oxide nanoparticles, 90% rapeseed oil & 10% butanol and 90% rapeseed oil & 10% butanol + 100ppm aluminium oxide nanoparticles were investigated. It was found that amongst all fuels, blend containing 90% biodiesel + 10% butanol + 100ppm aluminium oxide nanoparticles gave better spray characteristics; for example, the liquid sheet angle was 7.14% lower and the spray cone angle was 7.87% higher than the corresponding fossil diesel values. The study concluded that the spray characteristics and properties of biofuels could be improved by blending with both aluminium oxide nanoparticles and butanol.

Keywords: ***Nanoparticle, Butanol, Biofuel, Properties, Spray characteristics, CI Engine***

DOI: 10.1051/e3sconf/20172310001

INTRODUCTION

The increase in the energy demand, global warming, limited reserve of fossil fuels and refinery capacity have alarmed the world to harvest into green methods of producing energy. Scientists and researchers have been trying to find ways to reducing harmful emissions of internal combustion (IC) engines by modifying them, using alternative fuels, blending different fuels together and using additives [1]. Biofuels can be used in compression ignition (CI) engines and they may be produced domestically. However, they have their own setbacks as being highly viscous, less stable and incompatible with the existing fuel supply systems. One approach to improve the biofuels properties is by addition of nanoparticles; but currently there are concerns if the nanoparticles does not burn inside the engine cylinder and their effect on the exhaust emission gases, fuel pump, fuel filter and fuel injection characteristics [2]. Nanoparticles have a higher surface area to volume ratio, which acts as a highly reactive catalysts [3]. Aluminium oxide nanoparticles have the ability to donate their oxygen atoms from its lattice structure to the fuel which enhances combustion characteristics [4]. Lower NOx and soot emission can be achieved by addition of nanoparticles [2, 5]; Samuel and Shefeek [6] investigated the effects of cerium oxide nanoparticles addition in fossil diesel (FD) on performance and emission characteristics of a four cylinder CI engine. An ultrasonic shaker was used to mix the nanoparticles with FD. Four blends of FD were made with 10ppm, 20pmm, 30ppm and 40ppm of nanoparticles. It was observed that the specific fuel consumption was decreased as cerium oxide nanoparticles acted as an oxygen donating catalyst which enhanced combustion characteristics. CO and nitrogen gases reduced from 2% to 1.4% and from 83% to 81.6% respectively, with 40ppm cerium oxide nanoparticles blend when compared to neat FD [6]. Gumus et al. [7] assessed the impacts of aluminium oxide nanoparticles (Al_2O_3) and copper oxide nanoparticles (C_uO) on FD; they investigated the physicochemical properties, stability of nanoparticles in the FD with the use of various surfactants, and CI engine performance & emissions. Nanoparticles blends of 50ppm with FD for both nanoparticles were created using an ultrasonic mixer. The emulsion of nanoparticles in diesel was tested using two different surfactants (Sodium Silicate and Darvan-C). They reported that 2% Darvan-C showed the best emulsion layer [7]. Santhanamuthu et al. [8] investigated the performance and exhaust emissions of a CI engine with blends of polanga seed oil, FD and iron oxide nanoparticles. They observed best optimum engine performance with 100ppm iron oxide nanoparticles. D'silva et al. [9] added titanium oxide nanoparticles (T_iO_2) in FD to investigate stable dispersion, engine performance and emission characteristics. They reported that the flash point, fire point, kinematic viscosity, density and calorific value were increased with the addition of T_iO_2 by 41.17%, 40.74%,6.17%, 0.43% and 0.59%

respectively [9]. About 21.28% reduction in brake specific fuel consumption (bsfc) was observed at peak load when T_iO_2 were added. Other studies also reported improvement in fuels properties when blended with nanoparticles [10-12]. However, in some cases decrease in the calorific value was also reported [6]. Kinematic viscosity increased with the addition of nanoparticles, but the use of surfactants allowed the nanoparticles to emulsify in the base fuel properly, hence decrease in the viscosity was also reported [7]. Spray characteristics of neat biofuels were investigated [13-14]. Wang et al. [13] investigated the spray characteristics of biodiesels produced from waste cooking oil, palm oil and FD at various injection pressures. Biodiesels gave longer spray tip penetration with injection delays, smaller projected area, less volume and smaller spray angle than FD. With increased injection pressure, the difference between the sauter mean diameter (SMD) of biodiesel and FD was decreased. Butanol was found to be a suitable additive for biofuels for improved engine performance and reduced emission gases [15-16]. In this study, the effect of butanol addition on physicochemical and spray characteristics of biofuel-nanoparticles blends will be investigated. The study will be conducted in three stages: (i) two nanoparticles (C_uO and Al_2O_3) will be used to blend separately with FD, waste cooking oil biodiesel (B100) and neat rapeseed oil (RSO100); physicochemical properties will be measured and compared to find the optimum nanoparticles blends (ii) butanol will be added into the optimum nanoparticles blends, and the effectiveness of butanol as surfactant will be assessed by comparing the properties of the butanol-nanoparticles-biofuel blends (iii) spray characteristics of the fuel blends will be measured and analysed. Finally, properties and spray characteristics will be compared to find the optimum blends for CI engine testing.

MATERIALS AND METHODS

Waste cooking oil biodiesel (B100) was acquired from a local company, and rape seed oil (RSO100) was acquired from supermarket. The nanoparticles were purchased from Sigma-Aldrich. Butanol (1 Butane) was purchased from Fisher Scientific Ltd. The study was carried out in three stages: (i) Firstly, to investigate the blends of FD, B100 (from WCO) and RSO100 with Al_2O_3 and C_uO nanoparticles to find out which of the two nanoparticles was more feasible to be used and also in what concentration, (ii) Secondly, to investigate if the Butanol could be used as an emulsifier when nanoparticles were mixed with the fuel, and (iii) Finally, to investigate the effects of nanoparticles and butanol on macroscopic spray characteristics of the fuel blends. Parr 6100 Bomb Calorimeter (ASTM-D240), Setaflash series 3 plus closed cup flash point tester (ASTM-D3278), and Cannon Fenski u-tube viscosity meter (ASTM-D130) were used to measure the heating

DOI: 10.1051/e3sconf/20172310001

value, flash point temperature and kinematic viscosities of the fuel blends. Kinematic viscosity was measured at @40°C using a constant temperature water bath. The density was measured using hydrometer and volume-weight method. Fuel blends samples for all the three base fuels (FD, B100 and RSO100) with 25ppm, 50ppm, 75ppm and 100ppm concentrations of both nanoparticles (Al_2O_3 and C_uO) by mass were prepared. Physicochemical characteristics were measured. The results were analysed and compared, improved nanoparticles blends were taken for phase 2 study, where butanol was introduced – biofuel and butanol was mixed together and after that 100ppm the nanoparticle was added in the blend. The blends were shaken for 15 minutes so the nanoparticles could be dispersed into the blends properly. Macroscopic spray characteristics of fuel injection (liquid sheet angle and spray cone angle) were measured using optical and mechanical methods. Sealey VS2058 diesel injector nozzle POP tester and Lister Petter fuel injector (P751-62090) manufacture by Delphi were used (Figure 1). The maximum pressure of the POP tester was 600 bar and the opening pressure was 125bar. The nozzle had a single hole through which the fuel was sprayed as a full cone (Figure 1). Liquid sheet angle is defined as the angle of the fuel as it comes out of the nozzle (Figure 2), before it turns into spray particles. An optical technique was used to measure the liquid sheet angle. Nikon D3320 camera was used to take the photographs of the fuel. All the photographs were taken from a specific point marked on the ground for accuracy. The spray cone angle is the angle at which the fuels starts dispersing from liquid into tiny molecular particles when injected from an injection nozzle. A mechanical technique of using a 445×570 blotting (absorbent) paper at a distance of 302mm below the nozzle was used to find the indentation diameter on the blotting paper. The distance was selected in such a way so the fuel was allowed to form into tiny molecular

 DOI: 10.1051/e3sconf/20172310001

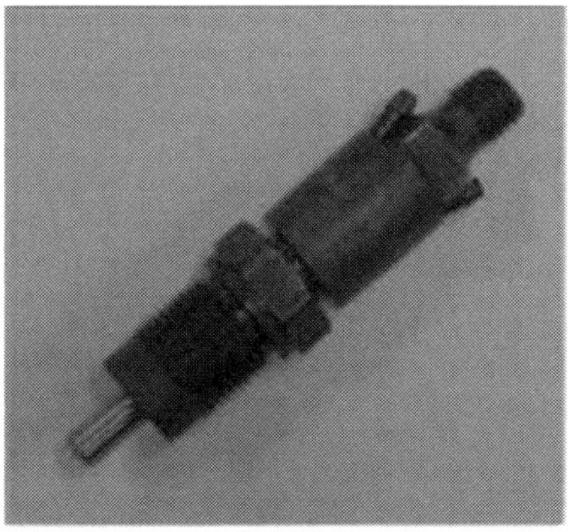

(a) Delphi injector

(b) POP tester

Figure 1 - The injector nozzle and POP tester setup

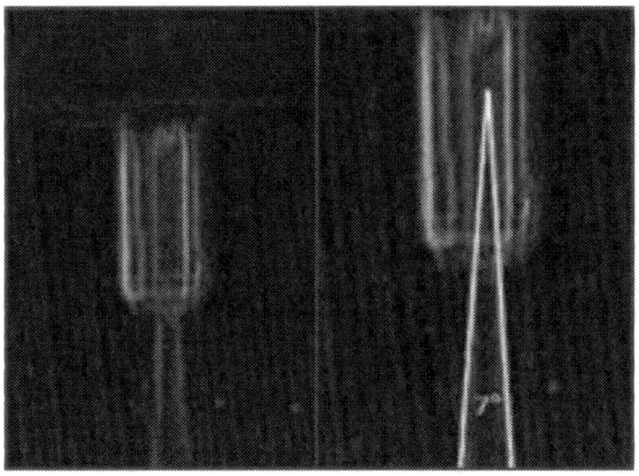

Figure 2 - Left: The edges sharpened of the magnified image, and Right: manually addition of lines on the edges to measure the liquid sheet angle, these images are for FD

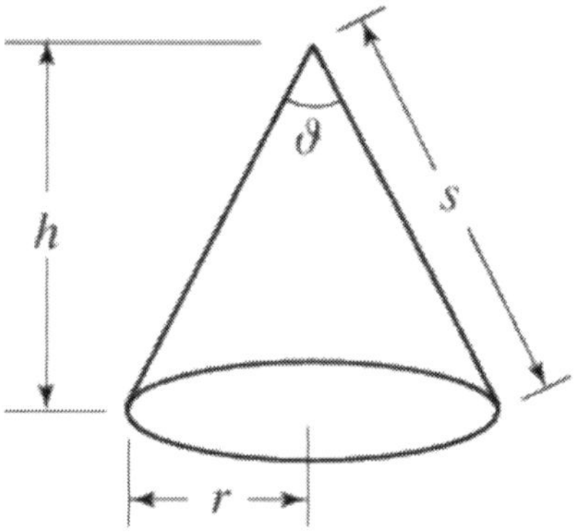

Figure 3 - Spray cone angle parameters [17]

particles. Platform was manually constructed so the blotting paper was as horizontal as possible, ensuring that indentation of the spray was accurate. The test was performed 5 times and the average value was taken. For each of the blends, the

DOI: 10.1051/e3sconf/20172310001

formula below was used to find the spray cone angle. Figure 3 shows the parameters of a cone used to estimate the spray cone angle using following equation:

$$Spray\ cone\ angle\ (°) = 2 \times \tan^{-1}\left(\frac{r}{h}\right)$$

Where r is the radius (of the indentation) and h is the height (distance from the nozzle tip to the blotting paper).

RESULTS AND DISCUSSION

Figure 4 illustrate the physicochemical properties of the neat fuels and blends when CuO and Al_2O_3 was added. Figures 4(a-c) shows comparison of the density [@15°C] of nanoparticles blends with respect to the neat FD, B100 and RSO100 fuels. The density of CuO and Al_2O_3 are $6.4\left(\frac{g}{m^3}\right)$ [18] and $3.7\left(\frac{g}{m^3}\right)$ [19]; hence, density increase due to C_uO addition was higher than Al_2O_3. In general, density of all blends were increased in the range of 0.07% to 0.13% as the nanoparticles concentration was increased from 25ppm to 100ppm. The flash point of neat RSO100 fuel was maximum; flash point for blends of Al_2O_3 and C_uO with FD increased as the concentration of the nanoparticles increased up to 75ppm. The flash point temperature stayed constant from 75ppm to 100pm (Figure 4(d)). Almost similar trend was also observed for B100 fuel (Figure 4(e)). In the case of RSO100 and blends, for Al_2O_3 nanoparticle, the flash point stayed constant up to 75ppm and then increased by 0.19% when the concentration was increased to 100ppm. However, for C_uO, the trend was not clear possibly due to the non-homogeneous mixing of nanoparticles (Figure 4(f)).

DOI: 10.1051/e3sconf/20172310001

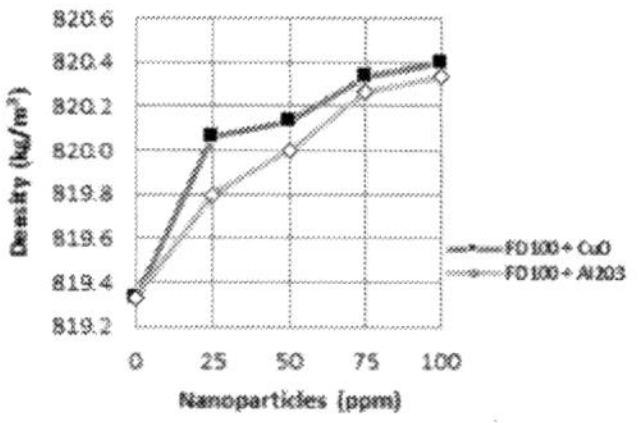

(a) Density of Fossil Diesel & blends

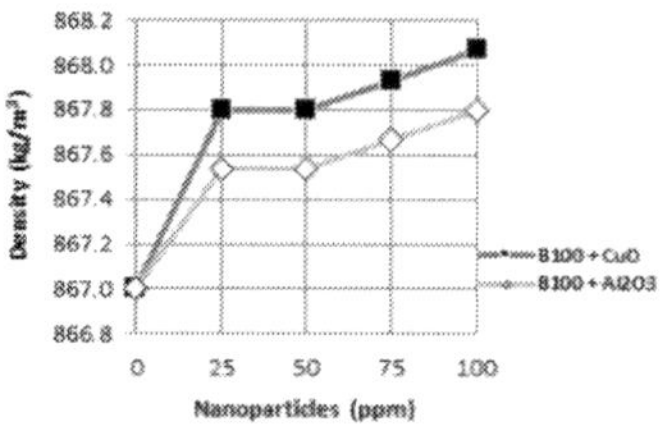

(b) Density of B100 & blends

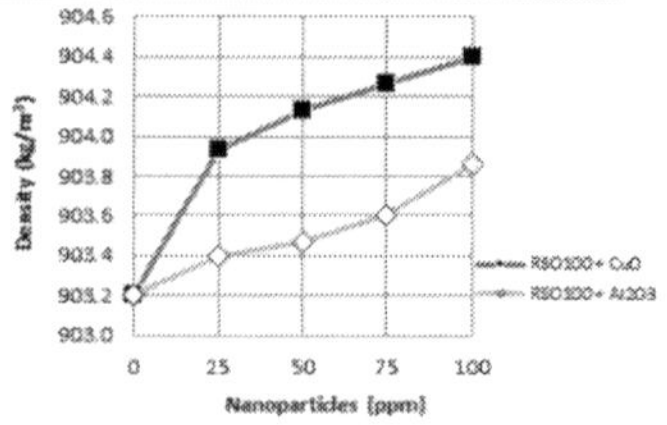

(c) Density of RSO100 & blends

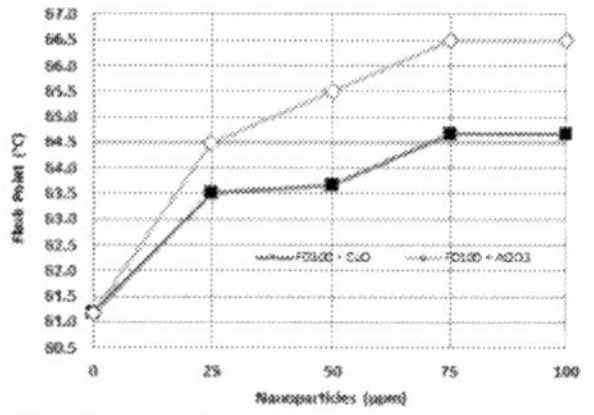

(d) Flash point of Fossil Diesel & blends

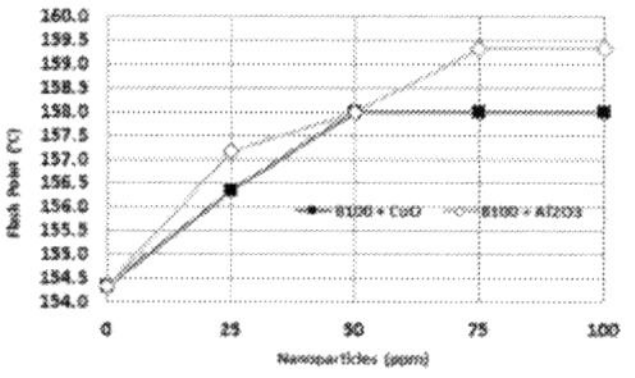

(e) Flash point of B100 & blends

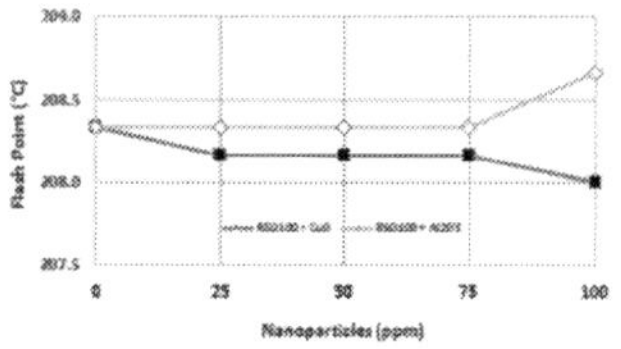

(f) Flash point of RSO100 and blends

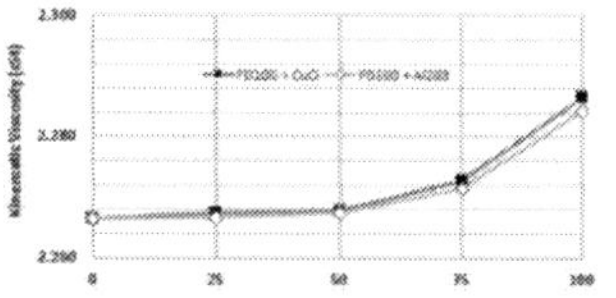

(g) Fossil Diesel & blends @40°C

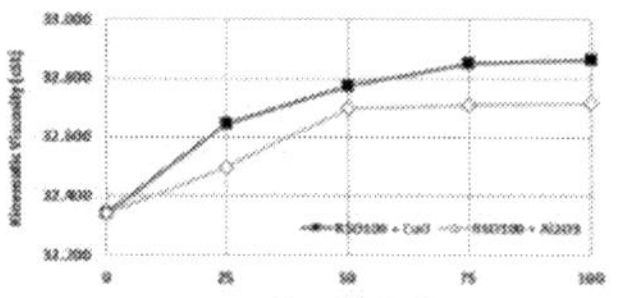

(h) B100 & blends @40°C

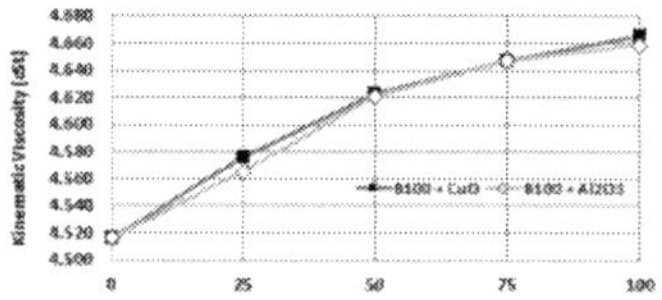

(i) RSO100 & blends @40°C

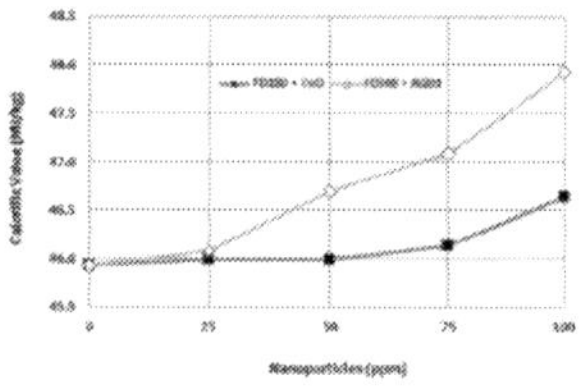

(j) HHV of Fossil Diesel & blends

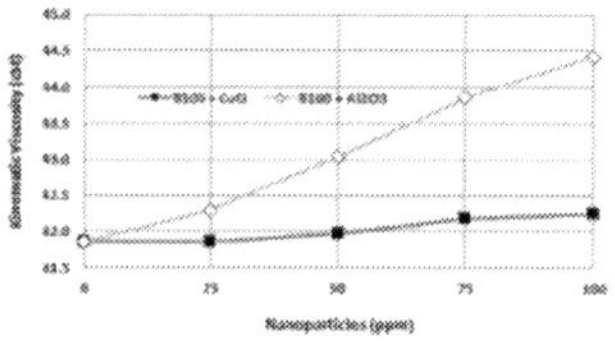

(k) HHV of B100 & blends

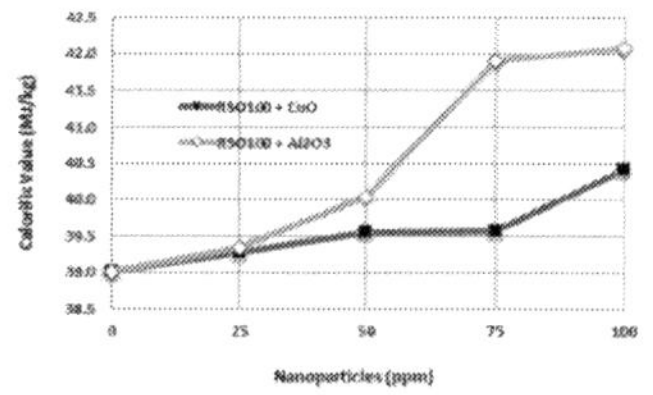

(l) HHV of RSO100 & blends

Figure 4 – Physicochemical properties of Al_2O_3 and CuO nanoparticle blends

For all nanoparticle blends of FD, B100 and RSO10O, the kinematic viscosity increased with increased concentration of nanoparticles (Figures 4(g-i). CuO caused higher kinematic viscosity in comparison to Al_2O_3 in all the cases. It was believed that this was due to the higher density of the CuO than Al_2O_3 blends. In addition, nanoparticles increases the surface tension in the fuel blends [20], with increased concentration of nanoparticles, the surface tension would also increase and hence the viscosity. The calorific value increased with increased concentration of nanoparticles (Figure 4(j-l)). For all blends and for same concentration of nanoparticles, Al_2O_3 gave higher HHV than CuO. The increase in HHV of RSO100 when 100ppm Al_2O_3 was added was 7.3%; whereas, these values were 5.8% and 4.1% for B100 and FD respectively (Figure 4(j-l)). Furthermore, it was found that calorific value of B100 with 100ppm Al_2O_3 is almost similar to that of neat FD. Higher oxygen content in Al_2O_3 caused higher HHV than CuO, as oxygen helps in combustion. On the basis of the above results, eight more blends were prepared using 100ppm Al_2O_3 nanoparticle and 10% Butanol : 90% B100 & 10% Bu (B90Bu10), 90% B100 & 10% Bu + 100ppm Al_2O_3 (B90Bu10 + 100ppm Al_2O_3), 90% RSO100 & 10% Bu (RSO00Bu10), 90% RSO100 & 10% Bu + 100ppm Al_2O_3 (RSO90Bu10 + 100ppm Al_2O_3), 80% B100 & 20% Bu (B80Bu20), 80% B100 & 20% Bu + 100ppm Al_2O_3 (B80Bu20 + 100ppm Al_2O_3), 80% RSO100 & 20% Bu (RSO80Bu20) and 80% RSO100 & 20% Bu + 100ppm Al_2O_3 (RSO80Bu20 + 100ppm Al_2O_3).

The emulsions of the nanoparticle blends were checked after 5 days of blend preparation and it could be seen that the Al_2O_3 particles had dispersed better with the use of butanol (Figure 5).

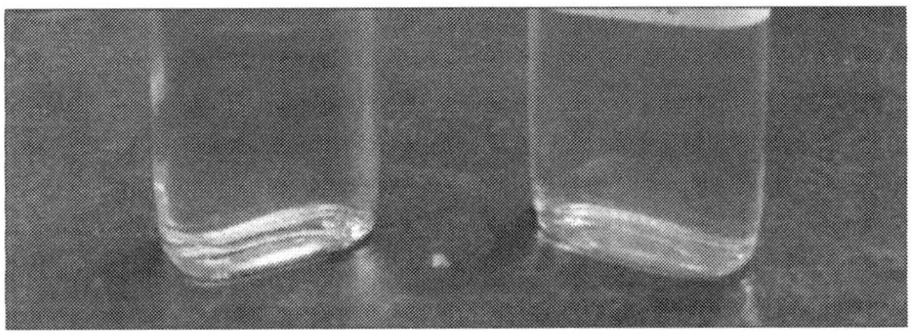

Figure 5 - Examples of a close up look of the blends after 5 days of Al_2O_3 nanoparticles addition

Figure 6 (a) shows the effect of butanol addition, for example, when 10% butanol was added to B100 (from B100 to B90Bu10) flash point was decreased by 58.1%.

DOI: 10.1051/e3sconf/20172310001

The flash point of Butanol is 35°C [21] which caused this behaviour. However, this is still 5.4% higher than FD. Similar behaviour was observed with RS100, the flash point was decreased by 66.7% when 10% Bu was added (ie. from RSO100 to RSO90Bu10).

Flash point temperatures were increased when 100ppm Al_2O_3 was added. For example, flash point was increased by 3.4% for B100 fuel when 100ppm Al_2O_3 was added, and in the case of B90Bu10 fuel this was increased by 7.6%. The increase in flash point for RSO100 when 100ppm Al_2O_3 were added was 0.2% and the increase rate was 5.14% for RSO90Bu10 fuel when 100ppm Al_2O_3 was added. Furthermore, it was observed that percentage increase for B100 and RSO100 fuels when 100ppm Al_2O_3 were added was much lower than B90Bu10 and RSO90Bu10. This proved that addition of butanol helped the nanoparticles to emulsify effectively. Kinematic viscosity of butanol @40°C is 1.0039cSt [22].

When 10% butanol was added to B100, viscosity decreased by 15.2% (Figure 6(b-c)). Similar characteristics were observed with RSO100 when 10% Bu was added - viscosity @40°C was decreased by 39.7%. The increase for B100 when 100ppm Al_2O_3 were added was 3.1%. On the other hand, viscosity was decreased by 0.4% for B90Bu10 fuel when 100ppm Al_2O_3 was added. The percentage increase for RSO100 when 100ppm Al_2O_3 were added was 1.1%, whereas for RSO90Bu10 viscosity was decreased by 3.3% when 100ppm Al_2O_3 were added. So, it was established that when Al_2O_3 nanoparticles were added to B100 and RSO100 the viscosity of the blends increased due to increased surface tension.

On the other hand, when 100ppm Al_2O_3 was added to B90Bu10 and RSO90Bu10 fuels, the viscosity decreased due to the fact that butanol acted as a surfactant and dispersed the Al_2O_3 into the blends properly. It was believed that the decrease in the viscosity of the biofuel + 10%Bu + 100ppm Al_2O_3 blends would give better spray characteristics than only butanol-biofuel blends.

The calorific value was decreased by 6.1% when 10% butanol was added to B100 – Figures 6(c) and 6(d). This was due to the low calorific value of butanol being $37.334\frac{\mathrm{MJ}}{\mathrm{kg}}$ [21]. The same happened with RS100 when 10% Bu was added (from RSO100 to RSO90Bu10), the calorific value was decreased by 2.8%.

In general, for all blends the calorific value increased when 100ppm Al_2O_3 were added. The increase in HHV for B100 and B90Bu10 fuels were 5.8% and 9.6% respectively when 100ppm Al_2O_3 were added. On the other hand, these values were 7.2% and 8.2% for RSO100 and RSO90Bu10 when 100ppm Al_2O_3 were added – Figure 6(d). These results proved that rate of increase in HHV were higher when nanoparticles were added in the biofuel –butanol blends instead of only neat biofuels. Hence, it was evident that the butanol was able to help catalyse the combustion process much better in the nanoparticles blends. Macroscopic spray characteristics with and without Al_2O_3 nanoparticles are shown in Figure 7.

When 10% Butanol was added to B100 (from B100 to B90Bu10), liquid sheet angle stayed the same. On the other hand, when 10% Butanol was added to RSO100 (from RSO100 to RSO90Bu10), the liquid sheet angle increased by 25% (Figure 7(a)). Furthermore, when Bu was added to the biofuels with Al_2O_3 blends, the liquid sheet angle increased by 23.1% for B90Bu10 + 100ppm Al_20_3 when compared to B90Bu10; and 11.1% for RSO90Bu10 + 100ppm Al_20_3 when compared to RSO90Bu10. The decrease for B100 when 100ppm Al_2O_3 were added was 40%, and this value was 16.7% for RSO100 when 100ppm Al_2O_3 were added (Figure 7(a)).

The results illustrated that overall, Butanol - biofuel - nanoparticle blends gave increased liquid sheet angle; this will eventually lead to improved combustion inside the engine cylinder and lower fuel consumption. The spray cone angle increased by 0.19% when 10% Butanol was added to B100 (from B100 to B90Bu10) – Figure 7 (b). Whereas, this was decreased by 21.4% in the case of RSO100. When Bu was added to the biofuels with Al_2O_3 blends, the spray cone angle increased by 19.5% for B90Bu10 + 100ppm Al_20_3 when compared to B90Bu10, and by 22.3% for RSO90Bu10 + 100ppm Al_20_3 when compared to RSO90Bu10. The spray characterisation results proved that butanol - nanoparticle blends gave better spray characteristics than only nanoparticles-biofuels blends. The increase in the liquid sheet angle and the spray cone angle would mean that the atomisation of the fuels would be better as more fuel particles will get in contact with air.

DOI: 10.1051/e3sconf/20172310001

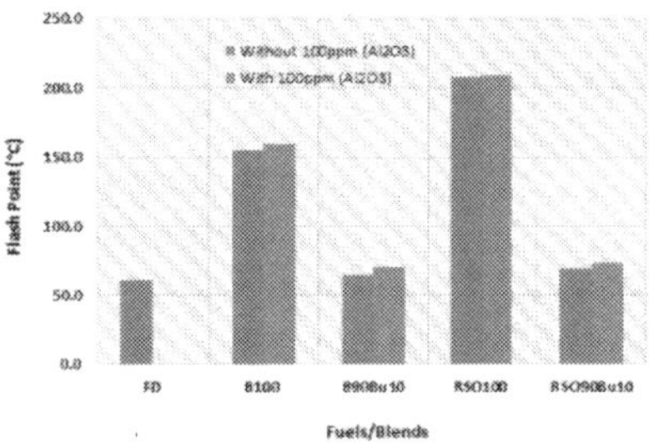

(a) Flash point with & without 100ppm Al_2O_3

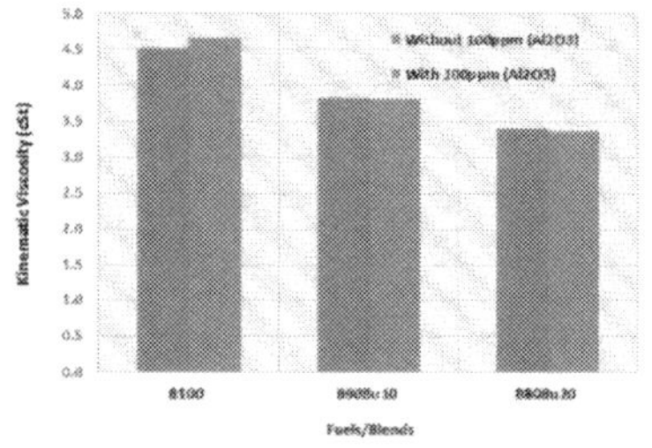

(b) Viscosity @40°C with & without 100ppm Al_2O_3

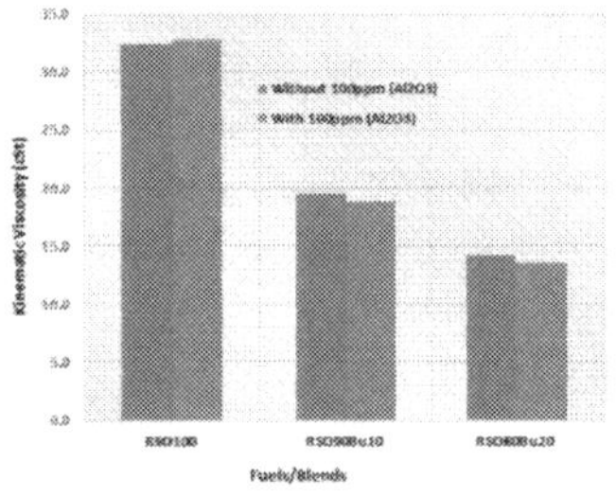

(c) Viscosity @40°C of RSO & butanol blends - with & without the addition of 100ppm Al_2O_3

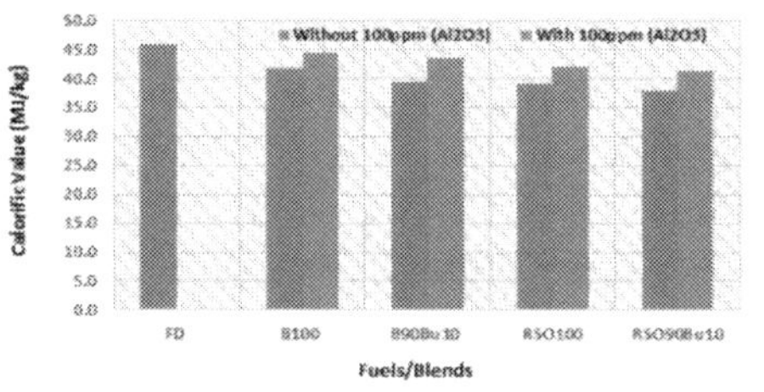

(d) Calorific value of blends with & without 100ppm Al_2O_3

Figure 6 – Comparison of biofuels (and blends) properties with and without Al_2O_3 nanoparticles

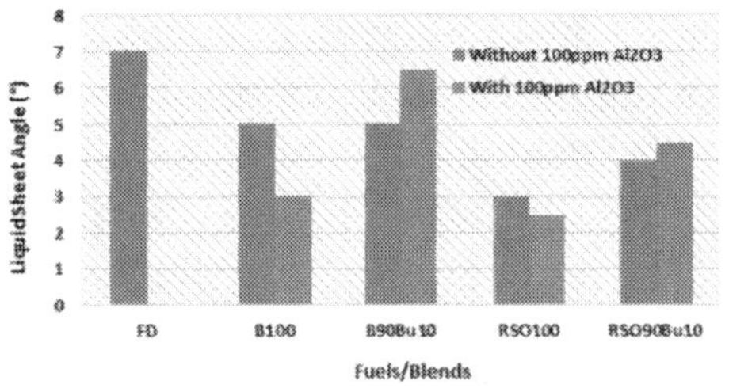

(a) Liquid sheet angle of blends with & without 100ppm Al_2O_3

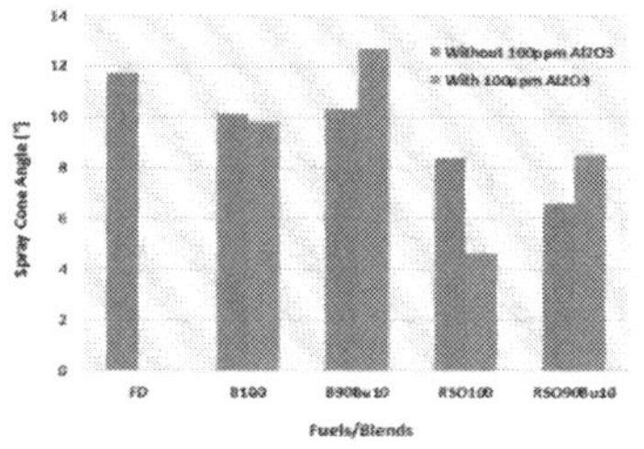

(b) Spray cone angle of blends with & without 100ppm Al_2O_3

Figure 7 – Spray properties of fuels with and without Al_2O_3 nanoparticles

DOI: 10.1051/e3sconf/20172310001

CONCLUSIONS

Two nanoparticles (Al_2O_3 and CuO) were used in this study to investigate the physicochemical properties and spray characteristics of the biofuel blends. It was found that the addition of nanoparticles increased the density, flash point temperature, kinematic viscosity and heating value of the fuels when mixed separately with neat FD, neat B100 and neat RSO100.

The density of the Al_2O_3 blends were lower than CuO blends. The flash point were much higher for Al_2O_3 blends in comparison to CuO blends. The viscosity was increased with both nanoparticles; however, Al_2O_3 gave smaller increase than CuO nanoparticles. It was proved that addition of Al_2O_3 gave better properties than CuO nanoparticles.

For internal combustion engines application, increase in the density and the viscosity would produce negative effects on engine fuel supply systems and combustion; whereas, on the other hand, increased flash point and the heating value would produce positive effects on fuel systems and combustion.

Hence, in the second stage of the study, the effect of butanol addition in Al_2O_3 nanoparticles-biofuels were assessed to see how butanol addition affect physicochemical properties and spray characteristics of the blends. It was observed that Butanol-biofuel- Al_2O_3 blends improved physicochemical properties and spray characteristics when compared to only biofuel-Al_2O_3 blends. The spray characteristic results showed that both spray cone angle and the liquid sheet angle parameters were improved significantly for Al_2O_3 - butanol - biofuel blends. The study concluded that addition of butanol helped to emulsify Al_2O_3 nanoparticles effectively in the blends; and butanol can be used as a surfactant for emulsification of Al_2O_3 nanoparticles in biofuels. Measurement of surface tension, engine performance and exhaust emission analysis are recommended as further work. Use of other nanoparticles and ethanol is another area of further investigation.

DOI: 10.1051/e3sconf/20172310001

REFERENCES

[1] Smith H, Winfield J, Thompson L. The market for biodiesel production from used cooking oils and fats, oils and greases in London. LRS consultancy, 2013.

[2] Challenges and opportunities-Understanding the Basic Properties of Nanoparticles. https://web.phys.ksu.edu/current/seminar/s09/mody-per-seminar.pdf

[3] AZO Nano. Copper Oxide (CuO) Nanoparticles - Properties, Applications; 2013. http://www.azonano.com/article.aspx?ArticleID=3395

[4] Gan Y, Qiao L. Combustion characteristics of fuel droplets with addition of nano and micron-sized aluminium particles. Combustion and Flame 2011; 158(2): 354-368.

[5] Soutter W. Nanoparticles as Fuel Additives. AZO NANO; 2012.

[6] Samuel N, Shefeek MK. Performance and Emission Characteristics of a C.I Engine with Cerium Oxide Nanoparticles as Additive to Diesel. Int. J. Sci & Res 2015; 4(7): 673-676.

[7] Gumus S, Ozcan H, Ozbey M, Topaloglu B. Aluminium oxide and copper oxide nanodiesel fuel properties and usage in a compression ignition engine. Fuel 2016; 163: 80-87.

[8] Santhanamuthu M, Chittibabu S, Tamizharasan T, Mani TP. Evaluation of CI engine performance fuelled by Diesel-Polanga oil blends doped with iron oxide nanoparticles. Int. J. ChemTech Res 2014; 6 (2): 1299-1308.

[9] D'Silva R, Binu KG, Bhat T. Performance and Emission Characteristics of a C.I engine fuel with diesel and TiO nanoparticle as additives. Mat. Today: Proc 2015; 2(4-5):3728-3735.

[10] George RO, James SJ, Franc S, John M. An Experimental Analysis on Synergetic Effect of Multiple Nanoparticle Blended Diesel Fuel on CI Engine. IJIRST 2015; 1(12).

[11] Aalam S, Saravanan CG. Effects of nano metal oxide blended Mahua biodiesel on CRDI diesel engine. Ain Shams Engg Journal 2015 (in press).

[12] Aalam CS, Saravanan CG. Performance Enhancement of Common-Rail Diesel Engine using Al2O3 and Fe3O4 Nanoparticles Blended Biodiesel. IRJET 2015;2(5): 1400-1410.

[13] Wang X, Huang Z, Kuti OA, Zhang W, Nishida K. Experimental and analytical study on biodiesel and diesel spray characteristics under ultra-high injection pressure. Int J Heat Fluid Flow 2010;31(4):659–666.

[14] Ejim CE, Fleck BA, Amirfazli A. Analytical study for atomization of biodiesels and their blends in a typical injector:Surface tension and viscosity effects. Fuel 2007;86(10):1534–1544.

[15] Şahin Z, Aksu O. Experimental investigation of the effects of using low ratio n-butanol/diesel fuel blends on engine performance and exhaust emissions in a turbocharged DI diesel engine. Ren Energy 2015; **77**: 279-290.

[16] Tücca, G, Özgür T, Aydın K. Effect of diesel–microalgae biodiesel–butanol blends on performance and emissions of diesel engine. Fuel 2014; **132**: 47-52.

[17] Wolfrom MathWorld. http://mathworld.wolfram.com/Cone.html

[18] Sigma-Aldrich.
http://www.sigmaaldrich.com/catalog/product/aldrich/544868?lang=en&region=GB

[19] Sigma-Aldrich
http://www.sigmaaldrich.com/catalog/product/aldrich/544833?lang=en&region=GB

[20] Chhetri AB, Watts KC. Surface tensions of petro-diesel, canola, jatropha and soapnut biodiesel fuels at elevated temperatures and pressures. Fuel 2013;104 (1):704-710.

[21] Rakopoulos DC, Rakopoulos CD, Giakoumis EG, Dimaratos AM, Kyritsis DC. Effects of butanol–diesel fuel blends on the performance and emissions of a high-speed DI diesel engine. Ener Con and Management 2010;51:1989-1997.

[22] Mužíková Z, Šimáček P, Pospíšil M, Šebor G. Density, Viscosity and Water Phase Stability of 1-Butanol-Gasoline Blends. Journal of Fuels 2014; 2014 p. 7, 2014.

AUTHOR INDEX